MEDICINE

THE DEFINITIVE ILLUSTRATED HISTORY

ANDREAE VESALII
BRVXELLENSIS, SCHOLAE
medicorum Patauinæ professoris, de
Humani corporis fabrica
Libri septem

MEDICINE

THE DEFINITIVE ILLUSTRATED HISTORY

STEVE PARKER

CONSULTANT
Steve Parker

CONTRIBUTORS
Alexandra Black, Philip Parker, Sally Regan, Marcus Weeks

DK LONDON
Senior Editor Kathryn Hennessy
Senior Art Editor Helen Spencer
Editors Alexandra Beeden, Polly Boyd, Anna Cheifetz,
Jemima Dunne, Georgina Palffy, Esther Ripley
US Editor Jill Hamilton
Managing Editor Gareth Jones
Senior Managing Art Editor Lee Griffiths
Senior Jacket Designer Mark Cavanagh
Jacket Design Development Manager Sophia MTT
Jacket Editor Claire Gell
Pre-production Producer Nadine King
Producer Mandy Inness
Associate Publishing Director Liz Wheeler
Publishing Director Jonathan Metcalf
Art Director Karen Self

DK DELHI
Senior Editors Dharini Ganesh, Bharti Bedi, Anita Kakar
Senior Art Editor Mahua Sharma
Project Art Editor Shreya Anand
Editors Arpita Dasgupta, Priyaneet Singh
Art Editor Anjali Sachar
Senior Editorial Manager Rohan Sinha
Managing Art Editors Sudakshina Basu, Anjana Nair
Jacket Designer Suhita Dharamjit
Managing Jackets Editor Saloni Singh
Picture Researcher Aditya Katyal
Manager Picture Research Taiyaba Khatoon
DTP Designers Vijay Kandwal, Pawan Kumar
Senior DTP Designers Harish Aggarwal, Sachin Singh
Pre-production Manager Balwant Singh
Production Manager Pankaj Sharma

First American Edition, 2016
Published in the United States by DK Publishing
345 Hudson Street, New York, New York 10014

Copyright © 2016 Dorling Kindersley Limited
DK, a Division of Penguin Random House LLC
16 17 18 19 20 10 9 8 7 6 5 4 3 2 1
001—283277—Oct/16

A catalog record for this book is available from the
Library of Congress.
ISBN: 978-1-4654-5341-9

DK books are available at special discounts when purchased in bulk for sales promotions,
premiums, fund-raising, or educational use. For details, contact: DK Publishing Special
Markets, 345 Hudson Street, New York, New York 10014
SpecialSales@dk.com

Printed in China

A WORLD OF IDEAS:
SEE ALL THERE IS TO KNOW

www.dk.com

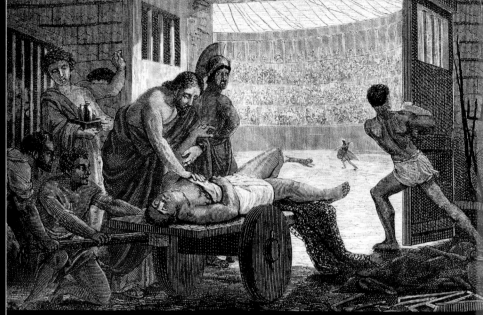

TREATING A GLADIATOR

CONTENTS

1
ANCIENT WISDOM TO 700

12 Timeline

14 Healers and Herbalists

16 Early Surgery

18 Shamanism

20 Medicine in Ancient Egypt

22 Secrets of Mummies

24 Medicine in Ancient Mesopotamia

26 Early Chinese Medicine

28 Acupuncture

30 Ayurveda

32 Medicine in Ancient Greece

34 The Four Humors

36 Hippocrates

38 Medicine in Ancient Rome

ANCIENT EGYPTIAN SURGICAL INSTRUMENTS

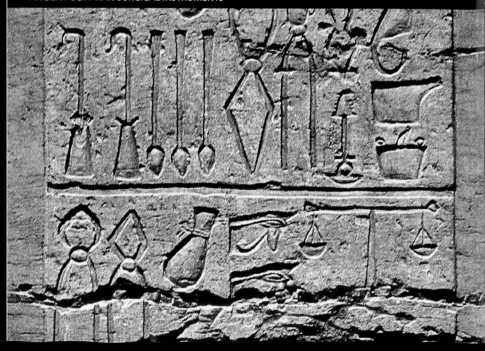

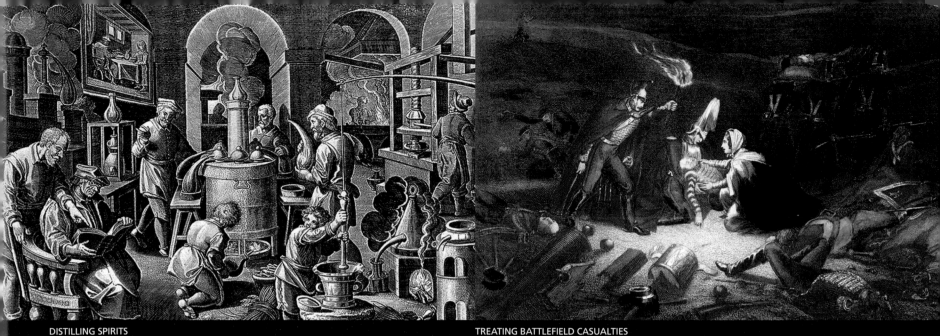

DISTILLING SPIRITS

TREATING BATTLEFIELD CASUALTIES

40 **Galen**

42 **Roman Surgical Tools**

2

REVIVAL AND RENAISSANCE 700–1800

46 **Timeline**

48 **The Golden Age of Islamic Medicine**

52 **Ibn Sina's *The Canon of Medicine***

54 **The First School of Medicine**

56 **Medieval Medicine**

60 **Anatomy Restored**

62 **Apothecary Store**

64 **Alchemy**

66 **The Black Death**

68 **Preventing Plagues**

70 **Alchemy, Chemistry, and Medicine**

72 **The Anatomy Revolution**

76 **Barber-surgeons**

78 **Ambroise Paré**

80 **Repair and Reconstruction**

82 **Discovering the Circulation**

84 **The Circulation Revolution**

86 **Cataract Surgery**

88 **Exchanging Epidemics with the New World**

90 **Thomas Sydenham**

92 **Early Microscopists**

94 **Evolution of Microscopes**

96 **The First Microanatomists**

98 **Scurvy**

100 **Smallpox: The Red Plague**

102 **The First Vaccination**

104 **Phrenology**

106 **The Modern Hospital**

108 **Homeopathy**

3

SCIENCE TAKES CHARGE 1800–1900

112 **Timeline**

114 **The First Stethoscope**

116 **Diagnostic Instruments**

118 **Resurrection Men**

120 **Miasma Theory**

122 **Cholera**

124 **John Snow**

THE MEDICAL SCHOOL AT SALERNO

INOCULATING A PATIENT

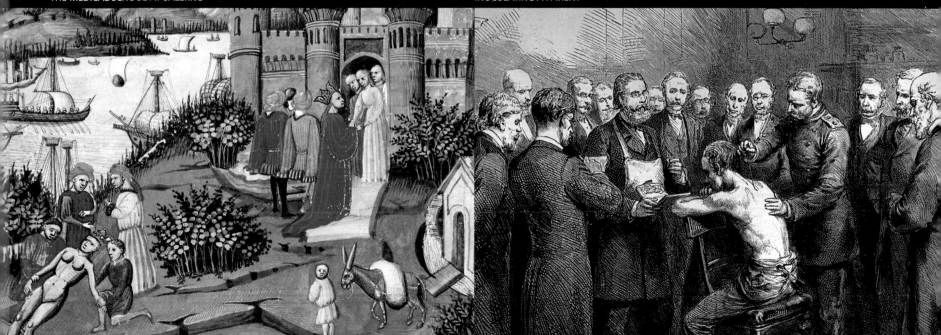

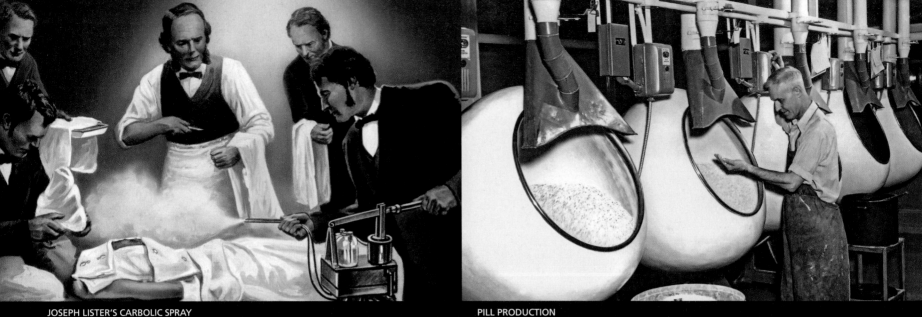

JOSEPH LISTER'S CARBOLIC SPRAY

PILL PRODUCTION

126 **Epidemiology and Public Health**

128 **Anaesthetics**

130 **Early Anaesthetics**

132 **Dentistry**

134 **Pregnancy and Childbirth**

136 **Midwives**

138 **Childbed Fever**

140 **Women in Medicine**

142 **Nursing**

144 **Medical Publishing**

146 **Microbiology and Germ Theory**

148 **Louis Pasteur**

150 **Cell Theory**

152 **Pathology and Medical Autopsy**

154 **The First Antiseptics**

156 **Tuberculosis**

158 **Vaccines Come of Age**

160 **Mysteries of the Brain**

162 **Mental Illness**

164 **Horror of the Asylum**

166 **Viruses and How they Work**

168 **Fighting Rabies**

170 **The Discovery of Aspirin**

172 **X-rays**

174 **The Struggle Against Malaria**

176 **Transfusion Breakthrough**

4

ERA OF SPECIALIZATION 1900–1960

180 **Timeline**

182 **Sigmund Freud**

184 **The Development of the ECG**

186 **A Cure for Syphilis**

188 **Minimally Invasive Surgery**

190 **Diabetes and Insulin**

192 **War and Medicine**

194 **Battlefield Medicine in World War II**

196 **Influenza and the Pandemic**

198 **The Discovery of Penicillin**

200 **Antibiotics in Action**

202 **The Evolution of Syringes**

204 **Women's Health**

206 **Heart Disease**

208 **Allergies and Antihistamines**

210 **Polio: A Global Battle**

GUESTS IN A TRANCE AT MESMER BANQUET

FIRST AID AFTER GAS ATTACK

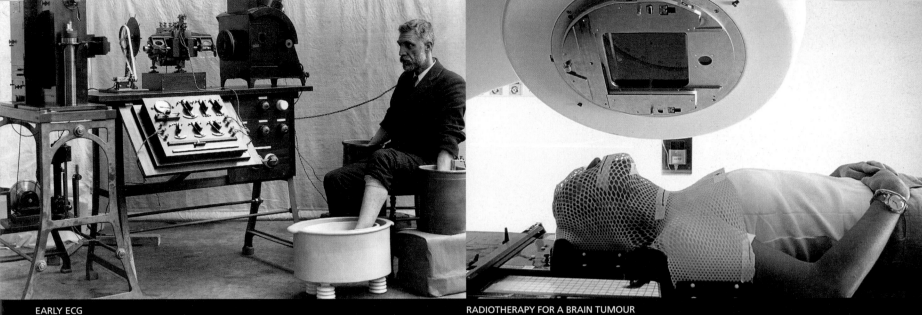

EARLY ECG

RADIOTHERAPY FOR A BRAIN TUMOUR

212 **The Structure of DNA**

214 **Inhalers and Nebulizers**

216 **Scanning Machines**

218 **The Pharmaceutical Industry**

5

PROMISES OLD AND NEW 1960–PRESENT

222 **Timeline**

224 **The Contraceptive Pill**

226 **Margaret Sanger**

228 **Cancers**

232 **Advanced Imaging**

234 **The First Heart Transplant**

236 **Implants and Prostheses**

238 **Artificial Body Parts**

240 **In Vitro Fertilization**

242 **HIV and AIDS**

244 **New Discoveries for Old Diseases**

246 **Genetic Revolution**

248 **Genetic Testing**

250 **Mental Health and Talking Therapies**

252 **Robots and Telemedicine**

254 **Robotic Surgery**

256 **Emergency Medicine**

258 **Antibiotic Resistance and Superbugs**

260 **Alzheimer's Disease and Dementias**

262 **End-of-Life Care**

264 **Nanomedicine**

266 **Global Medical Bodies**

268 **Ebola Virus Disease**

270 **Stem Cell Therapy**

REFERENCE

274 **Body Systems**

276 **The Skeletal System**

278 **The Muscular System**

280 **The Nervous System**

282 **The Cardiovascular System**

284 **The Respiratory System**

286 **The Endocrine System**

288 **The Digestive System**

290 **The Immune System**

292 **The Urogenital System**

296 **Sensory Organs**

300 **Who's Who**

305 **Glossary**

312 **Index and Acknowledgments**

NANOBOTS

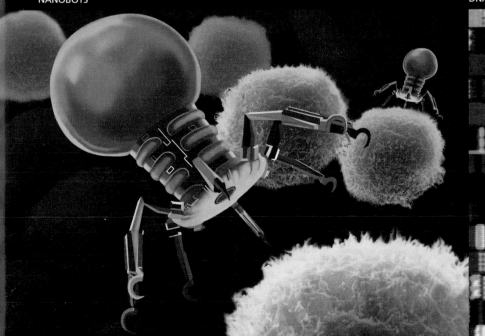

DNA SEQUENCING

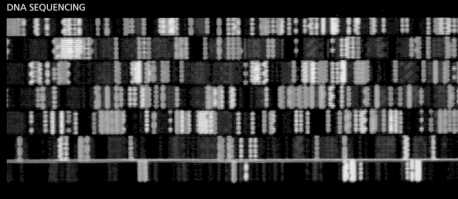

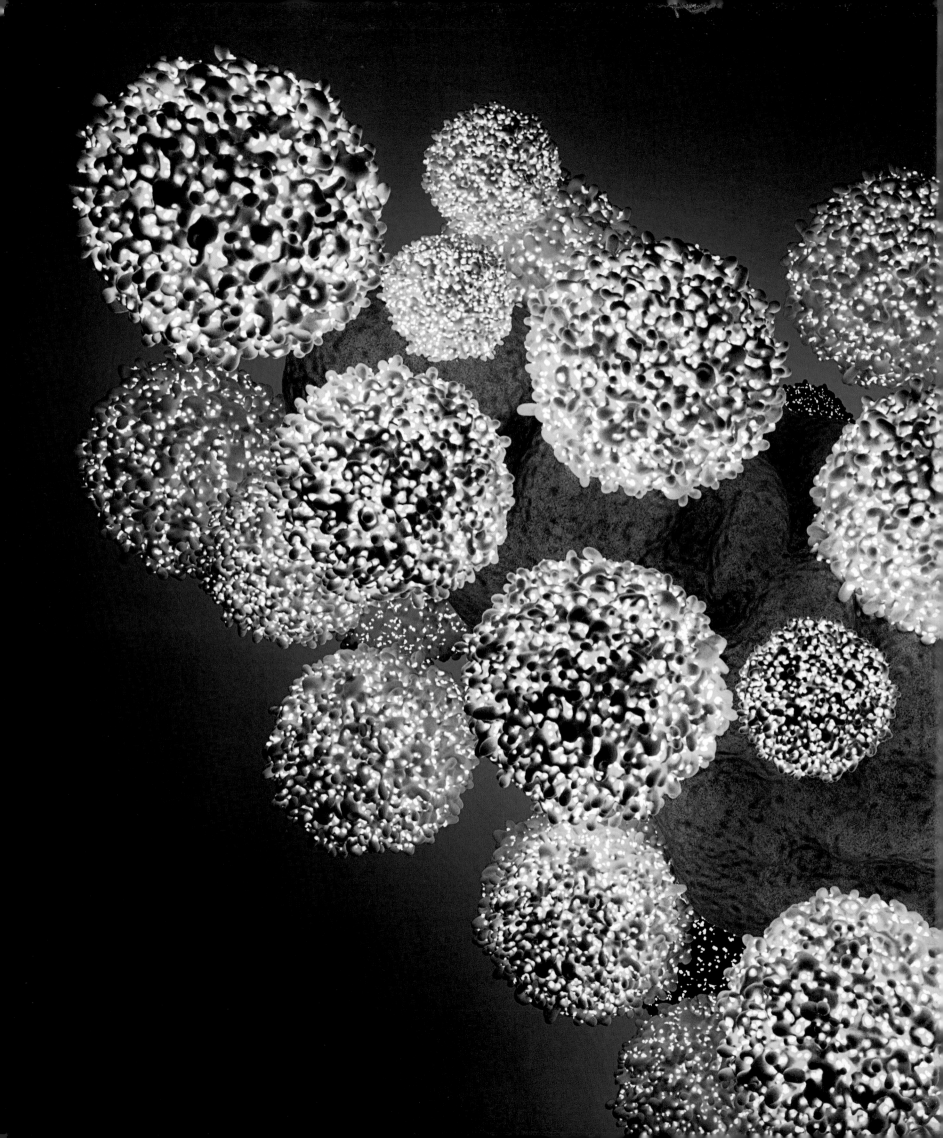

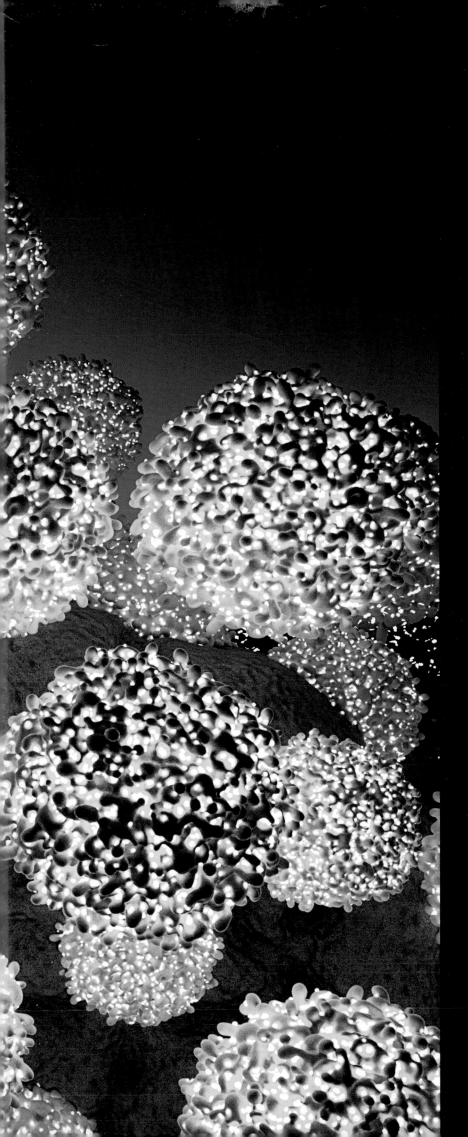

Introduction

One of the greatest figures in the history of medicine, Hippocrates of ancient Greece, believed that "A wise man should consider health as the greatest human blessing… To physicians: Cure sometimes, treat often, comfort always… And make a habit of two things: Help; or at least, do no harm." While these words are more than 2,300 years old, their sentiments still ring true today. Good health is a most precious commodity and in the modern world medicine has achieved towering status. Many nations spend more than one-tenth of their entire wealth on prevention and treatment of illness and allied health services.

The origins of medicine are hazy, but it is known that all great ancient civilizations had specialists in healing arts, as each region around the world developed knowledge and learning in the area. Varied traditions arose, some effective, but many bound up with spells and curses, spirits, demons, and other supernatural entities. Progress toward modern medicine quickened from about the 16th century, especially in Europe. Here, the Renaissance led to the rise of organized observation, recording, experimentation, analysis, and a rational, evidence-based approach, and medicine evolved from art to science.

The past two centuries have seen momentous advances— vaccination; antiseptics; anesthetics; the discovery of germs and the antibiotics to fight them; improved diet, hygiene, and sanitation; numerous uses for radiation; body imaging; transplants and implants; and progress against cancers. The average patient's experience has changed immeasurably since ancient times. But there are still abundant inequalities around the world and challenges to meet, such as malaria, HIV/AIDS, and other epidemic infections; chronic diseases of the respiratory and circulatory systems; and the provision of clean water, adequate nutrition, and comprehensive vaccination for all. The 21st century also sees major new treatments emerging, such as therapies exploiting genes and stem cells, and the prospect of tailor-made "personalized medicine."

All of these topics and more are covered in the following chapters. The history of medicine is a vast subject, but this book throws a spotlight onto what has been, the giant strides that medicine has achieved, and how the balance between health and illness looks set to improve for future generations.

◁ **Always something new**
The arrival of HIV/AIDS during the 1980s was a stark warning that new diseases will continue to emerge. Here, copious HIV (human immunodeficiency virus) particles (small bright spots) infect human white blood cells, quashing their immune defensive abilities.

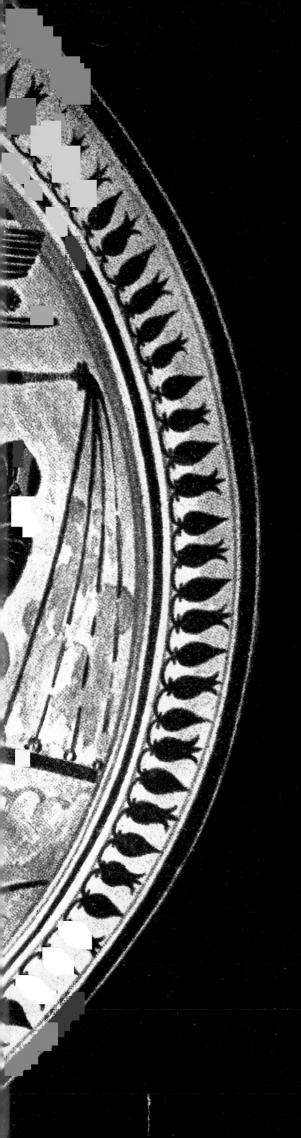

1
ANCIENT WISDOM
TO 700

ANCIENT WISDOM
TO 700

PREHISTORY	3000 BCE	1500 BCE

49,000 YEARS AGO
Neanderthals possibly use medicinal herbs, as evidenced by fossilized Neanderthal teeth.

7,000 YEARS AGO
A man undergoes a deliberate and successful arm amputation at what is now Buthiers-Boulancourt, France.

1500 BCE
The first reference to diabetes appears in an Egyptian papyrus.

500 BCE
The concept of the four humors, central to many medical systems for the next two millennia, begins to take shape in ancient Greece.

20,000 YEARS AGO
Holes are drilled into skulls—a procedure called trepanning—to treat medical conditions.

5,300 YEARS AGO
In the European Alps, Ötzi the Iceman suffers gut parasites, and painful bone and joint conditions.

1400 BCE
The Mesopotamian Gula Hymn includes: "I am a physician, I can heal; I carry around all healing herbs, I drive away disease; I give cures to mankind."

≫ Cupping vessel to treat humoral imbalance

10,000 YEARS AGO
Traditions of shamanism emerge on several continents.

3000 BCE
Egyptian mummies surviving from this time show broken bones, signs of tuberculosis, and other health problems.

≫ Stele of Hammurabi

1050 BCE
The landmark Mesopotamian Sakikku diagnostic handbook is completed by physician Esagil-kin-Apli of Borsippa.

500 BCE
Early versions of *Susruta Samhita*, an Ayurvedic compilation, appear in India.

2200 BCE
Per-Ankh, or Houses of Life, are built in ancient Egypt as places for creation and preservation of knowledge.

≪ Lord Dhanvantri, God of Ayurveda

2700 BCE
The tomb of one of the earliest known female physicians, ancient Egypt's Merit-Ptah, is inscribed "Chief Physician."

1755 BCE
The Code of Hammurabi, ruler of Babylon, includes several pronouncements on medical care, such as physicians are responsible for the success and failure of their actions.

≫ Mongolian shaman's decorated drum

1550 BCE
The Ebers Papyrus mentions medical use of willow bark, from which aspirin is derived.

7,000 YEARS AGO
Teeth of live patients are drilled, perhaps for abscess pain relief, in Mehrgarh, Pakistan.

2650–2600 BCE
In ancient Egypt, Imhotep becomes the leading priest-physician and is soon elevated to godly status.

Instincts for survival run deep. Our close cousins, chimpanzees and gorillas, respond to illness by self-medicating with herbs and clays. Early humans probably did the same. As civilizations evolved, individuals began to specialize in areas such as trade, warfare, and healing—and so medicine was born. The great ancient cultures of Mesopotamia, Egypt, China, and India developed their own medical systems, which were mostly entwined with gods, devils, and the spirit world. Around 2,500 years ago, ancient Greece and then Rome evolved their own styles of medicine, which focused more on the human body. However, progress stalled in the 5th century during Europe's "Dark Ages."

450 BCE

50 CE

400 CE

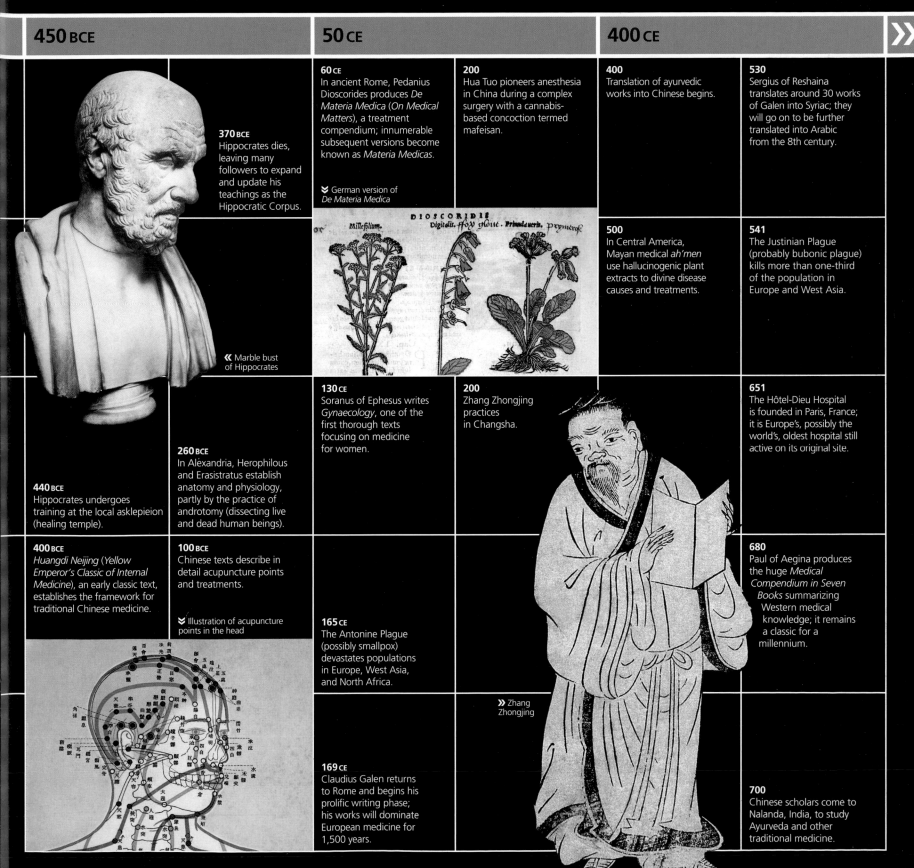

370 BCE
Hippocrates dies, leaving many followers to expand and update his teachings as the Hippocratic Corpus.

« Marble bust of Hippocrates

440 BCE
Hippocrates undergoes training at the local asklepieion (healing temple).

400 BCE
Huangdi Neijing (*Yellow Emperor's Classic of Internal Medicine*), an early classic text, establishes the framework for traditional Chinese medicine.

260 BCE
In Alexandria, Herophilous and Erasistratus establish anatomy and physiology, partly by the practice of androtomy (dissecting live and dead human beings).

100 BCE
Chinese texts describe in detail acupuncture points and treatments.

∀ Illustration of acupuncture points in the head

60 CE
In ancient Rome, Pedanius Dioscorides produces *De Materia Medica* (*On Medical Matters*), a treatment compendium; innumerable subsequent versions become known as *Materia Medicas*.

∀ German version of *De Materia Medica*

DIOSCORIDE

130 CE
Soranus of Ephesus writes *Gynaecology*, one of the first thorough texts focusing on medicine for women.

165 CE
The Antonine Plague (possibly smallpox) devastates populations in Europe, West Asia, and North Africa.

169 CE
Claudius Galen returns to Rome and begins his prolific writing phase; his works will dominate European medicine for 1,500 years.

200
Hua Tuo pioneers anesthesia in China during a complex surgery with a cannabis-based concoction termed mafeisan.

200
Zhang Zhongjing practices in Changsha.

» Zhang Zhongjing

400
Translation of ayurvedic works into Chinese begins.

500
In Central America, Mayan medical *ah'men* use hallucinogenic plant extracts to divine disease causes and treatments.

530
Sergius of Reshaina translates around 30 works of Galen into Syriac; they will go on to be further translated into Arabic from the 8th century.

541
The Justinian Plague (probably bubonic plague) kills more than one-third of the population in Europe and West Asia.

651
The Hôtel-Dieu Hospital is founded in Paris, France; it is Europe's, possibly the world's, oldest hospital still active on its original site.

680
Paul of Aegina produces the huge *Medical Compendium in Seven Books* summarizing Western medical knowledge; it remains a classic for a millennium.

700
Chinese scholars come to Nalanda, India, to study Ayurveda and other traditional medicine.

Healers and Herbalists

Preserved evidence in fossilized Neanderthal teeth shows that the history of medicine may stretch back almost 50,000 years, while modern anthropology reveals that many cultures weave ideas about health into their belief systems—believing in an invisible world of benign spirits, feared demons, lost souls, magic, and sorcery.

El Sidrón, an archaeological site in northwestern Spain, has yielded hundreds of fossilized bones and teeth from our closest cousins—the now-extinct Neanderthals (*Homo neanderthalensis*). Microfossils of plants including yarrow (*Achillea millefolium*) and chamomile (*Anthemis arvensis*) have been found in these Neanderthals' dental plaque—the hardened layer of debris on teeth. These herbs lack nutritional value and have a bitter, unpleasant taste. However, they are much used in traditional medicine. Yarrow is a tonic and an astringent, and chamomile is a relaxant and has anti-inflammatory properties. The fossilized teeth date to 49,000 years ago and are possibly the earliest evidence for the use of medications.

Each year, new evidence is being discovered, showing that prehistoric medicine was more advanced than once thought. Broken bones were

reset by smearing clay onto injured limbs; the clay then dried to form a supportive cast. Herb poultices were secured onto wounds with animal-hide bandages. Plant saps soothed burns while other constituents of

◁ **White Lady**
The "White Lady" cave painting in Brandenberg Mountain, Namibia, is probably more than 2,000 years old. Originally believed to depict a female, it may show the ritual dance of an African shaman or medicine man, with white minerals on his limbs.

plants were chewed for medicinal effects. For example, orchid bulbs were chewed for digestive problems and willow bark—the natural source of aspirin (see pp.170–71)—was chewed to ease fever and pain.

More than 7,000 years ago patients' teeth were drilled, perhaps to relieve abscess pain, while bow-operated drills were used to bore holes in the skull, a procedure known as trepanning (see pp.16–17).

Early healers

Prehistoric cave paintings and rock art of individuals wearing particular clothing and adornments suggest

◁ **Therapeutic herb**
For centuries yarrow has been a mainstay of herbal medicine across the Northern Hemisphere. Its astringent qualities stem bleeding, giving it local names such as woundwort and staunch-nose (for nosebleeds).

that they had a special role in their community as healers or therapists. These healing roles are still seen today in native cultures across the Americas, Africa, Asia, and Australasia. Spiritual, supernatural, and religious beliefs are all involved in their approach to illness and, as evil spirits and malicious demons are often blamed for ill health, treatments include offerings, spells, sacrifices, and exorcism, along with practical measures such as ointments made from herbs, minerals, and animal bones and blood.

An individual who conjures up supernatural powers and mediates with the spirit world is known as a shaman, medicine man or woman, soothsayer, or healer. He or she conducts ceremonies with chants, clapping, dancing, drumming,

25 **PERCENT of modern medicines that are made from plants were first used traditionally.**

burning aromatic plants, and taking potions to attain a trancelike state in order to communicate with the spirits. Modern analysis shows some of the herbs used in these rituals contain psychoactive, mind-altering, or hallucinogenic chemicals.

AUSTRIAN MUMMY c.33,000 BCE

ÖTZI THE ICEMAN

A 5,300-year-old, naturally preserved, mummified, frozen male found in the Ötztal Alps, Europe, in 1991 and named Ötzi, gives many clues about health and healing in prehistoric times. Ötzi was 45 years old when he died, and was found with a knife, ax, bow, arrows, bark containers, and what may have been a simple prehistoric medical

kit. Among his possessions were lumps of birch bracket fungus (*Piptoporus betuinus*), which has laxative as well as antibiotic properties. A detailed medical examination indicated the presence of whipworm parasite eggs in his large intestine. X-rays and scans of his skeleton

revealed that he had painful bone and joint conditions. Intriguingly, there are more than 50 skin tattoos on these painful areas. The tattoos, which correspond to known acupuncture points, were probably meant as symbolic "therapy" for pain relief.

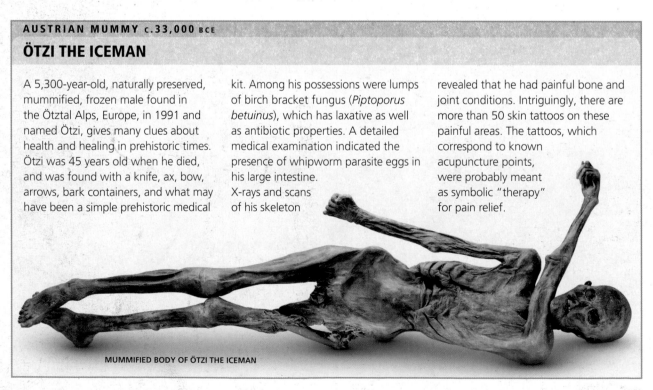

MUMMIFIED BODY OF ÖTZI THE ICEMAN

The practice of Shamanism (see pp.18–19) is seen especially in Africa and the Americas. Native American tribal groups all have distinct beliefs and medical practices, but they also have much in common, believing health is a balance among mind, body, and spirit. Healing involves restoring the balance in these three areas with the shaman's mediation, for example by reconsidering personal thoughts and emotions, receiving herbal remedies, and praying and making offerings to the spirits. Shamans are often apprenticed to a senior mentor, who imparts ritual practices using amulets, tokens, and charms. In divination, natural objects such as bones, feathers, and crystals are scattered to reveal a disease's cause and treatment.

"We return thanks to all **herbs,** which **furnish medicines** to cure our diseases."

TRADITIONAL IROQUOIS NATIVE AMERICAN OFFERING

Healers of the Maya civilization, in pre-Columbian Central America, were known as *ah'men*. They spent much time discussing their patients' personal life, habits, and worries—a practice that might today be termed psychotherapy or counseling.

Medicinal herbs

Herbs are still used in many cultures for medicinal purposes. In West and Central Africa, the root bark of the shrub iboga is used as both a stimulant in low doses and as a hallucinogen in larger quantities. The South African herb buchu is valued for its essential oils and as a traditional remedy for a number of digestive and urinary problems. In North America, smoking tobacco in a medicine pipe is a central part of prayer and healing ceremonies, and the shaman could choose from many other traditional herbal remedies.

Historically, the Aztec peoples also had a vast herbal medicine chest and they too believed that ill health was handed down from gods and spirits. One of their most important medicines was pulque or octli, an alcoholic drink fermented from maguey, a succulent plant. In South America, meanwhile, the herb ipecacuanha was used as an emetic, and the leaves of the coca plant were chewed as a stimulant—the source of cocaine, a drug that is globally much misused today.

▽ **Earliest herbal medicine**
Research has shown that the Neanderthals in El Sidrón had a gene that enabled them to taste bitter substances. This suggests that plants such as yarrow and chamomile were selected for reasons other than taste, such as medication.

Early Surgery

The first uses of surgery are unknown, but Stone Age scrapers and blades were certainly sharp enough to slice through flesh, and were perhaps used to remove growths. The earliest clear evidence of invasive surgery is trepanning—chipping or boring through the skull bones to the brain.

Trepanning, or trephining, involved making openings in the braincase, usually on the forehead or the top of the head. It may have been performed by early peoples for religious, ritual, or therapeutic purposes. In one large-scale survey of Neolithic skeletons—some dating back more than 7,000 years—about one in 10 skulls featured full openings or signs of attempts to make them. In these earliest examples, the holes had jagged, untidy edges from cutting with stone blades and scrapers, or perhaps chisel-shaped implements hit with a hammerstone. Hole shapes provide evidence that teeth from big cats and other predators were also used. In some cases, a circle of bone was chipped away and the freed part lifted out, perhaps to be kept as a memento.

A global phenomenon

Many surgeons in ancient Egypt, Greece, Rome, West Asia, and China were familiar with trepanning and wrote treatises on the subject. Evidence of its practice in Kashmir, India, was found in a 4,000-year-old skull with multiple trepanned holes. In China, the 2,000-year-old

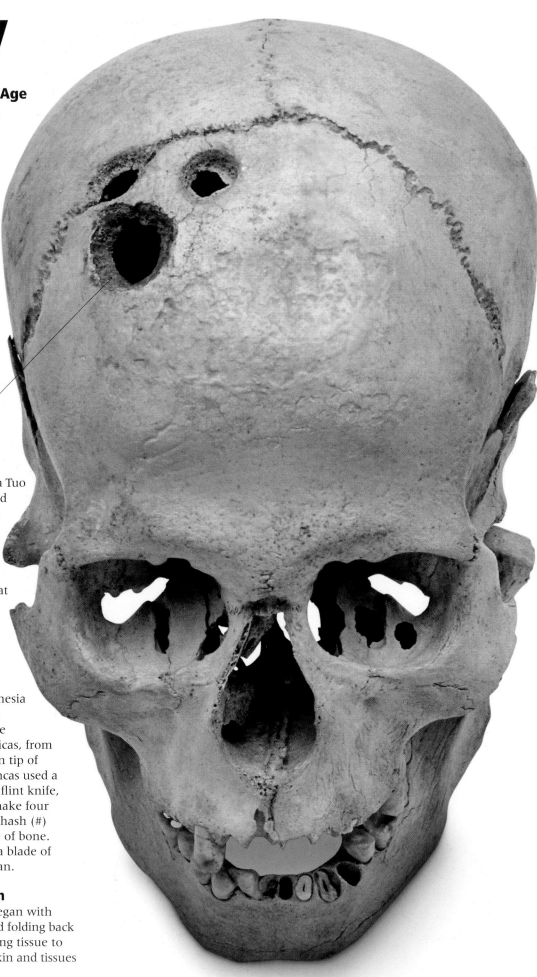

▷ **Multiple openings**
Dated to around 4,000 years ago, this multiple-trepanned skull was unearthed at Jericho (in modern-day Israel). The neat, circular holes of different sizes indicate that several drills were used.

Bone growth suggests healing

records of surgeon Hua Tuo show how he proposed to cure the headaches of teenage emperor Shao by "opening the skull"; the offer was declined.

There is evidence that by the 17th century, trepanning was being carried out on almost every continent, including in remote locations such as the Pacific islands of Polynesia and Melanesia. It was practiced widely in the pre-Columbian Americas, from Alaska to the southern tip of South America. The Incas used a ceremonial copper or flint knife, known as a *tumi*, to make four straight incisions in a hash (#) shape to free a square of bone. The Aztecs preferred a blade of the glassy rock obsidian.

Accessing the brain

Trepanning usually began with cutting, loosening, and folding back the skin and underlying tissue to reveal the skull; the skin and tissues

△ **Stone Age trepanning drills**
While the bottom two of these replica Neolithic drills are tipped with flint, the top example features a shark's tooth. The shafts were probably spun between the palms.

could be put back in place afterward. An opening was then made in the skull to reveal the brain's membranes, and in some cases the cortex, or gray surface layer of the brain. Some accounts describe patients being heavily intoxicated with alcohol, or given

52

The number of times French surgeon Jean-Jacques Bouestard trepanned one patient over a period of two months in the mid-18th century.

herbal or fungal sedatives and natural analgesics during surgery, but many were unanesthetized. Despite the high risk of infection, signs of bone healing after the operation indicate that many patients survived the procedure.

Tools of the trade

Mechanical trepans with drill-type rotation were being used in Europe by the medieval period. The string of a bow was wrapped several times around a metal- or stone-pointed stick, allowing the stick to be spun to and fro by the bow's sawing motion. In the late 1570s metal-geared woodworking drills were adapted to turn a variety of hard bits and burrs to give a round, neat-edged hole. However, this involved holding the trepan in one hand and turning it with the other—and it was hard to keep it steady. To stabilize the spinning mechanism, special frames that could be attached to the head were devised. The 1600s saw more developments, such as hand-cranked or clockwork-spring adaptations, small circular saws that could be turned around a central axle, and

"When an **indentation by a weapon** takes place in a **bone...** attended with fracture and contusion... it requires **trepanning.**"

HIPPOCRATES, FROM *ON THE INJURIES OF THE HEAD*, 4TH CENTURY BCE

△ **Painful procedure**
In the 17th-century painting *A Surgical Operation on a Man's Head*, Flemish artist David Teniers the Younger depicts a barber-surgeon, with female assistant, performing trepanning with a slim knife.

a saw-edged hole-cutter that freed a well-trimmed disk of bone. Another method involved boring a circle of small, closely spaced holes, then chiseling away the bone between them to free the middle section.

A radical solution

This painful and risky procedure may have been performed to address medical conditions that had no apparent external cause, such as severe headaches and migraines, epileptic seizures, encephalitis (inflammation of the brain tissue), and brain tumors and hemorrhage. Trepanning was also performed to treat deep wounds, and to heal skull bones fractured, depressed, or splintered in accidents or on the battlefield. The 16th-century French wartime barber-surgeon Ambroise Paré (see pp.78–79) described several trepanning techniques, and designed his own equipment.

In early South American cultures, trepanning may have been used in an attempt to revitalize someone

who had died—perhaps a powerful chief—by allowing a new, reviving life force to enter the head.

Trepanning was also used in medieval European cultures as a cure for mental conditions, such as paranoia, depression, and bipolar disorder—believed to be caused by demonic possession. A hole in the skull was thought to provide a much-needed exit for the demon during exorcism. The removed bone fragment could then act as an amulet or charm, worn by its owner to keep the demon at bay.

Trepanning began to fade from Western medicine in the 18th century. The growth of specialized treatments for conditions such as epilepsy and migraine, especially the development of new medicinal drugs, led to its decline as a surgical treatment. However, the procedure has its equivalent in modern surgery, where precision instruments and power drills are used to access brain tissue for a variety of conditions.

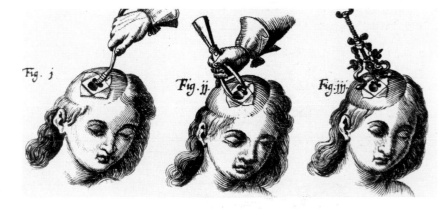

◁ **Trepanning instruments**
Three different types of trepanning instruments are being used in this 17th-century illustration, including a spinning mechanism stabilized on the patient's head by a frame with four legs.

1 TIBETAN TOOTH NECKLACE

2 AFRICAN HEALER'S NECKLACE

3 CONGOLESE HEALING DOLL

Teeth set in metal

4 INUIT SÉANCE CARVING

5 TANZANIAN DIVINING BOWL

6 TIBETAN RHINO HORN REPOSITORY

7 ZAMBIAN DIVINING BONES

Shamanism

The tradition of the shaman, who reaches into the unseen realm of spirits and souls to help and heal, is known in almost every part of the world (see pp.14–15). Shamans use a variety of objects, such as amulets and masks, to engage and direct their powers.

1 **Tibetan tooth necklace** Comprising many small teeth, this necklace is said to protect against evil spirits. **2** **African healer's necklace** The charms on this necklace include teeth, shells, claws, seeds, and a bird skull. **3** **Congolese healing doll** A Nte'va figurine made of wood, nuts, leather, bone, and cloth, this doll is used to ward off illness. **4** **Inuit séance carving** This carving depicts a shaman in a trance, with two fantasy helpers at hand. **5** **Tanzanian divining bowl** Items such as stones, bones, and teeth were swirled in the bowl. The position where they stopped were thought to reveal the answer to a particular question. **6** **Tibetan rhino horn repository** Despite being proven false, legends about the medicinal properties of rhino horns persist. **7** **Zambian divining bones** A shaman would throw these carved, fish-shaped bones onto a mat or into a bowl and interpret their arrangement. **8** **Sri Lankan exorcism mask** This mask of fearsome deity Maha Kola was used to scare demons from the body. **9** **Native American mask** Iroquois

shamans wore a mask with a broken nose, representing legendary healer Hado'ih. **10** **Native American fan** Plains Indians regarded the eagle as the most sacred bird. Its healing powers could be transferred by cooling the patient with a fan made of the bird's feathers. **11** **Tibetan headdress** The fiery skull on top of the headdress was said to frighten away evil. **12** **Malaysian shaman jacket** This garment is made of pangolin skin. Products derived from this scaly anteater are much used in traditional medicine. **13** **Native American soul-catcher** Amulets such as these were believed to retrieve an ill person's wandering soul. **14** **Mongolian decorated drum** This was used to create insistent rhythms to summon gods and spirits. **15** **Tlingit oystercatcher rattle** Tlingit shamans from coastal northwest North America carved their rattles to resemble birds, in this case an oystercatcher.

Heads of "spirits of affliction"

8 SRI LANKAN EXORCISM MASK

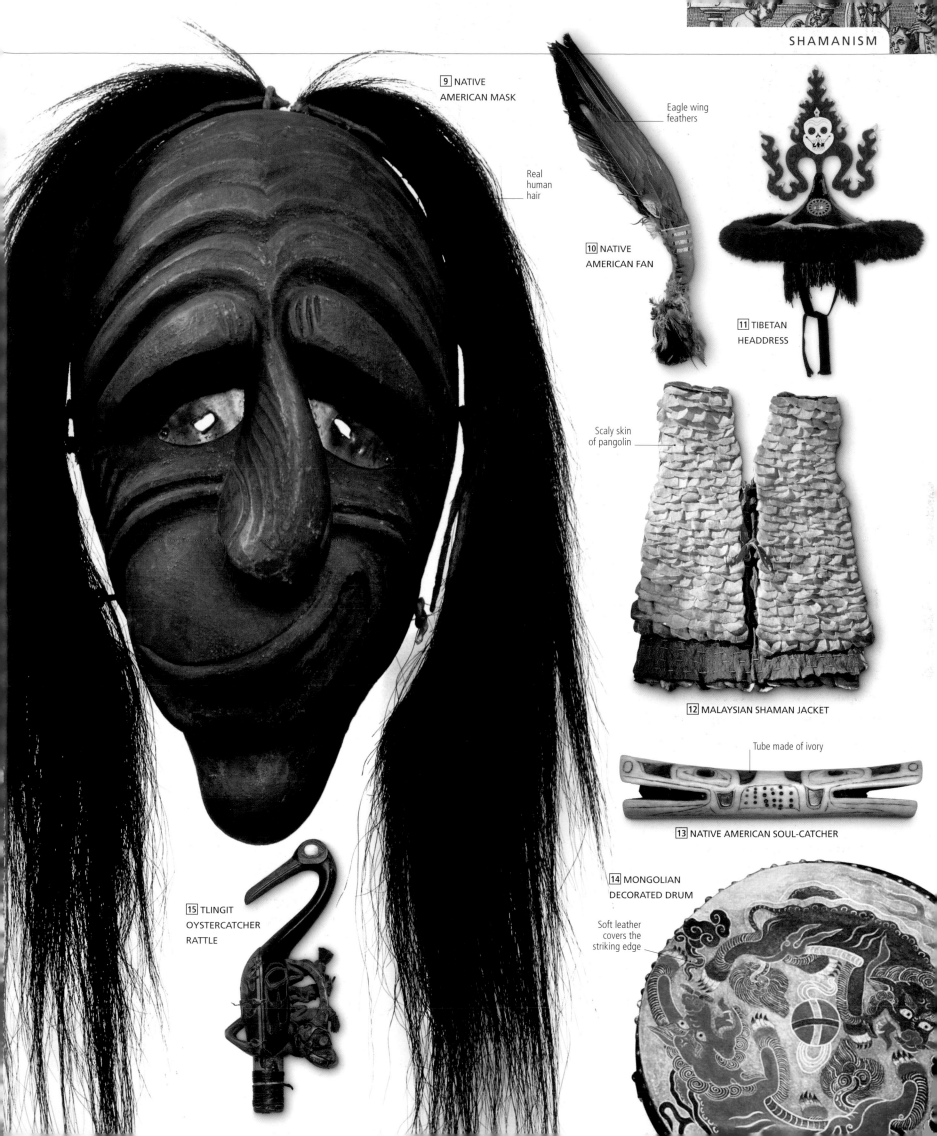

9 NATIVE
AMERICAN MASK

Real
human
hair

Eagle wing
feathers

10 NATIVE
AMERICAN FAN

11 TIBETAN
HEADDRESS

Scaly skin
of pangolin

12 MALAYSIAN SHAMAN JACKET

Tube made of ivory

13 NATIVE AMERICAN SOUL-CATCHER

14 MONGOLIAN
DECORATED DRUM

Soft leather
covers the
striking edge

15 TLINGIT
OYSTERCATCHER
RATTLE

Medicine in Ancient Egypt

For the ancient Egyptians, medicine and healing were inseparable from religious worship. Their physicians wrote manuals on human ailments and shared some surgical knowledge, but their treatments often revolved around magic, spells, and prayers to the gods.

The foremost figure in Egyptian medicine was Imhotep. Leader of a powerful cult of priest-physicians, he was active around 2630 BCE, during the early period of what is known as the Old Kingdom. Imhotep's origins are obscure, but he was probably an ordinary citizen rather than of royal descent. However, his fame grew so rapidly that even during his lifetime he came to be regarded as a god, believed to be the son of Sekhmet (goddess of healing) and Ptah (creator of the universe).

As a result of Imhotep's rapid deification, it is difficult to tell whether records of his life and achievements are factual or mythical. He may have been a practicing healer, dispensing herbs and potions to patients, but it is more likely that he was in charge of a team of physicians and took credit for their successes. His other roles included chancellor to the pharaoh, pyramid architect, and high priest to the sun god Ra. Even as Egypt's civilization faded some 2,300 years ago, Imhotep continued to

be worshipped, and in ancient Greece he became associated with Asclepios, the Greek god of healing (see pp.32–33).

Channels of the body

Influenced by Imhotep, other Egyptian priest-physicians worked toward developing theories of disease. They drew comparisons with the irrigation waterways dug between the Nile and crop fields, and conceived a system of up to 46 channels in the body, mostly emanating from the heart. They had only a vague knowledge of anatomy and may have viewed the arteries, veins, and intestines—and, possibly, tendons and nerves—as channels of the body. They believed that "flow" through the channels was important for good health, and that the body's channels could become blocked by evil spirits, which would cause sickness. Their remedy was to unblock these conduits by using various purges, laxatives, and emetics, and offering prayers and gifts to relevant gods to remove the root cause. The Channel Theory was an important turning point in medicine. Although it had a metaphysical basis, it was among the first attempts to link illness with the body's processes, and it resulted in the development of treatments that focused on the body rather than simply pacifying the spirits.

▷ **Lion-headed goddess**
Sekhmet ("powerful one") was the ancient Egyptian goddess of medicine and healing. Also the warrior goddess and a solar deity, she was usually depicted with the head of a lioness and a sun disk and cobra crown.

▷ **Mummy pathology**
Studies of mummies show that the average age of death in ancient Egypt was 40. Major causes included infectious and parasitic diseases, bacterial infections, and atherosclerosis leading to heart failure.

Medical papyri

Much knowledge of ancient Egyptian medicine comes from preserved papyrus documents. The most important of these are the Kahun papyrus—the earliest (c.1800 BCE), also known as the gynecological papyrus—and the Edwin Smith, Ebers, Hearst, Erman, London, Brugsch, and Chester Beatty papyri.

The papyri are generally named after the person who procured, financed, or translated them, or the place where they were stored. None can be ascribed to a particular physician, and many appear to be rewrites or updates of earlier versions. The longest of them is the Ebers papyrus (c.1550 BCE), which lists hundreds of magical chants and spells against bad

"Bandage him with alum and **treat him afterward [with] honey** every day until he gets well."

TREATMENT FOR A DISLOCATED RIB, FROM THE EDWIN SMITH PAPYRUS, c.1600 BCE

▷ **Edwin Smith papyrus**
The world's oldest surviving surgical text, the Edwin Smith papyrus was written in Egyptian hieratic script around the 17th century BCE. It is likely that the material was adapted from a series of earlier documents going back more than 4,000 years.

spirits, as well as mineral and herbal remedies. It describes a range of ailments too, including parasitic diseases, bowel disease, ulcers, urinary difficulties, female disorders, skin rashes, and eye and ear problems.

A more methodical approach

Dating back to around 1600 BCE, the Edwin Smith papyrus is much more systematic and explanatory—closer in approach to a modern medical text. It covers a total of 48 typical "case histories." The cases generally start at the head and work down the body, and each progresses in a logical manner, with a title and notes on examination, diagnosis, prognosis (prediction), and treatment.

> ## "[The heart] speaks at the tips of the vessels in all body parts."

ON THE HEART AND VESSELS, FROM THE EBERS PAPYRUS, 1550 BCE

For example: "Instructions for a split in his cheek. If you examine a man having a split cheek and you find that there is a swelling, raised and red, on the outside of his split. You shall say concerning him: One having a split in his cheek. An ailment which I will treat. You should bandage it with fresh meat on the first day. His treatment is sitting until his swelling is reduced. Afterward you should treat it (with) grease, honey, and a pad every day until he is well." Raw meat was believed to stop bleeding, and honey to counter infection.

The Edwin Smith papyrus was probably a teaching document. It covers mainly wounds, general

trauma, bone-setting, and minor surgery, which suggests that it may have been used by physicians tending to soldiers wounded in battle. Although examining a patient to make a diagnosis is an essential part of medical practice today, this method was new in ancient Egypt. More often, bad spirits were blamed for the ailment,

and treatment involved offerings and chants. Unusually for its time, the Edwin Smith papyrus focuses on practical advice not magic.

Surgical procedures

Evidence suggests that surgical operations in ancient Egypt were performed on the outside of the body only, and that truly invasive

procedures that involved cutting open the body were unheard of, except after death for purposes of mummification (see pp.22–23). One exception was trepanning (drilling or scraping a hole in the skull), which was probably performed to treat cranial trauma, migraine, epilepsy, and mental disorders, and to expel evil spirits.

▷ **Ancient surgical instruments**
Dating back to c.100 BCE, this relief from a temple in Kom Ombo, Egypt, shows a range of medical and surgical instruments including forceps, scalpels, and saws. The temple was used as a sanitorium in ancient times.

Secrets of Mummies

The study of Egyptian mummies today uses some of the most modern technology, such as medical imaging, when examining one of the most ancient methods of body preservation. Scans reveal details of health issues that afflicted even the most powerful people in ancient Egypt, from broken bones to gut worms and kidney tuberculosis.

The oldest Egyptian mummies date back about 5,000 years. They were preserved using a mix of sodium salts, substances containing elements such as arsenic and mercury—to dehydrate the body and prevent decay—and aromatic oils and resin. They were then wrapped in linen strips. These mummified remains preserve anatomical details in both their hard and soft tissues.

Current technologies such as X-rays and CT scans offer a way of studying some of the medical problems that afflicted the ancient Egyptians without disturbing their remains. Parasites such as tapeworms, roundworms, and the worms that cause elephantiasis—a disease that involves extreme enlargement of the legs or scrotum—have been detected in mummies. Dental decay, sinus infections, malaria, and tuberculosis also appear to have been prevalent. Dozens of mummies show atherosclerosis—the narrowing and hardening of the arteries due to the buildup of fatty deposits. Dispelling the idea that this is a modern disease resulting from a rich diet, in ancient Egypt it may have been caused by inherited factors in noble families, accompanied by long-term infection and parasites.

" Absence of malignancies in mummies…indicates **cancer-causing factors** are limited to…**modern** industrialization.**"**
PROFESSOR MICHAEL ZIMMERMAN, MANCHESTER UNIVERSITY, 2012

▷ CT scan of an Egyptian mummy
The 2,800-year-old coffin and mummified body of Egyptian priest Nesperennub were scanned at University College Hospital, London, in 2007. The 1,500 scans of the mummy reveal details of his age, lifestyle, and health, and how he was mummified.

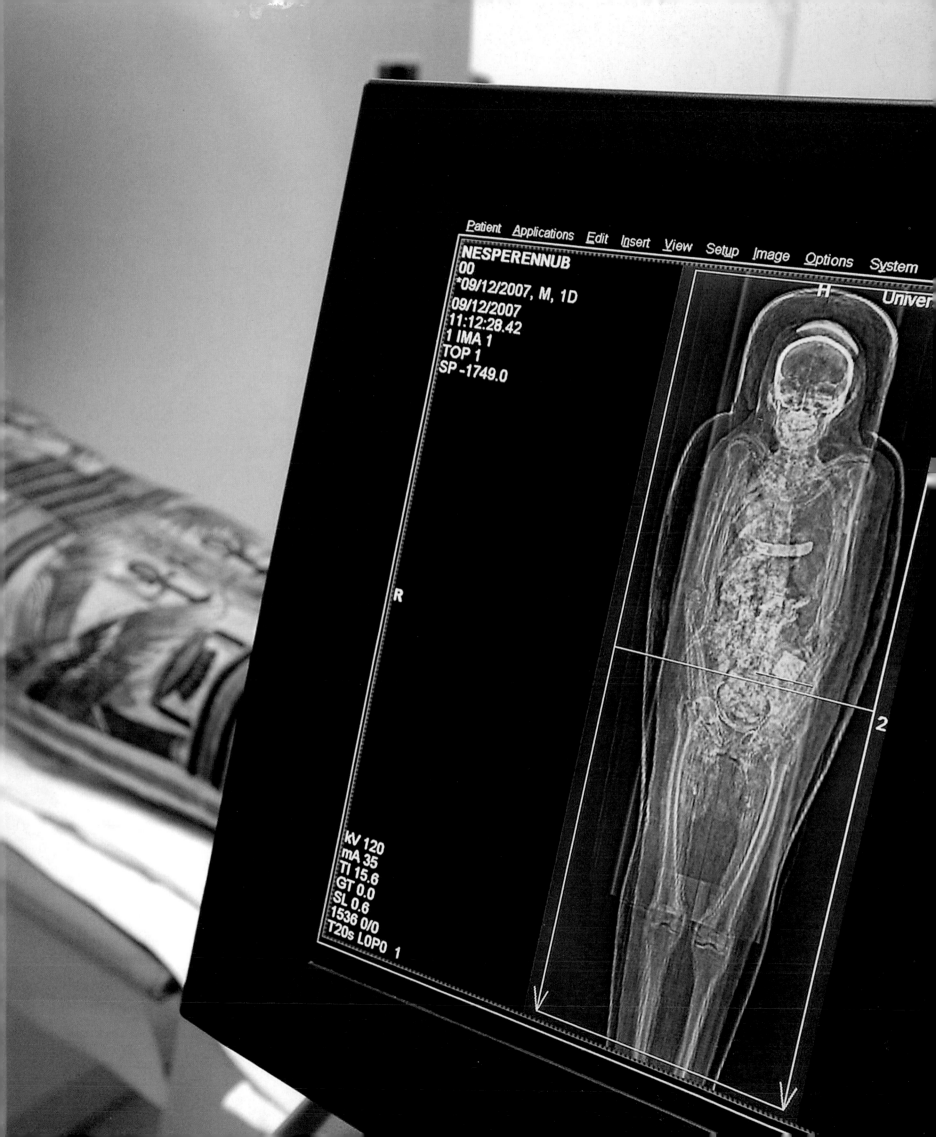

Medicine in Ancient Mesopotamia

Although healing practices in ancient Mesopotamia (roughly centered on modern-day Iraq) involved the use of magic, incantations, and divination, physicians had an extensive knowledge of diagnosis, a wide repertoire of drug treatments, and carried out basic surgery. They were also bound by a well-established, formal code of conduct.

△ **Symbol of Gula**
The goddess Gula, or "the lady of health," was the most important of the gods who had an influence on medical affairs. Her symbol was the dog, and canine figurines have been found at her cult temples in several Mesopotamian cities such as Isin, Nippur, Umma, and Babylon.

The first medical texts from Mesopotamia survive in the form of clay tablets that date back to c.2400 BCE. These give recipes for medicines, but the diseases for which these were intended as treatments are unclear. A much larger selection of diagnostic tablets from the library of the Assyrian King Ashurbanipal—who ruled in the mid-7th century BCE—gives a clearer impression of Mesopotamian medical practice.

The Mesopotamians believed that diseases were caused by a particular god or demon, so a person with venereal disease, for example, might be referred to as struck by "the hand of Lilith," a female demon. The primary job of a doctor was to chase out the disease-causing demon from the patient; the treatment of the symptoms was considered a secondary task. There were three types of doctor: the *masmassû*, or exorcist, who conducted rituals and made incantations to purify the patient; the *barû*, or diviner, who made predictions about the course of the illness, mainly through heptoscopy (reading the livers of sheep); and the *asû*, or physician, who made more conventional diagnoses and prescribed remedies.

Medical preparations

Mesopotamian physicians used around 250 medicinal plants, 120 minerals, and about 200 other substances. Some of the ingredients, such as mandragora, henbane, linseed, myrrh, and belladonna, were used by later physicians, while other more exotic ones, such as crushed gecko and raven's blood, soon fell out of use. Remedies were prescribed for specific diseases: for instance, fish oil and an extract of cedar were thought to treat epilepsy.
  Doctors were skilled in the treatment of wounds, applying bandaged poultices of sesame oil or honey and alcohol to prevent infection. They had a wide knowledge of the external symptoms of diseases, and were able to give accurate descriptions of afflictions, such as epilepsy and tuberculosis. They were also aware that some diseases spread by contagion, and they practiced a form of quarantine to prevent the spread of fevers.

Doctors in Mesopotamia could also perform surgery; a set of bronze needles meant for cataract operations dating from around 2000 BCE has been found, and an account survives of a surgeon cutting open the chest of a patient to drain pus from the lungs.

10 SHEKELS The fee paid to a doctor in Babylonia for performing successful surgery (with a scalpel) on an upper class patient—equivalent to more than a year's pay for the average tradesman.

Knowledge of anatomy, however, was limited, since human dissections were not carried out in the region.

Strict laws

The medical profession was strictly regulated by law, and the Law Code of Hammurabi, dating from around 1750 BCE, contains several clauses relating to doctors. They were paid a set fee: for example, a doctor was paid five shekels of silver for mending a broken bone (although this was reduced to three shekels if the patient was a commoner, and only two if the patient was a slave). Meanwhile, penalties for medical malpractice were severe: if a doctor caused a patient's death, the doctor's hand would be cut off.

◁ **Stele of Hammurabi**
Hammurabi, the ruler of Babylon in the 18th century BCE, is seen here receiving his law code from the sun god Shamash. The text contains more than 280 clauses, of which about a dozen deal with the regulation of the medical profession.

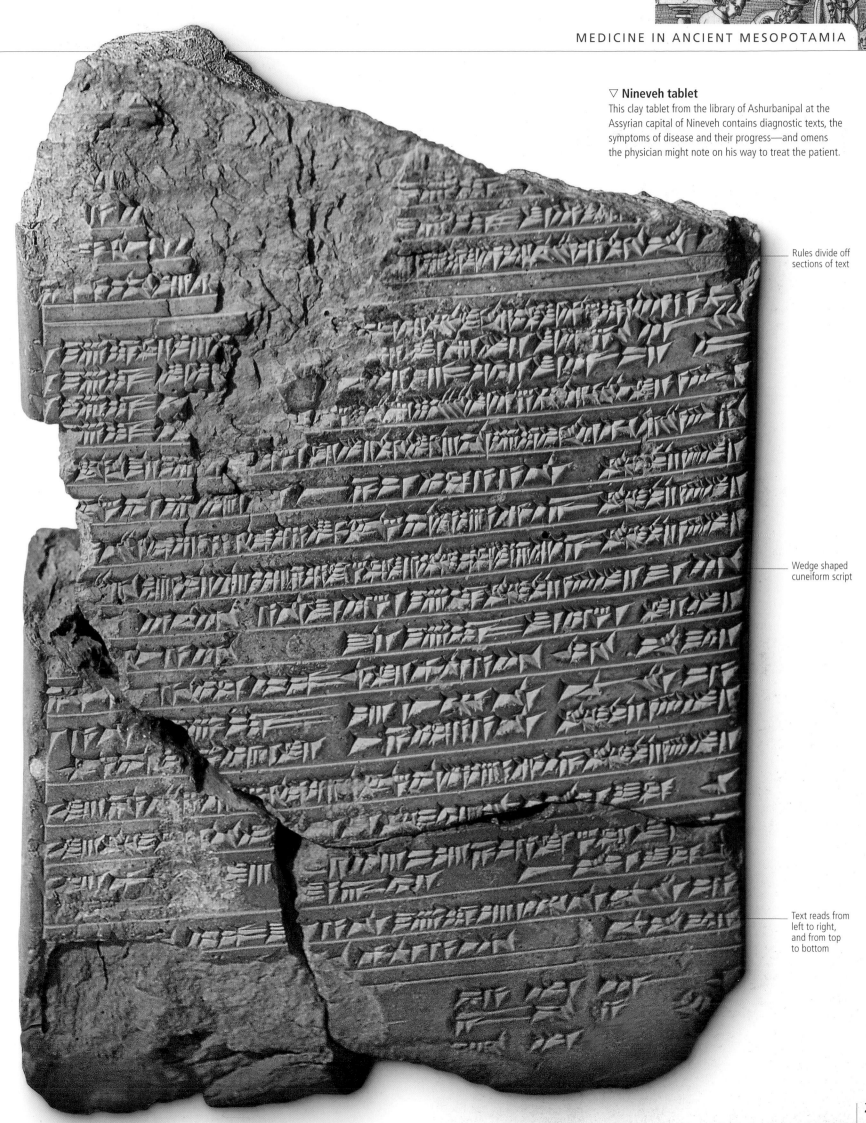

▽ **Nineveh tablet**
This clay tablet from the library of Ashurbanipal at the
Assyrian capital of Nineveh contains diagnostic texts, the
symptoms of disease and their progress—and omens
the physician might note on his way to treat the patient.

Rules divide off
sections of text

Wedge shaped
cuneiform script

Text reads from
left to right,
and from top
to bottom

Early Chinese Medicine

FIRE

WOOD

EARTH

WATER

METAL

The prime source of knowledge about early Chinese medicine is the 2,000-year-old _Huangdi Neijing_ (_Yellow Emperor's Classic of Internal Medicine_). Although it has been revised through the ages, it still remains central to traditional procedures and practices today.

The _Huangdi Neijing_, an ancient Chinese medical text, takes the format of question and answer discussions between the semi-mythical Yellow Emperor, Huang-di, and his advisors. Huang-di asks a question, which in turn is answered by his ministers, and through this process they cover an encyclopedic range of contemporary Chinese medical knowledge and practice. The work describes key traditional Chinese concepts such as yin-yang, zang-fu, the five phases, and the flow of _qi_ or "life energy" along channels known as meridians (see pp.28–29). It includes diagnostic procedures such as feeling the pulse, observing the tongue, and examining human excrement, as well as a range of treatments, including herbal and mineral concoctions, massage, special diets, bathing, meditation, and forms of physical exercise and ritualized movements.

▽ **Qigong massage**
One of the oldest and most adaptable therapeutics, Qigong focuses on relaxation, meditation, body postures, measured movements, and deep breathing techniques.

The concept of yin-yang has permeated Chinese philosophy, culture, and medicine for millennia. It represents the inherent duality—opposite yet complementary—in the universe. Yin is described as dark, watery, cool, passive, and feminine, while yang is bright, dry, hot, active, and masculine—and each cannot exist without the other.

Zang-fu is a system of assigning body parts as either yin or yang. The lungs, heart, liver, spleen, and kidneys are zang organs (and are assigned as yin); the stomach, intestines, gallbladder, and urinary bladder are fu organs (and assigned as yang).

Another concept is the five phases of energy, wu-xing: earth, water, fire, wood, and metal. The _Huangdi Neijing_ records: "The five elemental energies… encompass all the myriad phenomena of nature. It is a pattern that applies equally to humans." The five phases theory also incorporates the cycles in which the five elements interrelate: _sheng_ (generating); _ke_ (controlling); _cheng_ (overactive); and _wu_ (contradictory). It is believed that yin-yang, zang-fu, the phases (elements), and cycles interact to affect the flow of _qi_ (energy). An imbalance of the _qi_ results in disease; treatments aim to restore harmony and balance.

Influential physicians
One of the best known early Chinese physicians was Zhang Zhongjing (see panel, left).

△ **Yin-yang and the five phases**
According to traditional Chinese medicine, well-being incorporates the concepts of yin-yang, zang-fu, and the five elements, or "phases." The latter term reflects the belief that these entities are not fixed but rather, like energy states, undergo continuous change.

Surgery is not prominent in the history of Chinese medicine, and one of the few surgeons to gain fame was Hua Tuo, in the late Eastern Han Dynasty (25–220 CE), who also carried out acupuncture (see pp.28–29) and other forms of healing. He is reputed to have invented an anesthetic, known as mafeisan, probably based on a mixture of wine, cannabis, opium, and several relatively toxic herbs, which he used for open surgery, especially on the bowels.

Later, in about the 6th century, Sun Simiao compiled extensive texts listing thousands of remedies. He also practiced alchemy, and placed great emphasis on gynecology, pediatrics, and medical ethics. In _Qianjin Yaofang_ (_Prescriptions Worth a Thousand Gold_) he emphasized the significance of a careful approach, impeccable morality, and dignified attitude in a physician. His doctrine spread throughout China, and can be seen as the Chinese equivalent of the Hippocratic oath (see pp.36–37).

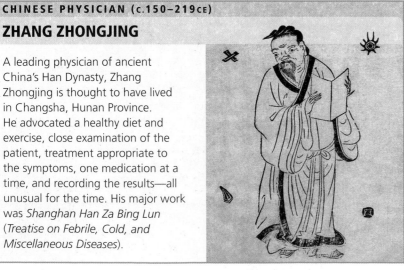

CHINESE PHYSICIAN (c.150–219CE)

ZHANG ZHONGJING

A leading physician of ancient China's Han Dynasty, Zhang Zhongjing is thought to have lived in Changsha, Hunan Province. He advocated a healthy diet and exercise, close examination of the patient, treatment appropriate to the symptoms, one medication at a time, and recording the results—all unusual for the time. His major work was _Shanghan Han Za Bing Lun_ (_Treatise on Febrile, Cold, and Miscellaneous Diseases_).

"If the authentic qi flows easily… How could illness arise?"

FROM _SUWEN_, THE FIRST PART OF THE _HUANGDI NEIJING_, 2ND–1ST CENTURY BCE

Rebalancing *qi*
This 10th-century Song Dynasty painting shows a doctor burning moxa (a powder made from the herb mugwort) on a patient's skin (a process known as moxibustion) to stimulate the acupuncture points and meridian channels, in order to rebalance the body's flow of *qi* (energy).

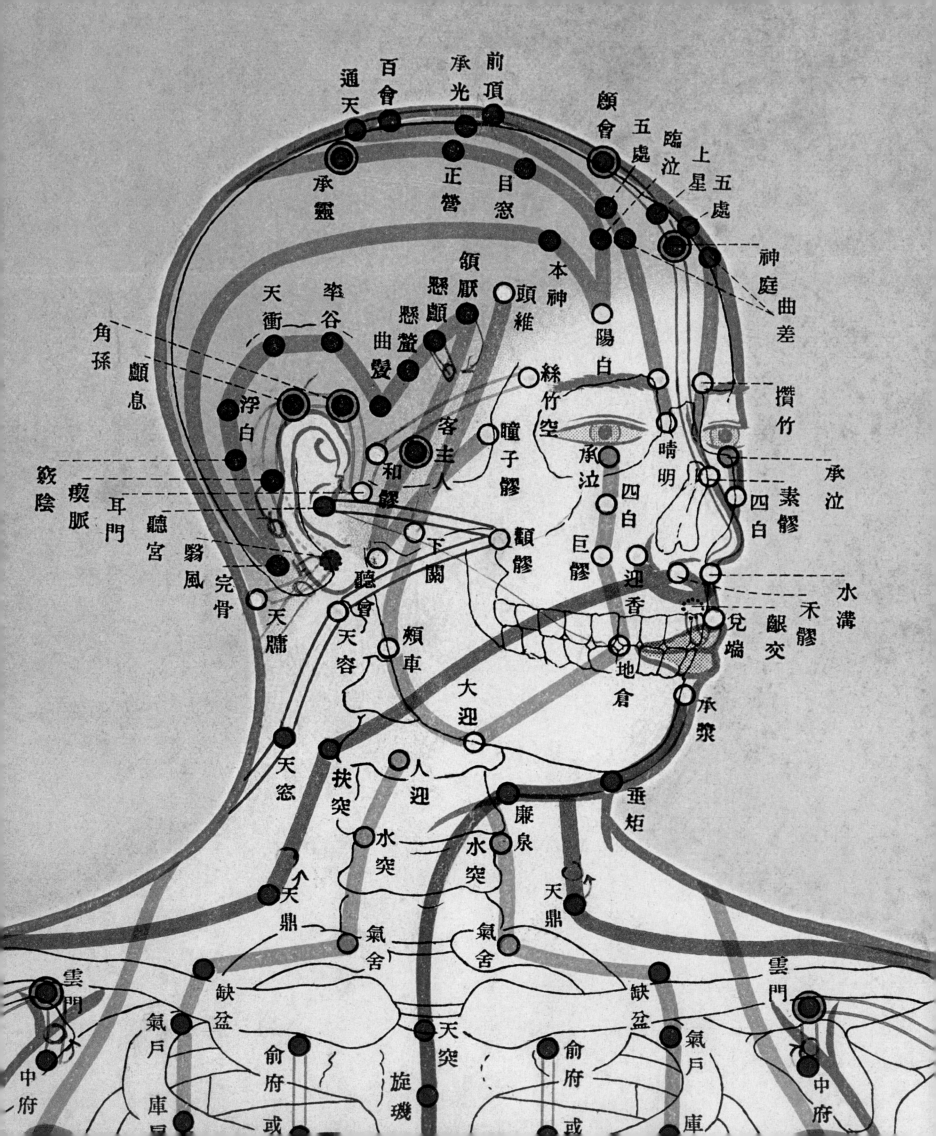

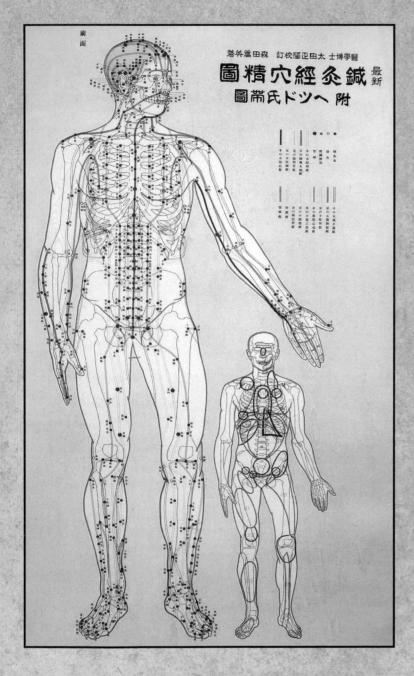

圖精穴經灸鍼 最新

森田蓮菜井咎　醫學博士田正圏校訂

附へツド氏帯圖

前面

肩髃

Acupuncture

Also known as needling, acupuncture is a traditional Chinese medical technique that has been used for perhaps four millennia. Along with moxibustion (see pp.26–27)—burning an herb called mugwort on the skin—it is one of the earliest known therapeutic systems with a logical theoretical basis.

As a method of alleviating pain, easing suffering, healing, and even curing a range of illnesses, acupuncture's origins may go back 4,000 years. The *Huangdi Neijing* (*Yellow Emperor's Classic of Internal Medicine*), a 2,100-year-old Chinese medical compendium, describes the techniques and uses of acupuncture in its second book, the *Ling Shu* (*Divine Pivot*). Widely used across East Asia in various forms, acupuncture has been found by modern Western studies to be effective in relieving certain forms of pain and discomfort.

According to traditional Chinese beliefs, health relies on a vital force, energy stream, or life flow moving through the body. Known as *qi*, this force flows along routes or channels called meridians. Problems such as pain and illness arise when someone's *qi* is disturbed. Acupuncture aims to correct the flow and restore the *qi* balance by inserting very thin needles into the skin and underlying tissues at specific sites called acupuncture points. These points may be located in parts remote from the problem area; for instance, some points for lower back pain can be found on the hand. Great skill and experience are needed when diagnosing and discerning relevant points, and when using the needles. The acupuncture points may also be stimulated by pressing (acupressure), or by using heat or strong light.

"Needling and moxa...
cure the corpse that is numb."
BIAN QUE, CHINESE PHYSICIAN, PROBABLY REFERRING TO A PERSON UNCONSCIOUS AFTER SEIZURE, 310 BCE

◁ **Acupuncture points**
This reproduction from an illustrated version of *Huangdi NeiJing* from 1000 CE shows the body's meridians and acupuncture points. The illustrated version itself derived from China's first great medical manual, of the same name, dating back 2,100 years.

Ayurveda

A traditional system for health, well-being, healing, and medicine, Ayurveda (meaning "life knowledge") has been prevalent in India and southern Asia for more than 2,000 years. It originated around the same time that the famed physician Hippocrates was developing the practice of medicine in ancient Greece.

Two major works form the basis of Ayurveda—the *Susruta Samhita* and the *Charaka Samhita*. However, both these ancient texts have been edited, reworked, and altered over the centuries, masking their original content. The *Susruta Samhita* is named after the celebrated Indian physician Susruta, who probably lived in Varanasi, India, in the 6th century BCE. The word *samhita* means a compendium, collection, or compilation. The *Susruta Samhita* contains information about *shalya chikitsa*, or Ayurvedic surgery, including a wide range of complex techniques for procedures such as tooth extraction, cyst drainage, cataract removal, repairing hernias, setting broken bones, and cauterizing hemorrhoids. It describes more than a thousand conditions and hundreds of herbal remedies.

The second work, the *Charaka Samhita*, is around 2,300 years old and is attributed to Charaka, who may have been a physician at an emperor's court. As with Susruta, the historical details of Charaka's life are unclear. The *Charaka Samhita* has more than 110 chapters divided into eight sections, and is written in verse to aid memorization. Like the teachings of Hippocrates (see pp.36–37), the treatise instructs physicians on how to examine a patient and make a diagnosis, and also recommends treatments. Most of the remedies emphasize lifestyle, hygiene, exercise, and diet, as well as herbal and mineral-based medicines.

Three further works contribute to the main body of Ayurvedic knowledge: the *Ashtanga Hridayam*, the *Ashtanga Sangraha*, and the Bower Manuscript. The *Ashtanga Hridayam* and the *Ashtanga Sangraha* date from around the 5th century CE and were written by the Indian physician and healer Vagbhata. The *Ashtanga Hridayam* has eight sections, including chapters on general surgery, internal medicine, gynecology, pediatrics, mental and spiritual problems, and sexual medicine. The Bower Manuscript (named after British officer Hamilton

143 OF THE 1,323 **VERSES** in the **Bower Manuscript deal with the origin and medical uses of garlic, demonstrating its importance in Ayurvedic medicine.**

Bower who acquired it in 1890) dates from about the same time as the *Ashtanga Hridayam* and the *Ashtanga Sangraha*. It contains a group of wide-ranging medical texts, with content adapted and updated from the earlier *Susruta Samhita* and *Charaka Samhita*, along with herbal recipes.

Elements of Ayurveda

While various forms of Ayurveda have developed over the centuries in different regions, most systems are based on the concept of five elements. These elements are *jala* or *ap* (water), *tejas* or *agni* (fire), *privthi* or *bhumi* (earth), *pavana* or *vayu* (air), and *akasha* (ether or space)—similar to the concept of the four elements and four humors developed in early European medicine (see pp.34–35). In each person the proportion of these elements varies over time and

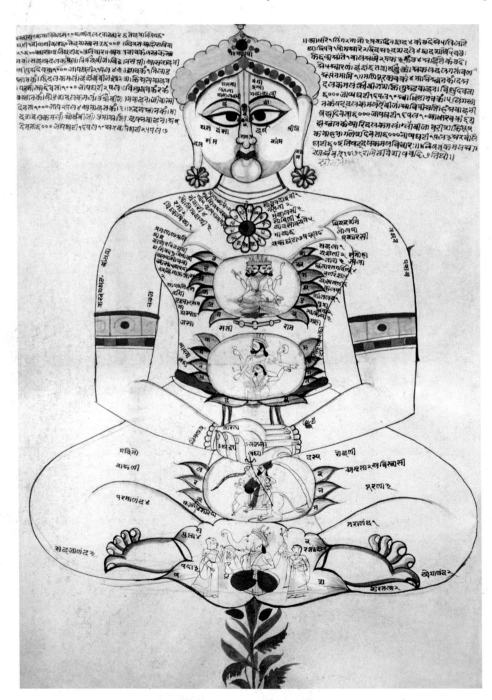

◁ **Human body *chakras***
The seven *chakras* are spinning centers of energy—part of the etheric realm—aligned along the middle of the body. If they whirl out of balance, they can upset other body systems, such as the *doshas*, and lead to illness.

AYURVEDA

IN PRACTICE

HERBAL HEALING

Ayurvedic medicine stresses the importance of preventing illness through good hygiene, exercise, and healthy diet, and of healing with natural herbal and mineral remedies. One of the main herbs used is *lahsun* or *lasuna*—garlic (*Allium sativum*)—which is viewed as a general stimulant. Different parts of the plant can be utilized for a range of ailments, including colds and coughs, digestive upsets, and skin problems such as sores, spots, bites, and stings. *Tulsi* or *thulasi*—holy basil (*Ocimum sanctum*)—is valued for its warming effect and soothes conditions caused by an excess of the *kapha* (phlegm) *dosha*, such as colds, coughs, and flu, as well as relieving bloating and indigestion.

▷ **Administering medication**
The ears are a traditional pathway or route into the body for Ayurvedic medications, which are administered as vapors, waxes, oils, and massage.

The *doshas* flow through the body along pathways and through pores known as *srotas*, rather like the meridian energy channels of acupuncture (see pp.28–29). Most Ayurvedic texts state that there are 16 *srotas*, which carry energy, nutrients, and waste, as well as learning and wisdom. Of these *srotas*, three are connected to the outside world: the *prana vaha*, which carries the *prana* (breath); *anna vaha*, which transports solid and liquid foods; and *udaka vaha*, which carries water. Another three *srotas* monitor and control the elimination of metabolic waste products: the *purisha vaha* for solid waste; *mutra vaha* for urine; and *sveda vaha* for perspiration. The *srota mano vaha* is associated with the mind and carries thoughts, ideas, feelings, and emotions. Two more *srotas* deal with menstruation (*artava vaha*), and lactation (*stanya vaha*). Seven *srotas* are linked to the Ayurvedic notion of *dhatus*—the seven tissues that make up the body. These *dhatus* are the blood (*rakta*), lymph (*rasa*), muscles (*mamsa*), bones (*asthi*), bone marrow (*majja*, which includes the brain and nerves), fat (*medas*), and reproductive organs (*shukra*). For example, the *mamsa vaha srotas* transport nutrients and waste for the *mamsa* (muscle) *dhatu*.

Another Ayurvedic concept is that of *agni*, or "digestive fire." This refers to the body's metabolism or ability to digest food efficiently, to process and assimilate learning, life experiences, and memories, and ability to prepare and burn off waste products for removal through the skin's pores, and from the mind. *Agni* can be affected by

▷ **God of Ayurveda**
Lord Dhanvantari is the god of Ayurvedic medicine, and physician to many other gods. It is believed that prayers and offerings to him help maintain health and ensure successful treatment.

the three *doshas* and other influences, such as the seven *chakras*, or "energy centers." These *chakras* are likened to spinning vortexes and are not part of the physical body but of the etheric, psychic, or "subtle" realm. While various forms of Ayurveda are common in the Indian subcontinent, Ayurvedic practice has also spread worldwide, especially among people interested in alternative and complementary therapies.

contributes to the three *doshas* (approximately corresponding to the European humors). The three *doshas* are *vata* (wind), *pitta* (bile), and *kapha* (phlegm). Good health and well-being occur when the *doshas* are well balanced. Imbalance brings unease and sickness, often related to the dominant *dosha*. For example, excessive *vata* can trigger indigestion, flatulence, and cramps. If *kapha* is dominant, it may result in problems linked to mucus and phlegm, such as lung ailments, coughing, and breathing difficulties.

2,000 herbs and mineral-based remedies are noted in the *Charaka Samhita*.

> " It is more important to **prevent** the occurrence of disease than to **seek a cure.** "

CHARAKA, INDIAN SCHOLAR, FROM *CHARAKA SAMHITA*, 1ST CENTURY CE

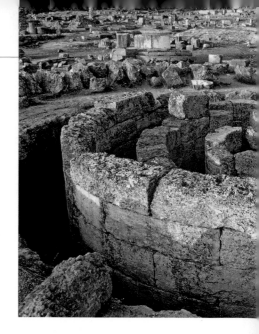

Early Greek medicine was influenced by and drew upon much from the ancient Egyptians (see pp.20–21) and their belief in the world of spirits and the supernatural. Diseases were regarded as punishments or even "gifts" from the gods, perhaps angered by sins and misdemeanors. Cures involved priests, prayers, offerings, and rituals to rid demons and lift curses. The Greek god of healing and medicine was Asclepios, and temples dedicated to him were called asclepeions. Here, the sick offered prayers and gifts to him. His sign was the Rod of Asclepios—a staff with a snake coiled around it—and this is still symbolic of medicine and the healing arts today. Although the origins of this sign are unclear, some historians trace the rod, serpent, and Asclepios himself back to Imhotep of Egypt, an architect and physician who was deified and worshipped as the Egyptian god of medicine.

Shift from mythology

As Greek medicine developed, its emphasis changed. Gradually, disease was seen more as a natural phenomenon or product of the earthly body, rather than a visitation from the gods, and symptoms, diagnosis, and treatment focused on the human,

Medicine in Ancient Greece

The most significant figure in ancient Greek medicine, and perhaps in all of medical history, is Hippocrates (see pp.36–37). However, many other physicians and healers helped to establish the Greek medical approach, procedures, and ethics that are still familiar today.

▽ **God of medicine**
In this stone-carved scene Asclepios treats a female patient. Women of a higher status had relatively good access to medicine.

△ **Sanctuary of Asclepios**
Temples devoted to the Greek god of medicine were places of refuge, rest, prayer, and healing. Asclepios was said to be born at Epidaurus, where his most famous temple—now a UNESCO World Heritage Site—was built in the 4th century BCE.

rather than on the supernatural and spiritual. This began a more scientific approach by which the physician made observations of the patient, recorded evidence, and assessed results.

Philosophers and thinkers such as Socrates, Plato, and Aristotle greatly contributed to the evolution of Greek medicine. Even before Socrates, Empedocles formulated the notion of the four classical roots or elements: air, fire, water, and earth. These were incorporated into Greek medicine as the four humors—blood, yellow bile, black bile, and phlegm (see pp.34–35). It was suggested by Greek thinkers that an imbalance in the humors caused illness. The concept of humorism was developed through the Classical Greek era (480–323 BCE) and was mentioned in the Hippocratic Corpus—a body of knowledge and written works

sometimes attributed to Hippocrates, but which was more likely compiled and built upon by his followers.

Developing theories
A century after Hippocrates, Greek physician Herophilus of Chalcedon worked in Alexandria, Egypt. He is often regarded as the first true anatomist because he dissected and studied human bodies. His writings were later taken up in Rome by the physician Claudius Galen (see pp.40–41) and others.

Herophilus made the first accurate descriptions of the brain, nerves, eye, arteries and veins, and digestive organs. His suggestion that conscious rational thought and intellect were based in the brain, not in the heart, were controversial at the time.

Herophilus worked with Greek physician Erasistratus of Ceos. Erasistratus is often seen as the first physiologist—he studied how the body works, or functions, and researched the brain, heart, and blood vessels. Like Herophilus, he believed the heart was not the center of thoughts, feelings, and emotions, but was a kind of pump with flaps, which could act as valves. Erasistratus suggested that air entered the body through the lungs, and went to the heart where it was transformed and distributed as a mysterious "animal spirit," or "pneuma," by the arteries. Veins carried blood, from the heart to the various organs. These early ideas on circulation were later

| **40–50** | **YEARS** The average lifespan of humans in ancient Greece. |

⊲ **Common treatment**
Greek physicians popularized blood-letting, or bleeding, as a treatment for many ailments. This was based on the concept that imbalance in the four humors caused illness. If the blood humor became too plentiful and dominating, it had to be removed.

extended by Galen and persisted until William Harvey accurately described circulation in 1628 (see pp.82–83).

As the Greek civilization faded and the Roman Empire expanded, many Greek physicians moved to the new regime. One of the best known was Asclepiades of Bithyni, in part because of his criticism of some classical Greek medical theories, including humorism, and the rational, observational, evidence-based

approach of Hippocrates. Asclepiades conceived a new theory of disease according to which tiny atoms, or corpuscles, moved around the body through minute holes or pores. Disturbances in the flow—caused, he thought, by pores too small, or atoms too numerous—led to ill health. His mainstays of treatment were exercise, massage, bathing, and diet, with few herbal potions. Despite his confidence, the theories of Asclepiades made little impact, and Roman physicians built on the mainstream aspects of Greek medicine (see pp.38–39).

▽ **Herophilus and Erasistratus**
These two eminent physicians were colleagues at Alexandria, Egypt, in the 3rd century BCE. Unusually for the time, relaxed regulations in the city allowed them to dissect human corpses. This led to the production of some of the earliest, realistic anatomical descriptions.

"When **health is absent,** wisdom cannot reveal itself, **wealth is useless,** and reason is powerless."

HEROPHILUS, GREEK PHYSICIAN, 3RD CENTURY BCE

The **Four Humors**

With its origins in ancient Greece, the concept of humorism is based on the balance of four humors (body fluids) in the human body—blood, yellow bile, black bile, and phlegm. This leading medical system thrived in Europe for more than two millennia before it began to lose prominence in the 18th century.

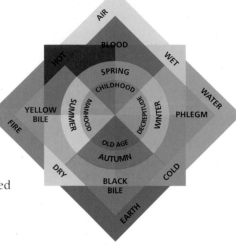

△ **The quartets**
This diagram shows the link between bodily humors and other quartet systems. For example, blood is associated with heat and wetness, spring, and childhood.

The theory of four humors with wide-reaching effects on the body and temperament was considered to be a well-rounded, wide-ranging, and highly integrated approach that offered insights into well-being and sickness. The concept fitted harmoniously with other foursomes in Greek scientific philosophy, such as the four elements (air, fire, earth, and water); the four attributes of matter (hot, cold, moist, and dry); and the four seasons (spring, summer, fall, and winter).

In the writings of Hippocrates (see pp.36–37) and his followers, the four stages of life were linked to the seasons, and four temperaments, or personality types,

24 **OUNCES (0.7 liters) of blood was let over four days when Charles II of England fell ill in 1685. He died shortly after.**

emerge from the humors. In Roman times, Roman physician Claudius Galen (see pp.40–41) formalized the system and added two further variables, namely

hot–cold and wet–dry, and four major organs that were each linked to a humor.

According to Galen, ideal temperament and health were the result of a balance of all four humors. This equilibrium was different for each individual, which is why people varied in their levels of health, fitness, personality, and susceptibility to ailments.

Personality and health

The humor of blood was associated with the heart, and an excess produced the sanguine temperament—social, optimistic, energetic, and easy-going. Blood was also linked with air, heat/wetness, and the spring season. Yellow bile was connected to the liver, and those with a choleric personality were deemed to be strong-willed, decisive, independent, and quick-tempered. Yellow bile was grouped with fire, heat/dryness, and the summer season. Black bile was allied to the spleen, an excess producing melancholic tendencies—quiet, private, cautious, and logical individuals. The humor was related to the earth, dryness/cold, and fall. Phlegm was associated with the brain: the phlegmatic person was calm, accepting, and slow to anger. Phlegm was grouped with water, cold/wetness, and winter.

When one humor became too strong, it was the likely cause of sickness. For example, an excess of phlegm caused illness

characterized by chills, shivering, coughs, and sneezes, which served to expel phlegm, mucus, and pus. Imbalances could also affect temperament: too much blood humor could lead to abandoning tasks, being forgetful and late, while an excess of yellow bile might cause over-assertiveness, disorganization, and depression. Surplus black bile could bring on worry, anxiety, and withdrawal. Signs of excessive phlegm might be laziness, carelessness, and fear of change.

The causes of humoral imbalance were numerous and ranged from stale vapor in the air and contaminated food and water, to offending the spirits, or a surfeit of emotions such as jealousy.

△ **Cupping vessel**
Dating back to 79 CE, this vessel from Pompeii, Italy, was used to restore humoral balance. The air inside was heated, and the cup placed on the skin to produce a vacuum to draw yellow bile to the surface.

△ **Public blood-letting**
An illustrated version of *Al Maqamat*, by Arab poet and scholar Ibn Ali al-Hariri, depicts a crowd watching the blood-letting of a patient in 13th-century Iraq.

Spread and decline

The principles of humorism, developed in Greece and Rome, made their way into Islamic medicine (see pp.48–51), were adopted by medieval practitioners, and also featured in Ayurvedic medicine in India (see pp.30–31). Renaissance physicians in Europe were drawn to Galen's teachings on humorism through new translations of his Greek texts.

Extensive tracts were written about the correct treatments to administer when the equilibrium

"[Humors] are the things that **make up the body's constitution** and cause its pains and health."

ATTRIBUTED TO POLYBUS, A FOLLOWER OF HIPPOCRATES, FROM *THE NATURE OF MAN*, 400 BCE

was disturbed. For example, the practice of blood-letting was thought to relieve an excess of the blood, a factor in many diseases. Cupping was believed to withdraw yellow bile, while emetics or purging potions removed yellow or black bile. Eccentric diets and herbs were often prescribed to damp down or restore the balance of a particular humor.

Throughout the 17th century, humorism was still practised widely in Europe; blood-letting, in particular, often had extreme consequences. From the late 18th century, it was swept away in a wave of methodical scientific research and a new understanding of human physiology that undermined its basic tenets.

▷ **Four temperaments**
This 1760s reproduction of the *Guild Book of the Barber Surgeons of York*, a 15th-century manuscript, shows the four temperaments—melancholic, sanguine, phlegmatic, and choleric—with clothes, facial expressions, and postures that contribute to the depiction of each one.

GREEK PHYSICIAN Born 460 BCE Died 370 BCE

Hippocrates

> **"** Sickness is not sent by the gods… find the cause, **we can find the cure."**

HIPPOCRATES, GREEK PHYSICIAN

One of the greatest names in the history of healing, Hippocrates elevated medicine into a respected profession with a scientific basis. He took Greek medicine and rid it of its supernatural elements, insisting on observation and accurate recording of case histories. By comparing these histories, he made the first systematic differentiation of diseases. He also set standards for doctors that are still admired and respected today.

Hippocrates was born on the island of Cos in Greece in around 460 BCE. His father was a doctor and Hippocrates learned medicine from him. He is known to have traveled widely, possibly going as far as Libya and Egypt, but very little is known about the man himself. The Hippocratic Corpus, a collection of around 60 works, some of which are ascribed to Hippocrates, marks Greek medicine as separate and distinctive from Egyptian (see pp.20–21) and Mesopotamian (see pp.24–25) medicine. However, there is no certainty that all the writings attributed to Hippocrates were actually authored by him.

△ **The Hippocratic Oath**
A professional code of conduct, the Hippocratic Oath is usually taken by all doctors and requires them to abide by ethical principles. Seen here is a medieval Greek copy of the oath.

Code of ethics
Although medical schools were flourishing in Sicily, southern Italy (see pp.54–55), and at Cyrene in North Africa, the school at Cos that Hippocrates founded became the most famous, and he came to be regarded as its greatest teacher. When entering this esteemed school, incoming students had to take an oath, now known as the Hippocratic Oath, in front of their elders and peers. The oath, with its code of ethics, set a high standard of expertise and etiquette, and established medicine as a profession that ordinary people could trust. It separated doctors from other "healers" and defined their practice.

◁ **Modernizing medicine**
This marble bust of Hippocrates celebrates him as the father of modern medicine. He turned away from divine notions of disease and healing and used observations of the patient as the basis of medical knowledge.

◁ **Ancient scene on marble**
This scene from the 4th or 5th century BCE shows a Greek physician attending to a patient. The doctor places great emphasis on the patient, using his hands to discern breathing and lung function.

surgeon and was interested in the study of orthopedics. Some of the principles found in the *Hippocratic Treatises On Fractures and On Joints* are still considered relevant today.

Ahead of time

Hippocrates believed that the body contained four basic humors (fluids)—black bile, phlegm, yellow bile, and blood (see pp.34–35). This system offered a rationale for understanding the human condition and for explaining illness. He believed that moods and disease result from an imbalance in the humors. He was probably the first physician to believe that diseases are natural occurrences and are not caused by supernatural forces or gods.

Hippocrates placed great emphasis on strengthening and building up the body's inherent resistance to disease. He prescribed diet, gymnastics, exercise, massage, hydrotherapy, and swimming in the sea. He also developed an understanding of the importance of hygiene and cleanliness, as well as that of rest and quiet.

When Hippocrates died, he was held in such high regard that it was believed that honey made from the bees living on his gravestone had special healing properties. Hippocrates put the doctor fully at the service of the patient, and his ground-breaking work has been a constant and enduring source of inspiration for doctors through the ages.

- **460 BCE** Born on the Greek island of Cos into a wealthy family. Hippocrates' schooling includes nine years of primary and two years of secondary education, during which he studies reading, writing, poetry, and music.

- **430–427 BCE** Helps fight the plague in Athens for three years. Recommends lighting fires to dry the atmosphere and boiling water before consumption.

- **431–404 BCE** Helps cure the injured in the Peloponnesian War. He excels at surgery, including that of the skull, and also at setting fractures and mending dislocations.

AN 11TH-CENTURY EDITION OF *HIPPOCRATIC TREATISES ON FRACTURES AND ON JOINTS*

- **420–370 BCE** Around 60 books including textbooks, lectures, and essays, are written during this period, and later collated in the Library of Alexandria. Written by Hippocrates and other authors, they are united in their focus on Hippocratic medicine. Hippocrates also writes *Hippocratic Treatises On Fractures and On Joints* during this time. Hippocrates promotes the concept of four humors and believes that an imbalance in the humors causes disease.

- **400 BCE** Sets up a school of medicine in Cos, Greece. In time he instructs his own sons, Thessalus and Draco, in the practice of medicine. His medical school produces many prominent scholars and pupils who add their experience and writings to the works of Hippocrates.

- **370 BCE** Dies in Larissa, Greece, at the age of about 90.

- **2nd century CE** Greek physician Soranus of Ephesus writes the first biography of Hippocrates. It becomes the main source for information about Hippocrates' personal life.

The oath included a promise to protect confidentiality, and not to "poison" patients. Hippocrates insisted that doctors be of "good appearance" and well fed because patients could not trust a physician who did not look capable of taking care of himself. According to the oath, the doctor must be calm and serene, honest, and understanding. A Hippocratic doctor visited his patient before noon, and enquired about what sort of night the patient had experienced, before performing a thorough examination of the body, and looking at the sweat and urine of the sufferer.

Father of modern medicine

Knowledge of anatomy and physiology was limited in Hippocrates' time because the Greek respect for the dead meant that dissection was not allowed. However, for the living Hippocratic

medicine, as shown in the Corpus, stressed three things: close observation of symptoms, being open to ideas, and a willingness to explain the causes of disease. The Corpus is full of case studies, which provide descriptions, for example of tuberculosis, mumps, and malaria. In it Hippocrates defined different categories of illness, such as epidemic, endemic, chronic, and acute—terms that have survived to this day. He was also a talented

"I will use my power to help the sick to the best of my ability… I will **abstain from harming** or wronging any man by it."

FROM THE HIPPOCRATIC OATH

Medicine in Ancient Rome

The civilization of ancient Rome is famed for its contributions to medicine. Founded largely on Hippocratic and Greek traditions, Roman physicians, surgeons, and pharmacists made many advances, and extensively recorded their medical theories and practices.

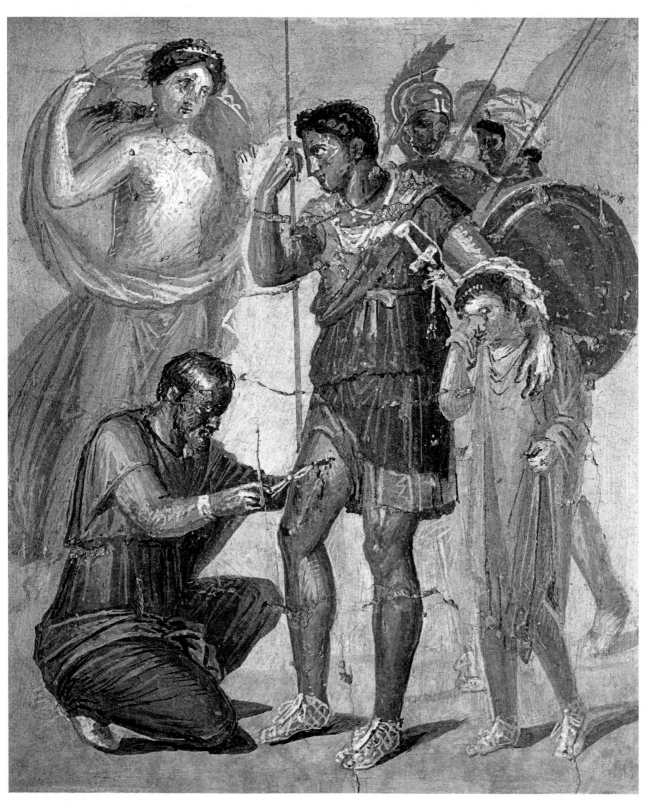

The civilization of ancient Rome rose to power around 1,500 years ago. The city gradually grew in influence to rule Italy and beyond, first as a republic and then as an empire, until its collapse in 410 CE. Roman writings, art, statues, surgical instruments, medicine jars, false teeth, and a host of other objects survive that provide a detailed picture of health, sickness, and healing in the "Eternal City" and the vast lands under its control.

The Romans were among the first to introduce public health measures, such as clean drinking water and organized sanitation, in their towns and cities. They also began spreading awareness about the

15 MILLION **The number of people in the Roman Empire who died in 165–85 CE in the Antonine Plague (probably smallpox).**

importance of general hygiene, including bathing. Exercise and diet, too, were a significant part of their lives. Often, at the first sign of illness, a physician would advise a change of foods and eating habits; for example, cutting down on rich meats and exotic spices in favor of more wholesome local bread and fruit.

Divine intervention

Roman philosophy and medical theories incorporated the belief that the gods wished sickness upon those who lapsed in their worship or morality. However, such divine

◁ **Mythical medicine**
A hero of Roman and Greek mythology, Aeneas is treated by Lapyx, the god of healing. Romans had many medicine-related gods who required prayers and offerings before physicians could effect a cure.

> # "People can **live without doctors, though not,** of course, **without medicine.**"

PLINY, *NATURAL HISTORY*, C.10 CE

◁ **Medicinal plants**

This German version of Dioscorides' *De Materia Medica*, published in 1543—almost 1,500 years after it was written—illustrates healing herbs such as yarrow, foxglove, and primrose, along with notes on their preparation and application.

PEDANIUS DIOSCORIDES

A Greek-born Roman physician, herbalist, and apothecary, Pedanius Dioscorides is best known for his five-volume compendium, *De Materia Medica* (*On Medical Matters*), see left. Dioscorides was attached to the Roman army, and traveled as a surgeon with the armies of Emperor Nero. His travels provided him with an opportunity to study the medicinal properties of a large number of herbs and minerals.

De Materia Medica is a five-volume work that details the features and uses of more than 600 substances, including plants, animals, oils, wines, and minerals. In the work, Dioscorides aimed to cover "the preparation, properties, and testing of medicines."

A landmark work of ancient Rome, *De Materia Medica* gained a great reputation over the following centuries and was regularly supplemented and revised. Its title, like the term pharmacopoeia, has since passed into general medical terminology to mean a database of collected information about a particular substance, whether a time-honored natural herb or the latest computer-designed chemotherapeutic compound.

abortion. His other works included *On Acute and Chronic Diseases, On Signs of Fractures*, and *On Bandages*.

The Empiric School held that experience was the key factor and that remedies should be familiar, tried, and tested. The Dogmatic School highlighted the traditions of Hippocrates and concepts such as the humors (see pp.34–35), which were to be followed as closely as possible. This school of thought was developed by the foremost physician of ancient Rome, Claudius Galen (see pp.40–41), who believed that illnesses were caused by an imbalance in bodily fluids. Rebalancing the humors to restore health included changes in diet and exercise as well as a wide array of herbal, mineral, and other treatments. Bleeding, cupping, and cauterization were common for many minor ailments.

For a civilization founded on military prowess, surgery became a leading medical discipline—both on the battlefield and during gladiatorial displays. Surgeons could treat many kinds of injuries caused on the battlefield and in everyday life, and they had extensive sets of equipment that included numerous knives; scalpels of various sizes and shapes; amputation saws with a range of tooth designs; rotary drills for procedures such as trepanning and tumor excision; hooks to extract foreign bodies such as embedded weapons; retractors to access inner parts; catheter tubes to insert into the urethra and bladder to remove stones and blockages; and various throat and vaginal speculums. There were many prostheses for the eyes, nose, teeth, arms, hands, legs, and feet, made of materials such as wood, iron, silver, and gold. Surgery was swift but careful and patients received alcohol, opium, and herbs for pain relief, and wound dressings of hot oils, herbal poultices, and vinegar.

Early hospitals

Late in the Empire's history, its organization spread to the medical system and the first dedicated hospitals were set up. These were largely reserved for eminent citizens like government officials and merchants, soldiers of high and medium rank, and sometimes, favored slaves. Medical units with physicians and caregiver slaves

3 MILLION Number of soldiers in Emperor Augustus' army.

2 THOUSAND Number of physicians to tend to the emperor's army.

were attached to the army. They set up mobile hospitals and medical rooms in forts. In the provinces a physician's role (apart from those who attended important people) was of relatively low status. Although some formal training and licensing was introduced, there were still no official qualifications and almost anyone could practice.

intervention was seen as less significant than in ancient Egypt or Greece. Chief among the medical deities was the adopted Greek god of healing, Asclepios. The Romans added others, including Vejovis, god of healing; Febris, goddess protecting against malaria and other fevers; Endovelicus for public health; Carna, for the heart and inner organs; and Bona Dea, goddess of women and fertility. Offerings and prayers to them were a routine part of many treatments.

Schools of thought

In ancient Rome, there were various approaches to medicine, known as schools. The Methodic School emphasized identification of the disease first, followed by treatment—it paid less attention to the individual patient. One of the most eminent followers of the Methodic School was Soranus of Ephesus (c.98– 140 CE), who moved from Greece to settle in Rome. He authored a number of books, including *Gynaecology*, which covered midwifery, baby care, and

Gold tooth holder

◁ **Roman dentistry**

Although dentistry had not emerged as a profession at the time, some Roman surgeons specialized in treatments for the mouth and teeth. A copy of a Roman original, these bridges were fitted over existing teeth to hold additional ones (real or ivory).

ROMAN PHYSICIAN Born c.129 CE Died c.216 CE

Galen

"The **best physician** is also **a philosopher.**"

CLAUDIUS GALEN, TITLE OF A TREATISE, ALSO QUOTED IN *PERI CHREIAS MORION, DE USU PARTIUM (ON THE USEFULNESS OF THE PARTS OF THE BODY)*, 165–175 CE

A physician who was elevated to godlike status, Claudius Galen was the foremost medical authority of the Roman Empire. Building on the work of Hippocrates (see pp.36–37) and other Greek physicians, he wrote a large number of works—more than 400 volumes, containing over 8 million words. His ideas and teachings on human anatomy, as well as the causes and symptoms of diseases, and their treatments, became, in effect, the laws of medicine for more than 1,300 years. Much is known about Galen's talents because he was a great self-publicist and regularly promoted his own work.

Brought up in Pergamon (now Bergama, Turkey), in a wealthy family and well educated, Galen was destined for a career in law or in the government, until his

◁ **Prolific medical writer**
More than half of Galen's written works were destroyed in a fire in 191 CE at Rome's Temple of Peace. Yet, the number of surviving volumes of his work still exceed those by almost any other medical author.

father dreamed that Asclepios—the Greek god of healing—asked his son to take up medicine. After his father's death, the 19-year-old Galen moved to Smyrna (modern-day Izmir, Turkey), where he was instructed by the physician Pelops and the philosopher Albinus. He then moved on to Corinth, Greece, and finally to Alexandria, Egypt, where he acquired knowledge from the great library. The young Galen was interested in the medicine of Hippocrates and the philosophy of Plato, and later analyzed their works in *On the Doctrines of Hippocrates and Plato*.

Illustrious career
In about 157 CE Galen returned to Pergamon, Turkey, and took up his first medical post as a physician-surgeon to the gladiators there, making notes on the variability of wounds sustained by them in the gladiatorial games. With his success at Pergamon, which saw death rates fall dramatically, his reputation and fame began to spread. The ambitious Galen then moved to Rome in 162 CE. Here, he was able to impress the Roman establishment with his medical abilities, speed of learning, and confidence. After treating the philosopher Eudemus in Rome, Galen was introduced to the government official Flavius Boethus, who encouraged him to begin to write and to give public lectures and demonstrations. However, he soon fell out with

colleagues, whom he claimed envied him, and decided to adopt a low profile. He eventually returned to Pergamon.

Galen went back to Rome in 169 CE after being summoned by Emperor Marcus Aurelius, and here began the most fruitful phase of his professional life. He began to write prolifically and continued to lecture and philosophize, while also attending to a series of five emperors as their personal physician, even accompanying them on their travels.

Discoveries and contributions

Galen's primary interest lay in anatomy, which he believed was the basis of all medicine, although he was constrained by laws that forbade the deliberate opening of the human body. Nevertheless, building on his experience with gladiators, he experimented on

▽ **Treating a gladiator**
This artwork from the 19th-century book, *Vies des Savants Illustres*, shows Galen treating a gladiator in Pergamon. As a physician, he studied human internal anatomy and regarded the physical body as a "vessel for the soul."

and dissected an array of animals, including Barbary apes (a type of Macaque monkey). His discoveries were numerous and accurate, and included finding the true identity and extent of many muscles and tendons, and he demonstrated the kidney's role in making urine by clipping the ureter of live animals and showing that it filled with urine. However, Galen's supreme confidence meant that he often

of his medicine, he acknowledged the achievements of Hippocrates. His extensive tracts on such themes included *On the Black Bile* and *On the Elements according to Hippocrates*.

Galen's writing style was diffuse, wordy, rambling, and contained subjective comment. His medicine, too, was interwoven with his very idiosyncratic beliefs. Over the centuries, while his philosophy

"In order to **diagnose,** one must **observe and reason.**"

MOTTO OF CLAUDIUS GALEN

took educated guesses, or clues derived from animals, as facts. For example, his study of the brain and the functions of its parts led to his assertion that the pineal gland helped support blood vessels, a belief that continued to be accepted through the Renaissance. Galen also developed the Greek idea of humors, or body fluids, into an extensive fourfold scheme (see p.34–35). In this, as in much

was discarded or superseded, Galen's medical teachings—complete with guesses and misconceptions—became, to many, undeniable.

It was not until the 16th century that challenges by Andreas Vesalius (see pp.72–75), William Harvey (see pp.82–83), and others began to dismantle the Galenic tenets of medicine, but even in the 1800s some Western medical doctors still referred to his works.

TIMELINE

- **c.129** Born into a wealthy family in Pergamon—in modern-day Bergama, Turkey—a major center of the region and Roman Empire.

- **148** Galen's father—Aelius Nicon—dies, leaving Galen financially well-off, and able to travel around Europe and North Africa to study medicine.

- **157** Returns to Pergamon and takes up a post as physician to the gladiators there, successfully treating their injuries and wounds. As the gladiatorial death toll reduces, his reputation spreads to Rome and reaches the senior medical fraternity who suggest that he moves there.

A 1561 EDITION OF GALEN'S WORK PRINTED IN BASEL, SWITZERLAND

- **c.162** Moves to Rome as a physician, but makes several enemies due to his attitude toward other physicians and their theories. He leaves the city occasionally, and returns to Pergamon for a time.

- **c.166** The Antonine Plague (probably smallpox or measles) sweeps across Europe. Galen writes extensively about the effects and possible treatments for this plague. A similar epidemic appears in 198 CE.

- **169** Recalled to Rome by Emperor Marcus Aurelius to become his personal physician, which he does until Aurelius dies in 180 CE.

- **170** Becomes physician to Emperor Aurelius' son and heir Commodus until his death in 192 CE.

- **191** A large number of his writings are destroyed in a fire at the Temple of Peace in Rome. Galen is devastated by the loss of his works.

- **193** Becomes physician to the new Emperor Septimius Severus. Although Galen starts fading from the spotlight, his writings continue to be widely circulated and remain immensely popular.

- **c.216** Dies in Rome, although some authorities say Pergamon or Sicily and put this date earlier, at around 200 CE.

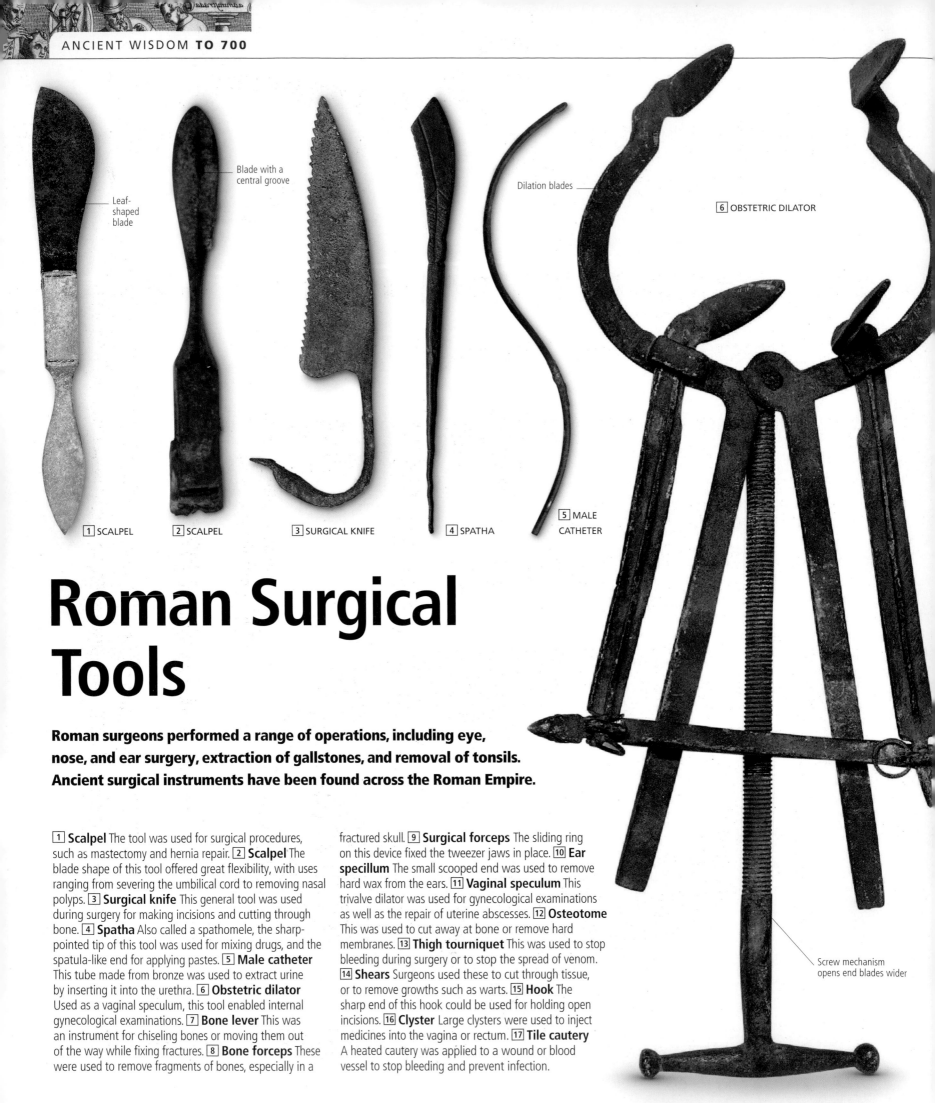

Leaf-shaped blade

Blade with a central groove

Dilation blades

6 OBSTETRIC DILATOR

5 MALE CATHETER

1 SCALPEL 2 SCALPEL 3 SURGICAL KNIFE 4 SPATHA

Screw mechanism opens end blades wider

Roman Surgical Tools

Roman surgeons performed a range of operations, including eye, nose, and ear surgery, extraction of gallstones, and removal of tonsils. Ancient surgical instruments have been found across the Roman Empire.

1 **Scalpel** The tool was used for surgical procedures, such as mastectomy and hernia repair. 2 **Scalpel** The blade shape of this tool offered great flexibility, with uses ranging from severing the umbilical cord to removing nasal polyps. 3 **Surgical knife** This general tool was used during surgery for making incisions and cutting through bone. 4 **Spatha** Also called a spathomele, the sharp-pointed tip of this tool was used for mixing drugs, and the spatula-like end for applying pastes. 5 **Male catheter** This tube made from bronze was used to extract urine by inserting it into the urethra. 6 **Obstetric dilator** Used as a vaginal speculum, this tool enabled internal gynecological examinations. 7 **Bone lever** This was an instrument for chiseling bones or moving them out of the way while fixing fractures. 8 **Bone forceps** These were used to remove fragments of bones, especially in a

fractured skull. 9 **Surgical forceps** The sliding ring on this device fixed the tweezer jaws in place. 10 **Ear specillum** The small scooped end was used to remove hard wax from the ears. 11 **Vaginal speculum** This trivalve dilator was used for gynecological examinations as well as the repair of uterine abscesses. 12 **Osteotome** This was used to cut away at bone or remove hard membranes. 13 **Thigh tourniquet** This was used to stop bleeding during surgery or to stop the spread of venom. 14 **Shears** Surgeons used these to cut through tissue, or to remove growths such as warts. 15 **Hook** The sharp end of this hook could be used for holding open incisions. 16 **Clyster** Large clysters were used to inject medicines into the vagina or rectum. 17 **Tile cautery** A heated cautery was applied to a wound or blood vessel to stop bleeding and prevent infection.

Gripping blade

7 BONE LEVER

8 BONE
FORECEPS

Sliding ring

9 SURGICAL FORCEPS

10 EAR SPECILLUM

Screw-operating device

11 VAGINAL SPECULUM

Trivalve dilator

12 OSTEOTOME

Bronze blade

13 THIGH TOURNIQUET

14 SHEARS

15 HOOK   16 CLYSTER

17 TILE CAUTERY

REVIVAL AND RENAISSANCE

700–1800

REVIVAL AND RENAISSANCE
700–1800

700		1100		1400	

750
Madhav Acharya compiles the 79-chapter *Rug Vinischaya*, also known as *Madhav Nidana*.

800
Varied works of Galen are translated into Arabic.

1000
Al-Zahrawi produces the immense surgical and medical classic *Kitab al-Tasrif* (*The Method of Medicine*).

1025
Ibn Sina (Avicenna) completes *Al-Qanun fi al-Tibb* (*The Canon of Medicine*).

1077
Constantine the African teaches at Salerno medical school, the first such teaching institution in Europe.

⌃ Salerno medical school

820
A Benedictine hospital is established in Salerno; the medical school will develop from it.

855
Zan Yin completes *Jingxiao Chanbao* (*Tested Prescriptions in Obstetrics*), the first Chinese text dedicated to gynecology and obstetrics.

1123
St. Bartholomew's becomes Britain's first truly medical hospital.

1144
Robert of Chester's *De Compositione Alchemiae* (*The Book of the Composition of Alchemy*) is one of Europe's first alchemical treatises.

1150s
Hildegard of Bingen produces *Liber Simplicis Medicinae* (*Book of Simple Medicine*, later called *Physica*).

⌄ Altarpiece depicting arrival of Hildegard at the Benedictine Abbey

⌄ Illustration from the 13th-century *Treatise on the Eye*

1200s
Treatments for eye conditions, such as bruising and infections, are regularly used.

1242
Ibn al-Nafis describes the pulmonary circulation from the heart's right side through the lungs to the left side.

1247
Song Ci produces *Xiyuanlu*, a collected record of medical jurisprudence, an early classic of forensic medicine.

1316
Mondino de Luzzi writes *Anathomia Corporis Humani* (*Anatomy of the Human Body*).

1347
The Black Death reaches Europe, in one of the greatest of all pandemics.

1363
Guy de Chauliac completes *Chirurgia Magna* (*Great Surgery*), which will be a standard anatomical, medical, and surgical work in Europe for three centuries.

1494–95
First reports of syphilis appear in Europe, the disease probably having been brought from the Americas.

1518
In Britain, the College of Physicians receives its royal charter.

1520s
Smallpox, brought from Europe, begins to take a toll on people in the Americas.

1529
Philippus Aureolus Theophrastus Bombastus von Hohenheim, who achieved both fame and infamy in various sciences and the occult, including alchemy, adopts the name "Paracelsus."

1530
The first text devoted to dentistry, *Little Medicinal Book for Diseases and Infirmities of the Teeth*, is published in Germany.

⌄ Early compound microscope

1537
During the Siege of Turin, Ambroise Paré tries an old recipe for a wound-healing balm, and begins a new era in battlefield medicine.

1543
Andreas Vesalius revolutionizes anatomy with *De Humani Corporis Fabrica* (*On the Fabric of the Human Body*).

1546
Girolamo Fracastoro suggests that epidemic diseases, such as rabies, spread due to some kind of communicable "spores."

1563
Garcia de Orta writes *Colóquios dos simples e drogas da India* (*Conversations on the Simples, Drugs and Materia Medica of India*), an early work in the field of tropical medicine.

1590
The compound microscope is invented, revealing a whole new world of tiny life forms that will impact hugely on medicine, but not for several decades.

From about the 8th century, the expanding Islamic world became the focus of progress in arts, architecture, sciences, and medicine. Al-Razi, Ibn Sina, and other great physicians of this "Golden Age" expanded and developed ancient knowledge, established hospitals, and returned Hippocratic humanity to medical care. Europe underwent its own Renaissance in arts, sciences, and medicine, which began in the 13th century. Pivotal developments included the anatomy of Vesalius, Harvey's description of circulation, the assimilation of the microscope into medicine, the founding of new-style medical schools and professional organizations, and Jenner's pioneering work in vaccination.

1600

1628
William Harvey publishes *De Motu Cordis* (*On the Motion of the Heart and Blood*)—a short report but monumentally significant due to its description of how the circulatory system works.

⌄ Harvey carrying out a postmortem

1665
Robert Hooke publishes *Micrographia*, a pioneering work in microscopy and one of the first science bestsellers.

1676
Thomas Sydenham publishes *Observationes Medicae* (*Observations of Medicine*), an extremely influential text in Europe for the next two centuries.

1630s
Cincona bark (the source of quinine) is brought from the New World to Europe to treat and prevent malaria.

1673
The Royal Society of Britain begins its publication of reports by innovative microscopist Antoni van Leeuwenhoek.

⌄ Antoni van Leeuwenhoek

1694
Zhang Lu's *Zhangshi Yitong* (*Chang's General Medicine*), a vast medical collection, describes inoculation against smallpox.

1661
Marcello Malpighi, founder of microanatomy, observes capillaries—the "missing link" between arteries and veins.

1700

1701
In Europe Giacomo Pylarini describes and practises variolation, a form of smallpox vaccination carried out in Asia.

1723
Pierre Fauchard establishes modern dental practices with *Le Chirurgien Dentiste* (*The Surgeon Dentist*).

1747
James Lind discovers how to prevent scurvy by carrying out one of the first organized clinical trials.

1748
Jacques Daviel pioneers a new technique to remove cataracts, greatly advancing their treatment.

1774
Prussian blue is one of the first stains (dyes) to color microscopic samples, advancing the area of histology.

1775
Percivall Potts describes how scrotal cancer is much more common in chimney sweeps—one of the first accounts implicating a carcinogen, and a landmark for occupational medicine.

1785
William Withering reports on his investigations into digitalis, the active substance in foxgloves used to treat dropsy.

1790
Samuel Hahnemann begins to devise therapies based on "like cures like," which becomes known as homeopathy.

» Homeopathic medicine chest

1793
Jean-Baptiste Pussin and his wife Marguerite, along with Philippe Pinel, begin improvements in the care and treatment of the mentally ill.

1796
Edward Jenner inoculates an 8-year-old boy against smallpox using cowpox material, establishing the principle of vaccination.

1796
Franz Joseph Gall writes his first main text on phrenology. It will flourish for a few decades, then disappear.

1799
Humphry Davy discovers that nitrous oxide acts as an anesthetic and wonders if it might alleviate pain during surgery.

⌄ Collection of model heads to explain principles of phrenology

Medical practice in action
Dated to around 1260, this "Europeanized" illustration of al-Razi treating a patient is from the *Recueil Des Traités de Médecine* (*Collection of Medical Treatises*), which was based on a Latin translation of his work by Gerardus Cremonensis, a noted translator of Arabic medical texts.

The Golden Age of Islamic Medicine

As Europe entered the "Dark Ages," the Middle East and western Asia saw a blossoming of culture and science, especially in the field of medicine. Building on knowledge from the ancient world, these advances eventually flowed back to Europe during the Renaissance.

The year 476 CE—when the last emperor, Romulus Augustulus, was deposed—is regarded as the end of the Western Roman Empire. After the collapse, Europe entered an era of social upheaval and disorder referred to as the "Dark Ages," during which little progress was made in the arts and sciences, including medicine.

In contrast, from around the 8th century the Muslim lands of the Middle East and western Asia experienced an Islamic "Golden Age." Spreading out from Baghdad (then capital of the Abbasid caliphate, now the capital of Iraq), academic and intellectual pursuits flourished in an atmosphere of tolerance. Part of an integrated approach to learning that viewed mathematics, astrology, literature, philosophy, alchemy, and the sciences as part of a unified truth, the field of medicine in particular saw unprecedented innovation.

A duty of care

Islamic teachings emphasize duties of care, both for the individual as regards aspects of self-care such as diet, exercise, hygiene, and mental and emotional matters, and care for others who are sick and needy. Medical treatment should be made available to all, and research into the prevention, treatment, and cure of illness should be sought. These attitudes led to much progress, not only in the skills of physicians, but also in the provision and organization of medical care.

Pioneering hospitals and medical schools funded by charitable individuals and wealthy rulers were established from the 9th century onward in Baghdad and other cities. Open to all, they had organized wards, inpatient and outpatient services, dedicated nursing care, and in many cases offered outreach services for rural areas. Most significantly, they also provided hubs for medical training and research.

A comprehensive system of medical education was established, with physicians undertaking basic scientific learning in subjects such as anatomy, physiology, and alchemy, followed by clinical training at hospitals that included instruction in conducting physical examinations, taking patient notes, and administering treatments.

Building on the past

The basis for these new advances in medical education and practice was knowledge drawn from the ancient world. Muslim physicians avidly translated, studied, and assimilated works from the scholars of the past—especially the texts of Greek physician Hippocrates (see pp.36–37) and Roman physician Galen (see pp.40–41), as well as traditional Chinese and Indian sources (see pp.26–27 and pp.30–31).

One of the greatest scholars to play a part in this process of synthesis was the physician al-Razi (also known as Rhazes).

△ **Medicinal substance**
Highly skilled Islamic pharmacists prepared a wide range of medicines using herbs and other substances, such as naturally occurring crystals and minerals. Sal ammoniac crystals, seen here on the black stone, were also used in alchemy.

Born around 865 CE in the city of Rey (now Tehran, Iran), al-Razi became chief physician in hospitals in Rey and Baghdad. He wrote more than 50 major texts and hundreds of minor commentaries that combined the principles and practices he had found in ancient medical works with his own clinical observations. His two most famous encyclopedic texts, *Kitab al-Mansouri fi al-Tibb* (*The Book on Medicine Dedicated to al-Mansur*) and *Kitab al-Hawi fi al-Tibb* (*The Comprehensive Book on Medicine*), were used for centuries after his death in 925 CE, in western Asia and, in Latin translation, in Europe.

Al-Razi's writings emphasized the importance of the relationship between doctor and patient. He revived the Hippocratic approach that regarded all patients as being equal and worthy of attention, and that charged physicians to do patients no harm through medical treatment. He also »

ARAB SCHOLAR AND PHYSICIAN (1213–88)

IBN AL-NAFIS

A Muslim medical scholar and polymath, Ibn al-Nafis attended the medical school at Nuri Hospital, Damascus (in modern-day Syria), before moving to Cairo in Egypt.

A prolific writer, he produced numerous texts on general medicine, ophthalmology, and surgery, as well as on the interaction of medicine with law, religion, and philosophy. However, al-Nafis may have invited controversy when he dissected corpses to study anatomy—a practice that was then forbidden. He came close to working out the body's circulatory system when he described, for the first time, the movement of blood around the pulmonary circuit, from the right side of the heart through the lungs to the heart's left side (see pp.82–83).

>> emphasized the importance of patient interviews in diagnosis, the need to amend treatments based on past experience, and the value of clinical observation in medicine in lieu of dogmatism and habit. These observations allowed al-Razi to advance theories on the nature of diseases and the importance of preventive medicine—the need to investigate the causes of ailments, not just provide cures—and the benefits of good diet and hygiene. Recording the symptoms of smallpox (see pp.100–01) and measles, for example, led him to propose the theory that blood froths like a fermenting drink with vapors that seep through the skin and create blisters and sores.

Age of discoveries

The advances in medical knowledge gained through meticulous record-keeping and

△ **Tools of the trade**
Traditional knowledge of chemists, alchemists, and apothecaries provided Arabic physicians with the skills needed to make medicines. This bronze mortar, from the 16th –18th centuries, would have held ingredients that were ground using a pestle.

▽ **Fighting smallpox**
This illustration is from a 17th-century Turkish edition of Ibn Sina's Canon of Medicine. The painting shows a man suffering from smallpox waiting for treatment while the apothecary weighs the ingredients for his medicine on a balance.

an emphasis on clinical observation led to progress in all medical fields as well as greater specialization. Physicians such as al-Zahrawi (also known as Albucasis), born in 936 CE, became renowned for their excellence in specific areas of medicine. Often referred to as "the father of surgery," al-Zahrawi pioneered new procedures and provided the first illustrations of more than 200 surgical instruments in his seminal encyclopedic work, *Kitab at-Tasrif* (*The Method of Medicine*). By the 13th century progress in the study of anatomy allowed the physician Ibn al-Nafis (see panel, p.49) to demonstrate an understanding of the body's circulatory system.

The introduction of new drugs and methods of testing, along with the development of processes such as dissolving and distillation, also fuelled advances in pharmacology. Many prominent physicians also translated ancient works and wrote their own texts on medicinal plants during

▷ **Return of medical knowledge**
During the Golden Age of Islamic medicine, physicians from the Middle East and western Asia continued to expand upon the medical wisdom of ancient Greece and Rome. From the 12th century, their writings were used in Latin translation in the new medical schools in Italy, Spain, and France.

this period, but in the early 13th century the Andalucian botanist Ibn al-Baytar produced a groundbreaking encyclopedia that was to become the authoritative text on herbalism for centuries. *Al-Kitab 'l-jami' fi 'l-aghdiya wa-'l-adwiyah al-mufradah* (*The Comprehensive Book of Foods and Simple Remedies*) alphabetically listed hundreds of herbal medicines and remedies—many of which were Ibn al-Baytar's discoveries.

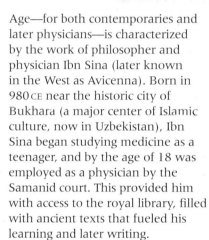

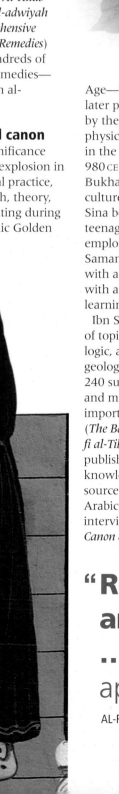

Medical canon
The significance of this explosion in medical practice, research, theory, and writing during the Islamic Golden

Age—for both contemporaries and later physicians—is characterized by the work of philosopher and physician Ibn Sina (later known in the West as Avicenna). Born in 980 CE near the historic city of Bukhara (a major center of Islamic culture, now in Uzbekistan), Ibn Sina began studying medicine as a teenager, and by the age of 18 was employed as a physician by the Samanid court. This provided him with access to the royal library, filled with ancient texts that fueled his learning and later writing.

Ibn Sina wrote on a wide range of topics, including mathematics, logic, astronomy, psychology, and geology, but is best known for his 240 surviving works on philosophy and medicine. Of these, the most important were *Kitab al-Shifa* (*The Book of Healing*) and *Al-Qanun fi al-Tibb* (*The Canon of Medicine*), published in about 1025. Collating knowledge from Greek and Roman sources, Ayurvedic, Persian, and Arabic works, and his own patient interviews and observations, *The Canon of Medicine* (see pp.52–53)

was translated into a number of languages, including Latin and Chinese, and became the standard medical textbook for physicians for the next few centuries.

Ibn Sina's influential writings promoted the development of a comprehensive medical system in which observation, methodical experimentation, and deduction were used to underpin medical practice. He found methods for testing the efficacy of drugs, established the importance of environmental factors (such as clean air and water) on health, and identified the contagious nature of infectious diseases.

These principles, and the great advances in medical science made during this dynamic period, began to filter westward from the middle of the 12th century. Primarily translated into Latin, the texts of Islamic physicians were copied (later printed), disseminated, and studied throughout Europe, eventually aiding the flowering of medicine in the West during the Renaissance of the 15th century.

> "**Restlessness, nausea,** and **anxiety** occur… with **measles** … **pain in the back** is more apparent with **smallpox.**"

AL-RAZI, IN *AL-JUDARI WAL HASABAH* (*CONCERNING SMALLPOX AND MEASLES*)

KEY

▭ GREATEST EXTENT OF ISLAMIC CONQUESTS

← ROUTE OF SPREAD OF MEDICAL KNOWLEDGE

PARIS

PADUA
MONTPELLIER
BOLOGNA
ROME
SALERNO
CORDOBA
CONSTANTINOPLE
ATHENS
BAGHDAD
ALEXANDRIA
CAIRO

Ibn Sina's *The Canon of Medicine*

Ibn Sina's masterpiece, *Al-Qanun fi al-Tibb* (*The Canon of Medicine*) had a vast influence on medical teaching in the West as well as in the Arab world. A definitive encyclopedia, it remained a standard medical textbook in Europe for 500 years—from the 12th to the 17th century—earning Ibn Sina the title of Prince of Physicians.

One of the most famous Arabic writers of medicine, Ibn Sina, later called Avicenna, was born in Persia in 980 CE. A precocious child, he could recite the entire Qur'an by the age of 10. He studied medicine at 16 and began to practice it at 18. He led a full life characterized by hard work, and alleged drinking and promiscuity.

Ibn Sina's *Canon*—a massive book containing a million words across five volumes—is a collection of all that was known at the time about medicine and surgery, including the doctrines of Hippocrates (see pp.36–37), Galen (see pp.40–41), and the Greek philosopher Aristotle. The first volume dealt with the origins of health and sickness and aspects of the body's anatomy and function. The second volume listed information on more than 700 drugs and medicines. The third volume covered the diagnosis and treatment of diseases specific to certain parts of the body, while the fourth focused on conditions that affect the whole body. The final volume discussed the preparation of medicinal remedies. The *Canon* was translated into Latin in the 1100s and consequently came to dominate approaches to medicine in the medieval period.

> "Therefore **in medicine** we ought to know the **causes** of **sickness and health.**"

IBN SINA, ON MEDICINE, c.1020

▷ *The Canon of Medicine*
Anatomical drawings of the heart, ear, brain, and other body parts from a 14th-century edition of the *Canon* are shown here. Human dissection was rare at this time, and Ibn Sina probably gained his anatomical knowledge from Galen and other ancient physicians.

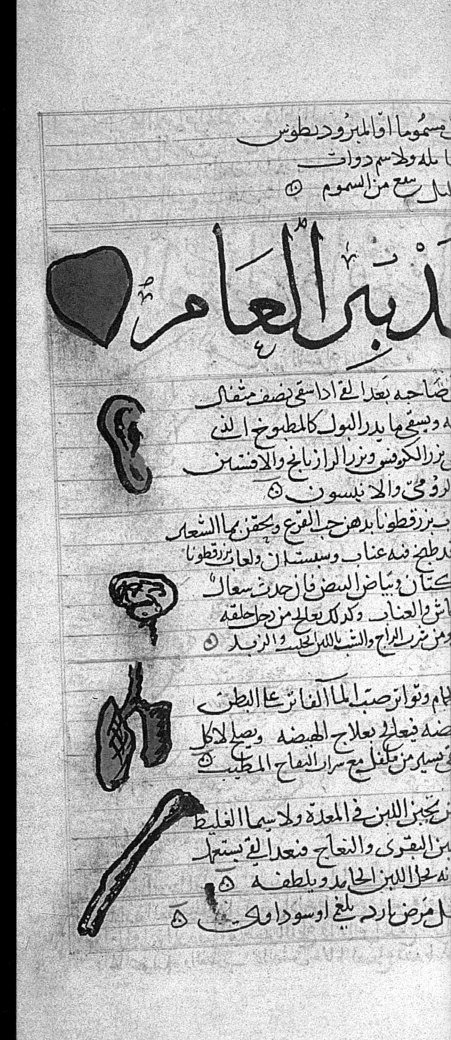

المحتوم فإن تعلّق ذلك إن نخرج الطعام
او ترياق الفاروق وان من سرّبه المستخرج
السموم ولا ذرعها ولا هشنها فيها

الأدوية الفتّالة وأصنافها ونفع

السبب	العلامة	الاستفراغ	النفع			الأعراض	الأعضاء	الأسباب
	مالك بالعقل							
	مالك بالحل							
	الحاد							
	والسطح							
	مالك بالماء							
	الحاد							
	السكّين							
	الغيا							
	حاد							

بل أحدها وبعد أحدها وهو العرض بي كيده ٥

The **First School** of **Medicine**

Although different approaches to medical training emerged around the world as early as the 23rd century BCE, the first formal facility for teaching medicine was the *Scuola Medica Salernitana,* in the southern Italian city of Salerno, which was founded in the 9th century.

According to ancient Egyptian scripts, medical schools were established by around 2200 BCE—when the first reference to *Per-Ankh*, or "Houses of Life," as places for the creation and preservation of written knowledge appears. Senior physicians taught students and worked with scribes to record information and produce copies of books on health practice.

Although some Egyptian medicine had its roots in logic and evidence, much of the thinking was based on religion and magic. Students from Greece and the Arab world studied in Egypt's medical schools, then returned home to integrate this knowledge with local practices.

Laying the foundation
Both the Greeks and Arabs built on the existing foundations of physician training established at

△ **Matthaeus Platearius**
Written in around 1470, by Salerno school physician Matthaeus Platearius, *De Simplici Medicina* (*The Book of Simple Medicine*) described 270 drugs in detail.

the Houses of Life, but they took medical learning to a new level based firmly on the principles of science rather than religion or superstition. This science-based approach reached a new height of sophistication hundreds of years later with the opening of the ground-breaking *Scuola Medica Salernitana*, the first modern medical school, in Salerno, Italy.

Founded on the site of a former monastery dispensary, the institute was unrivaled for four centuries in terms of both the scope of its teaching and in the production of medical textbooks, including translations of several important Arab works. The school's library was renowned, and its shelves were stacked with rare medical texts supplied by the Benedictine Abbey at nearby Monte Cassino, one of the great medieval centres of learning in Europe. The collection at the Salerno library represented the world's most extensive compilation of medical science knowledge. It included Latin translations of books by

▷ **The School of Salerno**
By the early 900s the Salerno medical school had become famous throughout Europe. In 1099 Duke Robert II of Normandy visited the school to seek treatment.

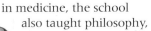

al-Razi and Ibn Sina, who were famous Arab practitioners of pharmacy and medicine (see pp.48–53).

One of the school's early supporters was the Archbishop of Salerno, Alfanus I, who was a talented physician himself. Fluent in several languages, he translated a number of medical books and worked to raise funds for the building of the Salerno school. Also instrumental in the success of the school was Constantine the African, a physician who arrived in Salerno from North Africa to study, but ended up staying on to teach. He shared his knowledge of Islamic medicine and translated several key Arabic texts that would become essential reading for medical students in Europe.

Well-rounded curriculum

The *Scuola Medica Salernitana* was a melting pot of different approaches to medicine and attracted many international students. The training methodology at the Salerno school fused Greek and Roman theory and practice together with Arab and Jewish traditions to create the most comprehensive curriculum available at the time. Courses were well-organized, with high standards and a strict policy of passing one level with the required marks before moving on to the next.

Typically, students would undertake three years of study, followed by four years of hands-on medical training with physicians, surgeons, medical herbalists, and other specialists. As well as preparing the students for careers

△ **Diagram of body showing muscles**
Treatise on the Human Body, published in England in 1292, illustrated numerous aspects of physiology as understood at the time. It included diagrams of the arteries, bones, and muscles of the body (shown here).

in medicine, the school also taught philosophy, religion, and law. Ethics was another important subject, as was physician-patient relations. According to one book, *The Coming of a Physician to His Patient*, "When the doctor enters the dwelling of his patient, he should… put the patient at his ease before his examination begins and the pulse should be felt deliberately and carefully."

Unusually for the time, women were welcomed as both students and staff. The school's most famous female faculty member was Trotula de Ruggiero, who wrote several books on gynecology. Once they had completed the relevant training, women were granted licenses for gynecology, obstetrics, midwifery, pre- and post-natal care, as well as for general practice. With its scope, its acceptance of women, and the volume of books generated there, the *Scuola Medica Salernitana* set the standard for the medical colleges of the future.

▷ **Anatomy lesson**
A woodcut from 1493 shows the practical anatomy instruction that was common at the medical school in Salerno. Initially only animals were dissected but human dissection was introduced at the school in 1250.

Medieval Medicine

In the early medieval period in Europe (the 5th to the 10th century), progress in medicine and science virtually ground to a halt. By the 12th century, however, the translation of ancient medical texts and circulation of new ideas were promoting greater knowledge.

When the Western Roman Empire finally dissolved around 476 CE, the orderly regime of hygiene, literacy, medical practice, and systematic agriculture also faded. Western Europe fragmented into small fiefdoms as Germanic tribes such as the Goths, Vikings, Saxons, and Huns swept across the continent, replacing the

1–2 PER YEAR The number of dissections that took place at medieval medical academies.

cohesive administration of Rome with independent regions that were organized according to the feudal system. Medical practices during this time were based largely on religious beliefs, folk tradition, and superstition. The progressive thinking of the Greek and Roman scholars, and the great Arabic texts on medicine and science, seemed all but forgotten.

Under the Roman Empire, Europe had benefited from an influx of Greek doctors, the Roman Army medical corps, good hygiene

practice, and information about herbal medicines. However, the new structure of Europe meant that there was little transfer of information and limited means of preserving existing medical knowledge other than in religious centers. Monasteries were one of the few places that did promote learning and book production—safeguarding a legacy of knowledge until interest in medicine revived in the mid- to late medieval period. Indeed, the one unifying element in Europe was the Catholic Church, which had become dominant in the power vacuum left after the fall of the Roman Empire.

The rule of religion

Ideas and practices relating to medicine—such as how the human body, sickness, and treatment were perceived—came to be dictated by the Church. Autopsy and dissection were banned, making it difficult to advance medical knowledge and understanding. The Church viewed spiritual intercession and prayer as the primary cure for disease, which was thought to be a punishment

for sin, and urged sick people to pray to the saints for help.

However, some devout Christians, in particular the Benedictines, considered it a Christian duty to care for and treat the sick on a more practical level. The use of natural medications and treatments (particularly herbs) was sanctioned on the basis that they had been provided by God to assist man, and so were spiritual in origin. Herbs were grown by monks and nuns to make remedies for their own use, and to treat sick members of the wider community. Historical documents stored in monastic libraries also provided monks with a degree of medical information and guidance on the use of natural remedies.

A number of hospitals across Europe were founded by religious orders in the medieval era, but most functioned like hospices or almshouses, providing general medical care, housing, and spiritual guidance for those in need.

542 The year when the first hospital in France was constructed.

30 The number of hospitals in Florence, Italy, at the end of the 14th century.

HILDEGARD OF BINGEN

Hildegard of Bingen claimed to have had religious visions from a young age and her parents offered her to the Benedictine monastery at Disibodenberg, Germany, where she eventually became abbess. Hildegard is renowned for her prolific writing and her diverse talents. During her lifetime and beyond, she earned a reputation as a mystic and prophet, a scientist, music composer, and writer, writing two monumental works on natural medicine and cures for illness.

Meeting medical needs

Surviving childhood—and for women, surviving childbirth—presented major medical challenges throughout the medieval period. Conception and childbirth were considered a priority as populations dwindled due to disease, but access to maternal care was limited and variable. Aristocratic women were generally attended by a physician familiar with the Greek and Roman texts on childbirth, but most of their knowledge was theoretical rather than based on practical experience of women's medicine. Other women managed childbirth with the help of a local midwife, who probably

◁ **Sacred reliquary**
Reliquaries, such as this one from 13th-century France, housed relics that were thought to be the bones or remains of saints. Christians believed that by touching a relic they would be protected from sickness.

▷ **Giving birth**
The *Cantigas de Santa Maria* (*Canticles of Holy Mary*) is a collection of illustrated poems set to music, written in Spain during the 13th century. One poem describes a Jewish women in labor who prays to the Virgin Mary, then gives birth to a healthy baby, and converts to Christianity.

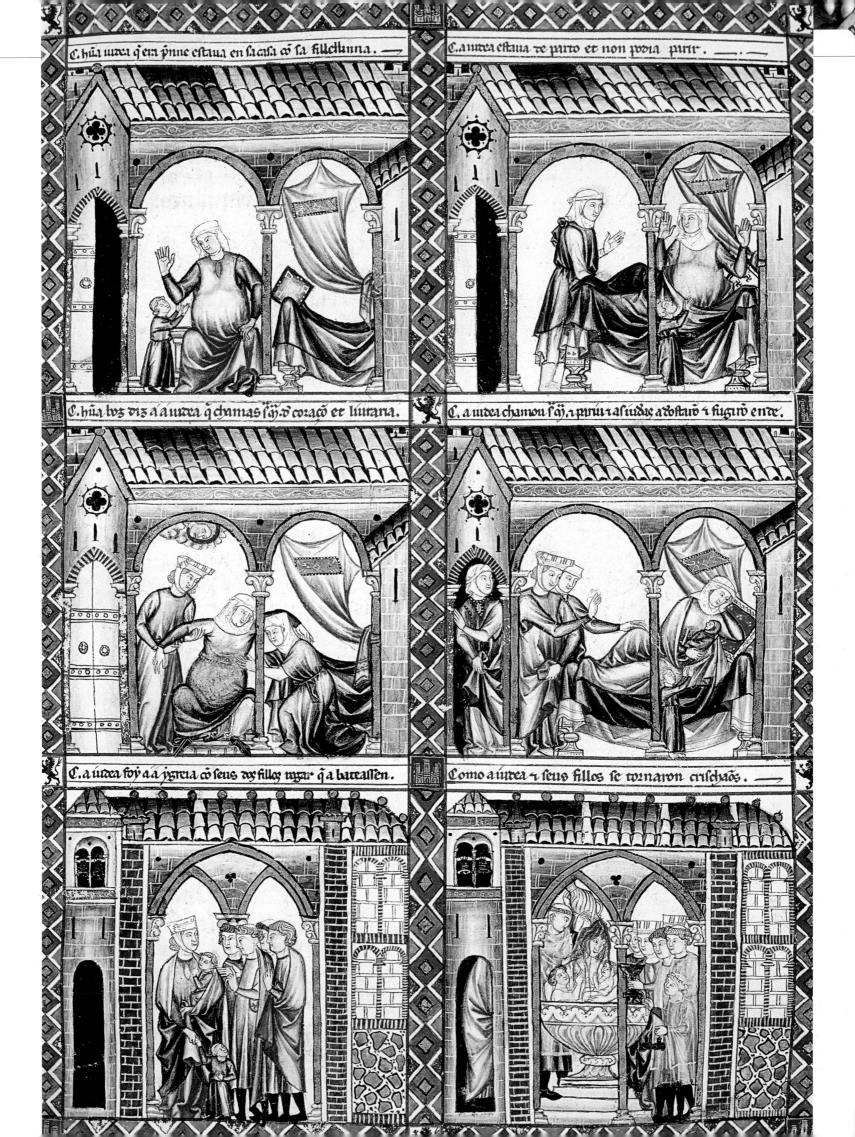

medieval period. Born in the German Rhineland-Palatinate at the end of the 11th century, Hildegard became one of the most important authorities of the 12th century on the subject of medieval pharmacology, and the beneficial properties of plants.

Living in a monastery, Hildegard had access to early translations of medical treatises from antiquity, (see pp.32–33 and pp.38–39) and also benefited from a boom in translations of Islamic medical texts (see pp.48–51) during the 12th century, as interest in the subject grew. She began to write her own books on the subject of sickness and treatment, all carefully set within a framework that placed God firmly at the top, as the divine creator of the natural world.

Some of Hildegard's publications became essential reading for medieval physicians and pharmacists. Her *Causae et Curae* (*Causes and Cures*), for example, was a massive work comprising almost 300 chapters on the causes of human diseases and their treatment. Perhaps even more impressive was the accompanying nine-volume *Physica*, which detailed remedies that could be made from plant and animal extracts. Both works took a well-organized, encyclopedic approach that made them very user-friendly.

Central to Hildegard's view was the use of herbs and botanical tonics as both preventive measures and cures for specific conditions— many still valued in modern medicine for their pharmaceutical properties. To promote brain and nervous system function, for example, she recommended chestnut; today, nutritionists know

that chestnuts are high in folates, which are essential for brain and nervous system development. To aid the heart, Hildegard advocated a tonic of parsley and honey-wine; parsley—rich in folic acid and essential oils—is today championed as a heart-healthy herb.

The four humors

Like other medical writers and practitioners of the time, Hildegard believed in the four humors (see pp.34–35), a theory promoted by Hippocrates in ancient Greece. The four humors were identified as blood, yellow bile, black bile,

△ **Leper with bell**
Early medieval physicians diagnosed leprosy as an excess of "black bile," and prescribed regular blood-letting as well as a drink containing gold, which was thought to be purifying. They wrongly believed that leprosy was easily spread, and forced lepers to ring a bell as a warning not to approach.

⟩⟩ learned her skills through an apprenticeship, but had little or no scientific training (see pp.140–41). This traditional type of medicine was often the main recourse for ordinary people without access to a physician. Focusing on herbal remedies, potions were typically dispensed by women, who had learned from older generations how to make folk remedies. Alternatively, a patient could visit an apothecary, who would concoct a tonic or remedy from herbs, spices, and wine.

300 **The number of plants with medicinal properties listed in the 12th century manuscripts of Hildegard of Bingen.**

Acquiring knowledge

One author who could claim some authority on the subject of both women's health and plant-based medicine was Hildegard of Bingen (see panel, p.56). Hildegard represents the reawakening of interest in medical knowledge, and the increase in its dissemination, that began in the mid- to late

▷ **Leeches**
Following principles first written down in ancient Greece, physicians in the medieval period would place leeches on a patient's skin to draw out blood that was supposedly bad. In modern medicine, leeches are sometimes used during reconstructive surgery to drain congested blood.

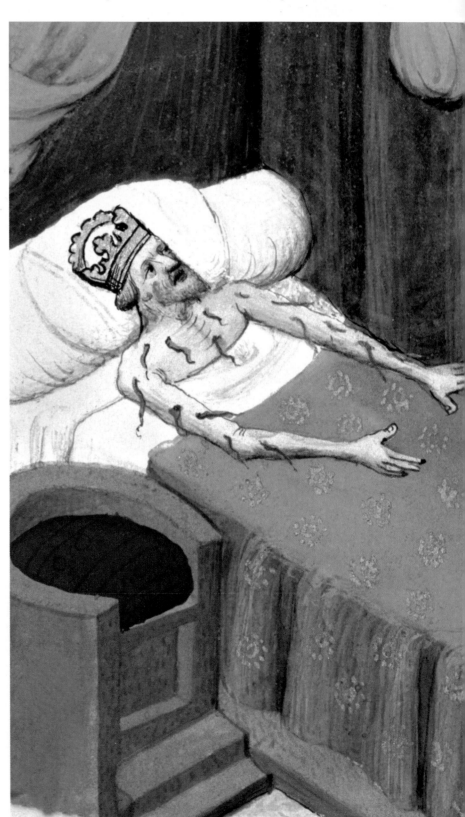

and phlegm, and were thought to directly affect the health of the body and emotions. All conditions were considered to stem from either an excess or a lack of one of the humors. Menstruation, for example, was of great interest to medieval scholars and physicians, who believed that the monthly discharge of blood was essential to keep the humors in balance. Following this line of thought, they believed that post-menopausal women were in great danger, since they were no longer able to get rid of "excess" blood.

Blood-letting

Reducing excess humors was one of the main medical procedures in medieval times—through blood-letting, intestinal purging, and induced vomiting. Blood-letting was the most severe of these treatments and was prescribed for many types of illness, including smallpox, epilepsy, and gout. Two main methods of blood-letting were used: leeching and the cutting of veins. Leeching (the milder of the two options) involved placing live leeches on the skin and leaving them to suck the patient's blood. The alternative was to open a vein with a lancet or pointed wooden stick, and let the blood flow into a basin.

If a doctor was not available to carry out blood-letting, monks and priests were authorized to step in and perform the procedure instead. In 1163, however, a church edict forbade the clergy from carrying out blood-letting, and barbers spotted this opportunity to expand their businesses. Barbers began to function as medical practitioners—offering blood-letting treatments, tooth extractions, lancing of boils, and even amputation, as well as the usual haircuts and shaves. Barber-surgeons (see pp.76–77) not only worked from their shops— identifiable from the blood-soaked towels drying outside—but also traveled around the countryside performing surgical procedures, and setting up temporary operating rooms on battlefields. Anesthetics were used, made from herbs or alcohol, but some of these were so potent that they could kill the patient before the operation had even begun.

1140 The year when King Roger II of Sicily forbade anyone from practicing medicine without a licence— the first regulation of its kind.

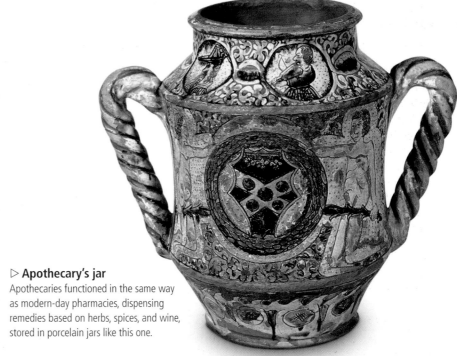

▷ **Apothecary's jar**
Apothecaries functioned in the same way as modern-day pharmacies, dispensing remedies based on herbs, spices, and wine, stored in porcelain jars like this one.

"Every day we see **new instruments** and new methods being **invented** by clever and ingenious surgeons."

THEODORIC OF LUCCA , SON OF HUGH OF LUCCA, MEDIEVAL SURGEON, 13TH CENTURY

Anatomy Restored

The origins of the modern study of anatomy are usually dated from Flemish anatomist Andreas Vesalius' 1543 text *De Humani Corporis Fabrica* (*On the Fabric of the Human Body*) (see pp.72–75). However, Vesalius and his colleagues owed much to the early anatomists who had developed their knowledge at the medical schools of Europe.

Toward the end of the early medieval period, Europe witnessed a revival in medical learning from ancient Greece and Rome. Medical knowledge from the Islamic world also flowed west. There was a renewed interest in anatomy, dissection, and autopsy, partly facilitated by new laws that permitted the dissection of human bodies for educational purposes. A prominent practitioner, Italian physician Mondino de Luzzi reintroduced public dissections for the benefit of students in about 1315, and wrote *Anathomia Corporis Humani* (*Anatomy of the Human Body*) in 1316. Mondino's pupil Nicola Bertuccio continued the practice and also produced works on how the body is affected by diseases, diets, and poisons.

In turn, Bertuccio's most famed student French physician Guy de Chauliac (see p.69) wrote *Chirurgia Magna* (*Great Surgery*), which went on to become a standard anatomical, medical, and surgical text in Europe for three centuries. In it, Chauliac urged all surgeons to study anatomy, and acknowledged the work of the physicians who had helped advance the field before him, including Hippocrates (see pp.36–37) and Galen (see pp.40–41), and their Islamic colleagues al-Razi (see pp.50–51) and Ibn Sina (see pp.52–53). It was almost two centuries later that Andreas Vesalius (see p.75) took the study of anatomy to the next level.

> **"A surgeon** who does not know his **anatomy** is like a **blind man** carving a log."
>
> GUY DE CHAULIAC, FROM *CHIRURGIA MAGNA* (*GREAT SURGERY*), 1363

◁ Anatomy class
This scene from an illustrated version of Guy de Chauliac's *Chirurgia Magna* shows the physician-surgeon identifying parts of the body while referring to a book. Assistants (center) carry out the actual dissection as students crowd in to observe.

Apothecary Store

The profession of apothecary—the formulator and dispenser of drugs to the sick—dates back to at least 2500 BCE. Skilled medics in their own right, apothecaries prepared medical remedies with the herbs stored in their shops.

1 **Pot marigold** Also called calendula, the flower is used to treat wounds and swelling, and as an infusion to calm fevers. 2 **Vervain** This plant was used to treat jaundice and gout, and to stimulate lactation in new mothers. 3 **St. John's wort** A strong anti-inflammatory, this plant is useful as a wound balm and to treat back pain. 4 **China rose** A tropical plant, China rose helps treat arterial and menstrual disorders. 5 **Saffron** Ground into a paste, this spice can be used as a sedative or a diaphoretic—to induce sweating. 6 **Cloves** These dried flower buds were once, and are sometimes still, used as an anesthetic and antiseptic in dentistry. 7 **Hops** The flowers of the hop plant were used as a sedative, useful for insomnia, anxiety, and stomach pain. 8 **Pestle and mortar** These were used to grind pharmaceutical ingredients into powders. This ivory example dates to 1500–1700. 9 **Opium** This container held Thebaic opium, a reference to its place of origin in the ancient Egyptian city of Thebes. In small quantities, opium worked as a calmative, sedative, and an expectorant to treat coughs. 10 **Pill silverer** This pill silverer from the UK dates to c.1860. It was a device used to coat pills in silver, or sometimes gold; the pills dropped inside, and the apparatus rotated to form the coating. 11 **Galangal** A type of ginger, galangal is used as a remedy for colic, flatulence, and respiratory problems. 12 **Garlic** Used as an antiseptic and against parasitic stomach infestations, garlic was also employed as a remedy for leprosy and smallpox. 13 **Ginger** This root is helpful in alleviating nausea, vomiting, and indigestion. 14 **Wild celery** Commonly employed as a diuretic (to promote urine production) it is also used to treat rheumatism and arthritis. 15 **Fresh mint** Used to ease indigestion, colic, and flatulence, this is either chopped on food or used as an infusion. 16 **Rosemary** Said to improve memory and banish bad dreams, this herb is also used to calm headaches. 17 **Aloe vera leaves** Taken internally, these cure constipation. When applied to the skin they soothe rashes and itches. 18 **Apothecary jar** Jars such as this Italian one from the 1500s were used to store drugs in apothecaries' shops.

Leafless spike

1 POT MARIGOLD

Pale lilac flower

2 VERVAIN

3 ST. JOHN'S WORT

4 CHINA ROSE

8 PESTLE AND MORTAR

9 OPIUM

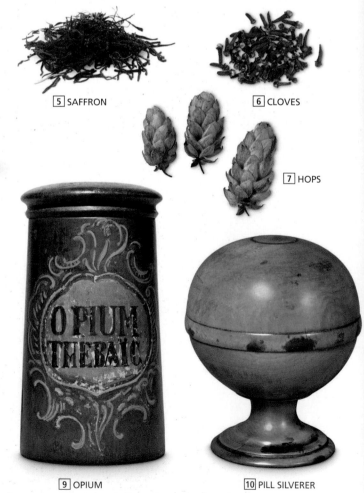

5 SAFFRON

6 CLOVES

7 HOPS

10 PILL SILVERER

Long stem appears
above ground

11 GALANGAL

Tuberous root
grows below
the ground

12 GARLIC

13 GINGER

14 WILD CELERY

15 FRESH
MINT

16 ROSEMARY

17 ALOE VERA
LEAVES

18 APOTHECARY JAR

AGISTRALE

Alchemy

A peculiar mix of science and magic, alchemy had various lofty aims that ranged from changing ordinary metals into gold to curing all illnesses. Dating back 4,000 years in Asia and Africa, alchemy enjoyed a golden age in Europe from the 12th to the 18th centuries.

The ancient civilizations of Egypt, India, and China all had long traditions of alchemy. The aims of early alchemists varied, but the underlying thread was always change or transmutation for the sake of improvement—physically, to alter a common substance into a precious one; spiritually, to bring light to darkness; medically, to give good health to the sick; or preferably all three.

There was a tendency toward esotericism among alchemists— restricting knowledge to a few privileged practitioners who could thereby mystify ordinary people. Yet alchemists also helped develop many real-world skills, such as extracting ingredients from plants, animals, and rocks; mixing, boiling, condensing, and purifying elements; and other procedures still practiced today.

Alchemy flourished during the "Golden Age" of Islamic medicine (see pp.48–51) and then journeyed west. Englishman Robert of Chester's 1144 translation of Persian polymath Jabir ibn Hayyan's (also known as Gcbcr) *Kitab al-Kimya* (*The Book of Composition of Alchemy*) encouraged alchemical practices across Europe. Among the alchemists' medical aspirations were to find a universal panacea to cure all ills and an elixir of youth. Swiss physician Paracelsus was a celebrated practitioner, whose free spirit, lengthy wanderings, contradictory statements, and yet practical talents, embodied the alchemic tradition. However, by the 1700s, faced with the rigorous application of the scientific method and the young subject of chemistry, alchemy faded into an occult pursuit.

"An **alchymist** is either a **physician** or a **soap boiler.**"

CORNELIUS AGRIPPA, GERMAN POLYMATH, FROM *THE VANITY OF THE ARTS AND SCIENCES*, 1530

◁ Seeking the elixir of youth
In the 13th century, English friar, philosopher, and alchemist Roger Bacon experimented with finding the elixir of youth. His reputation grew during the following centuries and inspired many others to turn to medicinal alchemy.

The **Black Death**

In 1347 a devastating epidemic struck Europe. Over the next five years, the infection—a plague characterized by black blotches across the victim's skin—killed approximately 60 percent of the continent's population, causing massive socioeconomic disruption.

Europe had experienced terrible plague epidemics before. The Great Plague of Athens, described by Greek historian Thucydides in 430 BCE, and the Plague of Justinian, which devastated the Byzantine Empire in 542 CE, both resulted in large-scale mortality, and may have been caused by the same organism responsible for the Black Death; however, these earlier outbreaks of plague affected a much smaller geographical area.

Recurring ravages

The "Great Pestilence"—as contemporaries called the Black Death—seems to have begun in Central Asia in the 1330s, before reaching Crimea in 1347, from where it rapidly spread westward along maritime trade routes. Venice and other Italian towns were struck that fall, and by the summer of 1348 France, Spain, Portugal, and England had been infected, with Germany and Scandinavia falling victim the following year.

The Black Death's vector, or spreading agent, was infected fleas harbored by the black rat (*Rattus rattus*), which thrived in the unsanitary conditions prevalent in medieval cities, where rubbish and human waste were omnipresent, and animals lived in the houses. The first symptoms of the disease were swellings in the lymph nodes of the groin, armpits, or neck, known as buboes – giving the Black Death its other common name, the bubonic plague. Black blotches then appeared on the skin, and death soon followed.

The Black Death caused panic throughout Europe. There was no cure. Ineffective treatments (see pp.68–69) included avoiding foods that were hard to digest, and purifying the air with attar (essential

▽ **The Great Plague of Marseilles**
These victims of the 1720 Marseilles plague show the swellings characteristic of the bubonic plague. The outbreak killed almost 100,000 people in Marseilles and its hinterland, and caused panic in other European countries, which feared a recurrence of the Black Death.

▷ The spread of the plague

The Black Death is believed to have reached Europe in 1347 through the port of Kaffa (today Feodosiya) in the Crimea, from where it spread throughout the Mediterranean on ships. By 1351 it had reached northern Scandinavia and Russia. Only a few regions, such as Poland, escaped.

KEY

	1347		1350
	1348		1351
	1349	→	ROUTE OF PLAGUE SPREAD

oil) of roses, cinnamon, and cloves (one theory maintained that the plague was spread by "miasmas" or noxious vapors). Doctors tried prescriptions of elixirs, such as *Theriaca Andromachi*—a concoction of herbs with up to 70 ingredients. Nothing worked, and only very remote communities escaped the epidemic. After it had killed around 50 million people, the first pestilence died out. It recurred in further waves, in 1360–63, 1374, and 1400, as new generations who lacked the immunity acquired from a previous infection fell victim.

Socioeconomic repercussions

The social and economic effects of the plague were devastating. Amid the terror of the first epidemic, thousands of Jews were slaughtered in Germany because they were blamed for poisoning wells and thereby causing the plague. As the population in Europe declined, laborers became scarce and land became vacant, allowing peasants to demand higher wages. Despite attempts to control wage levels, they rose inexorably, particularly in England.

Periodic epidemics of plague became a feature of European life for more than three centuries. England experienced its final outbreak in London in 1665, when 68,000 people died, and Marseilles, France, became the last European city to suffer, in 1720, when an infected ship carried the bubonic

plague into its port. Elsewhere, the disease remained endemic and a new wave of the plague began in 1894 in Canton in China, spreading the next year to India, where it claimed over a million lives.

23 DAYS The average time from first introduction of plague contagion among rats in a human community until first person dies from the disease.

Finding a cure

In 1894 the plague-causing bacillus was discovered by Japanese bacteriologist Shibasaburo Kitasato and French bacteriologist Alexandre Yersin; it was eventually named *Yersinia pestis*. Although early attempts to produce a vaccine against the plague failed, the rat flea was identified as the vector in 1898, leading to successful efforts to curb the spread of the disease by controlling the rat population.

By 1896 Yersin had produced an antiserum that was successful in about half of the cases, and the introduction of the antibiotic streptomycin in the 1940s increased the cure rate to about 95 percent.

While the Black Death can no longer decimate populations unchecked, it has not been entirely eradicated. In 1910 researchers realized that wild rodents, such as marmots (in Central Asia) and prairie dogs (in North America), act as reservoirs of the disease, and human contact with these

▷ Plague doctor

In order to avoid becoming infected, physicians called on to treat plague victims wore elaborate costumes, including masks with birdlike beaks, to reduce exposure to the "miasmas" believed to be the cause of the disease.

species causes periodic outbreaks. In 2013 a boy died in Kyrgyzstan after eating a plague-infected marmot, while in the US there were 15 cases of plague infection, including four deaths, in 2015.

"Its **earliest symptom...** was the appearance of... **swellings** in the groin or the armpit, some of which were **egg-shaped,** while others were roughly the **size of the common apple.**"

GIOVANNI BOCCACCIO, ITALIAN WRITER, IN *THE DECAMERON*, 1350

In medieval times, the term "plague" was used to refer to any epidemic. These plagues were often what we now know to be diseases such as malaria, typhoid, cholera, measles, syphilis, and smallpox. The Black Death (see pp.66–67), however, was the worst of all plagues, unprecedented in its virulence and destruction of human life. These devastating epidemics evoked a variety of responses, not the least of which was fear and panic.

Prayer or flight

With no idea of what caused such diseases or how they spread, some people simply fled them. However, in the Islamic faith fleeing was not an option: plague was viewed as an act of God, so had to be endured.

▷ **Spreading fragrance**
This spherical, eight-sectioned pomander was used to carry flowers, herbs, and spices such as nutmeg and musk that were thought to cleanse the air and ward off infection by the plague.

Preventing Plagues

Plagues were nothing new, but the arrival of the Black Death in the 14th century was one of the most devastating pandemics in human history. Medicine was powerless to treat it, but over time, organized responses were developed to prevent the spread of such diseases.

Dealing with the disease
During the Great Plague of London (1665–66), fires burned day and night to purify the air. Bell-ringers chimed for people to bring out their dead, and infected houses were sealed and marked with a red cross.

Many Christians believed that God was punishing humanity for its sins, so only prayer and penitence could end the plague. As a result, self-flagellation became increasingly popular, which led to thousands of penitents traveling through towns and countryside, flogging themselves with a three-tailed lash and praying that the Lord would take pity on their suffering and end the plague.

Over time, less responsibility was placed on God for inflicting such punishment. Plagues such as the Black Death and mass outbreaks of disorders such as St. Anthony's Fire (a gangrenous condition caused by ergot fungus poisoning), and St. Vitus' Dance (which presented as manic dancing), were thought to be the work of the devil, using his human agents—heretics, Jews, or witches. In turning fear and anger outward, thousands of innocents were scapegoated and massacred.

Attempts at prevention

Public officials, state rulers, and individuals all took action to try to prevent the spread of disease. Some thought that the air was filled with disease-causing noxious vapors or

giorni, meaning 40 days. Quarantine gradually became an accepted measure for treating outbreaks of plague. In 1374 the Duke of Milan drew up an edict insisting that all those suffering from plague should be taken outside the city walls to a field or forest, until they either recovered or died.

a theory contested by Muslims who believed that Allah directed such plagues. Failure to prevent the spread of the Black Death was only explained centuries later, when it was discovered that fleas were carriers of the plague (see p.67).

In the following centuries, systems for isolating the sick were greatly improved. In the early 1600s a law was passed forbidding travelers from entering Paris without a medical examination. By 1650 this

ethos had made its way as far as America, where thousands of travelers to the New World were stopped at Boston Harbor to be checked, or risk a hefty fine of $100. By the time the Great Plague ravaged London in 1665–66, all London-bound ships were made to drop anchor at the mouth of the Thames River for 40, sometimes 80, days. Sick Londoners were forced to stay in their homes, which were often boarded up. Those who could afford to, fled to the countryside.

In the 18th century, the arrival of another plague—yellow fever—in the Mediterranean ports of France, Spain, and Italy forced governments to introduce strict quarantine rules. The first major American yellow fever epidemic hit Philadelphia in July 1793, but politicians resisted quarantines because they were reluctant to limit trade. It was only the continued outbreaks of this disease over the next few decades that finally prompted US Congress to pass federal quarantine legislation in 1878.

> "Such **terror** was struck into the hearts of men and women by this calamity, that **brother abandoned brother...** fathers and mothers refused to see and tend their children."
>
> GIOVANNI BOCCACCIO, ITALIAN WRITER, ON THE PLAGUE AS IT RAVAGED FLORENCE, 1348

"miasmas" (see pp.120–21), which could be removed by lighting fires. People also began carrying sweet-smelling pomanders in an effort to cleanse the infected air.

In some places, the authorities reacted by isolating the sick. The cities of Venice and Milan refused entry to anyone suspected of being infectious. In 1348 ships arriving in Venice from infected ports were required to sit at anchor for 40 days before landing. The name for this practice—quarantine—was derived from the Italian words *quaranta*

The first permanent plague hospital (lazaretto) was opened by the Republic of Venice in 1423 on the small island of Santa Maria di Nazareth, away from the heart of the city. This concept spread to other parts of Europe as a way of containing the sick. Public officials also used disinfection procedures, such as fumigation and the burning of infected clothing and bedding. The nature of the contagion itself was not yet understood, but these measures suggested a belief that the disease was spread by people,

FRENCH PHYSICIAN (1300–1368)

GUY DE CHAULIAC

Born in Auvergne, France, Guy de Chauliac (see p.72) was a physician and surgeon who studied at the oldest university in Europe, the University of Bologna. In 1342 he was appointed by Pope Clement VI as his private physician. He attended the pontiff during the Black Death that came to France in 1348. A third of the cardinals at Avignon died, but Clement survived. Chauliac was also infected, but lived to record the experience and, unlike many physicians, he stayed and cared for the victims. In 1363 he wrote about it in graphic detail in his book *Chirurgia Magna* (*Great Surgery*), which became the most influential surgical text for more than 200 years.

Alchemy, Chemistry, and Medicine

For centuries, people investigated the properties of substances, how to purify them, and how they reacted when mixed. This field of science eventually became chemistry, but its mystical forerunner, alchemy (see pp.64–65), had a much greater influence on European medicine from the 12th to 18th centuries.

The ancient Greeks began to offer explanations about the structure of physical substances, or matter, as early as 380 BCE. The Greek philosopher Democritus believed that all matter was made up of invisible components called atoms that could not be broken down any further. Around the same time, the Indian philosopher Kanada came up with a similar proposal. However, neither of these theories were based on physical evidence.

A major step forward came in the 8th century, when Persian polymath Jabir ibn Hayyan examined the properties of materials using very basic laboratory equipment and processes such as crystallization distillation. Through his work, Hayyan developed an early chemical classification of matter: spirits, which vaporized when heated; metals, including iron and lead; and nonmalleable substances such as stone, which could be powdered. His breakdown is remarkably close to the modern

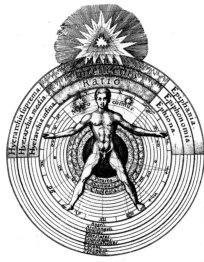

classification system of physical chemistry. Hayyan's texts describe processes familiar in chemical and drug research laboratories today. Hayyan produced hundreds of concoctions which, as a physician, he was able to test on patients, but he was not systematic about recording and analyzing his results.

More popular at the time was alchemy—a mix of mystical, philosophical, religious, and pseudoscientific approaches to

◁ **Understanding the world**
In *Utriusque Cosmi Historia* (*History of the Two Worlds*), published in 1617, physician Robert Flood illustrated his ideas on how the world worked, divided into physical, celestial, and spiritual dimensions.

explain the composition of, and changing states of, matter. Among its primary aims were finding a way to convert common materials into gold and silver and creating an elixir for everlasting life. However, the secretive and often intentionally baffling work of many alchemists, who jealously guarded their materials and methods, eventually led to skepticism from the general public and the wealthy patrons who funded their work.

Alchemical contribution

Nevertheless, medieval alchemists did make useful contributions to the field of medicine. One of the most influential was the 16th-century Swiss physician Philippus Aureolus Theophrastus Bombastus von Hohenheim, also known as Paracelsus. Although he adhered to some of the spiritual dimensions of alchemy, as well as various folk beliefs, Paracelsus also introduced many useful elements of chemistry to medical practice. He advocated that doctors should study nature and conduct experiments to understand the body's workings. He believed that metals were key elements, and he connected certain minerals to particular illnesses. For example, he found that goiter was caused by the presence of certain minerals in drinking water. He

◁ **At work**
Physician Theophrastus Aureolus Bombastus von Hohenheim called himself Paracelsus after the ancient Roman writer Celsus, who wrote the important early medical book *De Medicina* (*On Medicine*).

△ **Distilling spirits**
An engraving by Mannerist artist Jan van der Straet shows distilling equipment from the late 1500s and early 1600s. Medical alchemists used distillation to purify minerals and herbal extracts for use as drugs.

wrote: "Many have said of alchemy, that it is for the making of gold and silver. For me such is not the aim, but to consider only what virtue and power may lie in medicines." One of his beliefs—that which makes a man ill can also cure him—is the premise on which most modern vaccines are based. Gradually, during the course of the

Hamburg-born alchemist Henning Brand searched for the philosopher's stone. In the process he discovered a new chemical element in 1669. After reading a recipe claiming urine could be turned to silver he heated the residue from boiling down 60 buckets of urine, and isolated a white, waxy, glow-in-the-dark substance—called phosphorus after the Greek for "light-bearing." This was a new material for alchemists to exploit and some found that its compounds helped patients suffering from muscle weakness and lack of energy—a condition known as hypophosphatemia.

◁ **Alchemy in the Middle East**
This illustration from *Five Arabic Treatises on Alchemy* shows the distillation process. A large number of natural substances were discovered by Islamic alchemists using equipment such as this.

" The alchemists in their **search for gold** discovered many other things of **greater value.** "

ARTHUR SCHOPENHAUER, GERMAN PHILOSOPHER, 1780–1860

16th and 17th centuries, alchemy's focus became less supernatural and more rational, and alchemists were seen less as sorcerers and more as serious practitioners. Inspired by the ideas of Paracelsus, English physician Robert Flood wrote and illustrated *Utriusque Cosmi Historia (History of the Two Worlds)* (1617), which mixed medicine with

mysticism, and attempted to identify the materials of the universe, picturing God as an alchemist in a laboratory.

Switch to chemistry
The individual approaches of the alchemists, and the persisting spiritual and mystical dimensions, meant that alchemy could not

progress in a scientific way. The popularity of alchemy began to wane in the later part of the 17th century. In his textbook *The Sceptical Cymist* (1661), Anglo-Irish chemist Robert Boyle proposed that scientific investigation was the key to understanding chemistry. By the 18th century, chemistry had become a fully fledged science.

The **Anatomy Revolution**

One of the most important publications in medical history, Andreas Vesalius's anatomical masterwork *De Humani Corporis Fabrica* (*On the Fabric of the Human Body*), 1543, was pivotal in jolting medicine out of the stagnation of the medieval period.

As the power of Rome faded in Europe during the 4th and 5th centuries, the arts and sciences declined, along with many other intellectual pursuits (although progress continued in the Islamic world; see pp.48–51). Medicine relied on the great works of ancient Greece and Rome, although these gradually came to be distorted as new findings were introduced. Medicine is built on the twin foundations of anatomy and physiology—the structure and the workings of the human body. However, the study of anatomy almost disappeared, and surgeons, physicians, and others relied on the teachings of Claudius Galen (see pp.40–41). At a time when new attitudes and the quest for fresh knowledge were regarded as threats, Galen's works gained godlike status and were accepted without question.

Beginning around the 13th–14th century, the European Renaissance gave fresh impetus and a new questioning approach to art, architecture, and literature, allowing room for innovation and invention. However, medicine, and science in general, lagged behind. Although some advances were made by practitioners such as the Italian physician Mondino de Luzzi and French physician and surgeon Guy de Chauliac, the influence of Galen, Hippocrates, and other ancient physicians was so great that most medical authorities saw no need to follow the new Renaissance trends, and any challenges to the accepted traditions were suppressed.

The breakthrough

In 1543 Flemish physician and anatomist Andreas Vesalius produced *De Humani Corporis Fabrica Libri Septum*. It is now considered to be the first major anatomical work of the modern era, yet at the time it was ridiculed by some members of the medical establishment, who not only refused to understand what they saw with their own eyes, but even refused to look. Vesalius had studied medicine in Paris but he had to leave when his homeland, now part of Belgium, was caught up in a war between the Holy Roman Empire and France (see p.75). In 1536 he made his way back to Belgium via the University of Leuven (Louvain), before moving on to Venice and then to Padua in northeast Italy, where he studied for his doctorate in medicine (Padua had an exceptional reputation as a seat of learning). On qualification as a physician in 1537, Vesalius was immediately appointed professor of surgery and anatomy at the age of just 22 years.

Vesalius soon began to show his independent attitude, adopting a hands-on approach rather than following the established method. He focused on the demonstration of anatomy by dissection, believing it was fundamental to medical knowledge and surgical practice. Following the example of his mentor in Paris, Jacques Dubois (also known as Jacobus Sylvius), he opened up bodies himself during anatomy lessons. He and his students peered inside and they studied what they saw. Vesalius illustrated the actual anatomy in front of them, using his own skills and the guidance he obtained from his artist colleagues. This observational, empirical

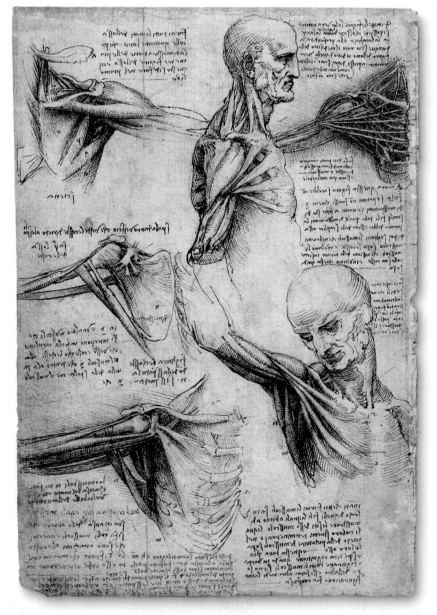

◁ **Leonardo's anatomy of the shoulder**
Vesalius was inspired by the works of artist and scientist Leonardo da Vinci, who had also produced anatomical illustrations. Both were interested in the way form (shape) reflected function in the body.

" Aristotle… says **men** have **more teeth** than **women**… no one is prevented from counting…. "

ANDREAS VESALIUS, FROM THE CHINA ROOT EPISTLE, 1546

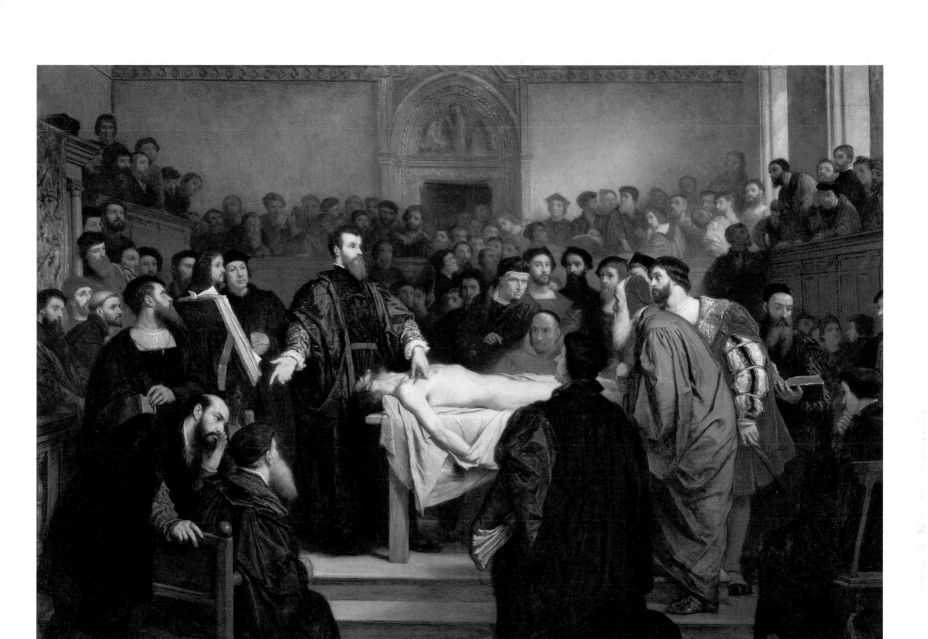

△ Vesalius in Padua

This 1859 work is by Belgian artist Edouard Hamman, who specialized in portraying famous people. Vesalius is shown lecturing and demonstrating at Padua, while reading from a traditional text (perhaps by Galen) held by an assistant.

approach was very unusual for the time. Traditionally, an assistant or barber-surgeon (see pp.76–77) carried out dissections and the corpse received only a brief survey, as the demonstration was deemed secondary to the professor reading out texts by Galen and others.

Around 1840 Vesalius began to notice discrepancies between Galen's time-honored works and what he was seeing with his own eyes. He realized that Galen had only been allowed to dissect animals and had then made assumptions based on their anatomy about the human body. Vesalius also studied animal anatomy, but unlike Galen he could compare it directly with »

▷ Padua's anatomy theater

Part of Vesalius's legacy was the promotion of anatomy to an essential medical subject for physicians and surgeons. This anatomy theater was built in his honor at Padua, and opened in 1595 to allow students a close view of the proceedings.

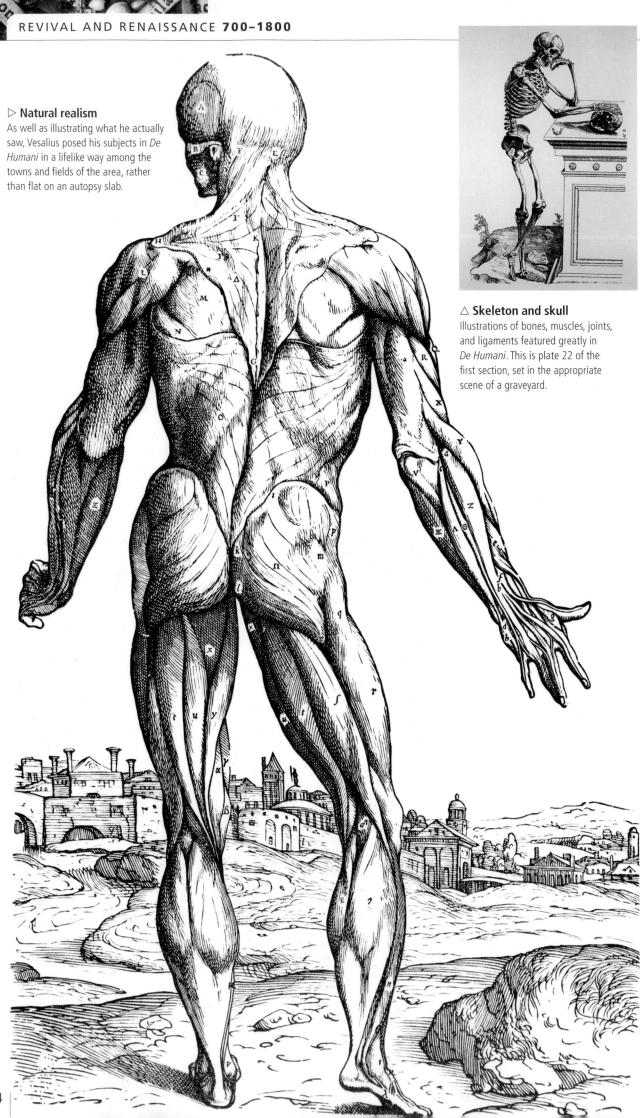

▷ Natural realism
As well as illustrating what he actually saw, Vesalius posed his subjects in *De Humani* in a lifelike way among the towns and fields of the area, rather than flat on an autopsy slab.

△ Skeleton and skull
Illustrations of bones, muscles, joints, and ligaments featured greatly in *De Humani*. This is plate 22 of the first section, set in the appropriate scene of a graveyard.

his knowledge of human anatomy gained by dissection. Vesalius's view began to clash with that of other anatomists at Padua. However, a local judge was interested in Vesalius's approach and he agreed to provide him with the bodies of executed criminals, thereby allowing him much more freedom to dissect, scrutinize, and analyze at great length.

The medical tome
By 1538 Vesalius had published six posters of anatomical illustrations, *Tabulae Anatomicae Sex* (*Six Anatomical Plates*), for his students. He decided that a full-length book derived solely from his own anatomical studies was needed. *De Humani*, published in 1543, was an immense and ground-breaking work in many ways.

400 ILLUSTRATIONS in 260 scenes depicted in *De Humani*, which had a page size of 16.5 x 11 in (42 x 28 cm).

A huge book, with more than 600 pages, it comprised seven sections: bones and ligaments; muscles and tendons; blood vessels; nerves; digestion; heart and lungs; and brain and sense organs. The images were based on observations and studies of real dissections. They were beautifully executed with contours and shading to give a three-dimensional effect. The artist is unknown, but is unlikely to have been Vesalius himself. The illustrations may instead have been drawn by noted painter Jan van Calcar, whom Vesalius had met in Venice and who probably also contributed to *Tabulae Anatomicae Sex*. In *De Humani*, the bodies are shown in inventive lifelike poses, many set in the Italian countryside. Vesalius chose the eminent Joannis Oporini of Basel for the printing to ensure that the book was of the highest quality and used the latest technology. The size, scale, clarity, and content of the work astounded everyone in the medical profession, and despite its high price, it soon sold out. In his work, Vesalius took the standpoint that, as in nature, as

◁ **Colored frontispiece**
The first printings of *De Humani* and *Epitome* were in black and white. Special presentation copies and later editions were hand-colored, like this frontispiece from *Epitome*.

FLEMISH PHYSICIAN (1514–1564)

ANDREAS VESALIUS

Vesalius was born in Brussels into a well-educated family—his father being an apothecary to both the Holy Roman Emperor Maximilian I and his successor Charles V. After publishing *De Humani* at the age of 28 years, Vesalius's fame spread to the court of the Holy Roman Emperor Charles V, who invited him to join him as imperial physician in 1544. This may have been in part because Vesalius had presented to Charles a specially dedicated, bound, and hand-colored copy of *De Humani*. That same year Vesalius married and the couple had a daughter in 1545.

Vesalius traveled widely as a practicing court physician and from 1556, when Charles stood down in a series of abdications, he continued as physician to Charles's son Philip II, King of Spain. Again, Vesalius had already dedicated to Philip a condensed edition of *De Humani*, usually called *Epitome*. Vesalius and his family continued to enjoy the privileges of the royal court, but in 1564 he left Spain, perhaps to avoid rumored accusations of heresy by the Spanish Inquisition. His wife and daughter went to Brussels while he journeyed on a pilgrimage and medicinal plant-hunting expedition to the Holy Land. In Jerusalem he received a request to return to Padua, but on the way his ship was wrecked. Vesalius was stranded and died in obscurity on the Greek island of Zante (Zakynthos).

well as technology and mechanics, form and function are closely linked. He made many corrections to traditional beliefs—for example, showing that men and women have the same number of ribs; that the mandible (lower jaw) is a single bone, not two; that the liver has two lobes, not five; that nerves run from the organs to the brain, and not between organs; that the kidneys do not produce urine through the filtration of blood (though this was later proved to be true); and that the heart's central dividing wall, the septum, does not have visible pores and so blood cannot pass from one side to the other (see pp.82–83).

Observations and results
Some medical experts were horrified at the way *De Humani* contradicted the knowledge of Galen and others. Vesalius was also accused of being antireligious. However, more progressive members of the medical profession soon recognized that they could not deny what was in front of them. In 1555 Vesalius produced a revised edition of *De Humani* that corrected some of his own errors and extended the scope to include more on female anatomy and pregnancy.

In establishing the modern science of anatomy, the bold and independent Vesalius corrected long-held misconceptions and introduced new theories. He also inspired a fresh breed of anatomists, physicians, and surgeons, including the Italians Gabriele Falloppio and Bartolomeo Eustachi, both famed anatomists in their own right.

> "I am not **accustomed** to saying anything with certainty after only one or two **observations.**"

ANDREAS VESALIUS, FLEMISH PHYSICIAN, FROM *THE CHINA ROOT EPISTLE*, 1546

ANDREAS VESALIUS DEMONSTRATES MUSCLE DISSECTION

Barber-surgeons

The 11th and 12th centuries saw the birth of a new profession in Europe—that of barber-surgeons. Less well bred and educated than doctors, barbers—with their haircutting and shaving tools of sharp blades and potions, as well as their knowledge of skin and blood— were well equipped to take on medical challenges.

Doctors in the medieval period were wealthy and educated. They were well versed in the works of Hippocrates (see pp.36–37) and Galen (see pp.40–41), but they did not undertake hands-on activities such as blood-letting, administering enemas, wound dressing, and callus and worm removal. This was where barber-surgeons came in. Originally apprentices to the doctors and physicians, barber-surgeons gradually gained importance as indispensable medical practitioners in their own right. They moved up from their local barber shops to more official medical premises, rubbing shoulders with the medical elite. The scope of their work widened from setting broken bones to dressing wounds; soon they were appearing on battlefields across Europe, where their practical skills and pragmatic approach saved many lives.

In the 16th century, ambitious practitioners, such as Ambroise Paré (see pp.78–79), helped the barber-surgeon community gain legitimate recognition. However, the role of barber-surgeons faded by the 1700s, when medical training became more formal and organized. Specialized surgeons with university training and hands-on experience came to dominate the field of surgery and barbers went back to hair and beards.

> **"**At this, I resolved **never again cruelly to burn** poor people who had suffered gunshot.**"**
>
> AMBROISE PARÉ, FRENCH BARBER-SURGEON, AFTER SUCCESSFULLY APPLYING A WOUND DRESSING OF EGG WHITE, ROSE OIL, AND TURPENTINE, 1537

▷ **All in a day's work**
This painting from the 1670s by Flemish artist David Teniers II shows barber-surgeons busy at work. The chamber is cluttered with instruments, jars, and other equipment—distinctly different from the elegant consulting rooms of physicians of the time.

FRENCH BARBER-SURGEON Born 1510 Died 1590

Ambroise Paré

"I dressed him, and God healed him."

AMBROISE PARÉ'S MOTTO

French barber-surgeon Ambroise Paré started a quiet revolution in surgery in the mid-1500s. The changes he brought about were the result of his harrowing battlefield experience, which led him to question many established surgical practices.

A key moment for Paré came in 1537, when he was serving as an army surgeon during the Siege of Turin. Paré ran out of the boiling oil concoction used at the time to cauterize (sear and seal) wounds involving gunpowder, a process that allegedly "detoxified" the body of poisons believed to be carried by gunpowder and projectiles. In need of an immediate alternative, Paré recalled an ancient treatment. He mixed a potion of egg yolks, rose oil, and turpentine and applied it to the soldiers' wounds. The next day, Paré saw that the injuries were beginning to heal. Moreover, the horrific pain caused by the boiling oil treatment had been avoided. In light of this experience, Paré resolved to change his attitude toward

medicine and surgery. He decided to observe carefully, use his own judgment, try new ideas, and assess the results. This experimental approach went against the blind acceptance of age-old methods used by most physicians and surgeons at the time.

Humble beginnings

Born into a working-class family in France, Paré was apprenticed to his elder brother, a barber-surgeon (see pp.76–77) in Paris, when he was a teenager. At the age of 22, Paré was accepted as an apprentice barber-surgeon at the Hôtel-Dieu in Paris, which was linked to the forward-looking Faculty of Medicine at the Paris University. Unlike in other such institutions, the apprentices here attended lectures, and received extensive training in medical theory, diagnosis, and complex surgical procedures. They often worked alongside the highly qualified surgeons and physicians, rather than as assistants. The Hôtel-Dieu also introduced examinations and qualifications, giving barber-surgeons professional recognition for the first time.

Paré progressed well toward his exams, but when his funds ran low he joined the army as

◁ **Father of modern surgery**
In the great Hippocratic tradition, Paré believed that his role was to ease suffering rather than increase it, and to assist the body's natural curing powers rather than challenge them.

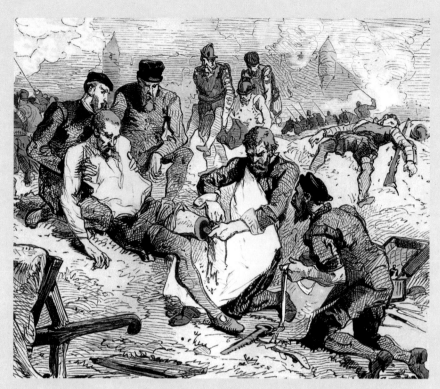

◁ Army surgeon
Paré's experience with amputations during his time as a battlefield surgeon led him to investigate the use of ligatures—strings or threads tied around the stump or vessel to stop blood loss.

obstetric procedure of repositioning an unborn baby to increase the chances of successful delivery. He disproved the myth of the antidotal power of the bezoar stone—a lump found in the intestines of various creatures. He tested a poison on a royal cook who had been sentenced to death, on the condition that should he survive, his life would be spared. The cook died seven hours after receiving the poison despite being given the bezoar stone.

Paré wrote at length about his experiences in French rather than the usual Latin of medical texts. This allowed less-educated barber-surgeons to learn from his experiences. With this readership in mind his books were also highly illustrated—yet another of Paré's innovations.

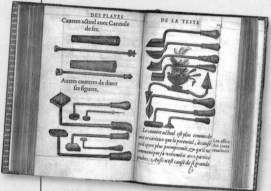

ILLUSTRATION OF SURGICAL INSTRUMENTS FROM PARÉ'S *LA MÉTHODE CURATIVE DES PLAIES ET FRACTURES DE LA TÊTE HUMAINE* (*TREATMENT METHOD FOR WOUNDS AND FRACTURES OF THE HUMAN HEAD*)

a regimental surgeon to raise money (he would pass the exams later, on his return).

Novel methods
Emergency amputations were usually followed by cauterization. Paré noted that this method was ineffective in containing blood

Paré secure his finances and allowed him more time to experiment. He devised several new forms of prostheses, including hands, arms, and legs—some with working mechanics—as well as false eyes and noses. In obstetrics, Paré is credited with reviving podalic version—the

> " See how **I learned** to treat gunshot wounds; **not by books.** "

AMBROISE PARÉ, FROM *LES VOYAGES FAITS EN DIVERS LIEUX* (*JOURNEYS IN DIVERSE PLACES*), c.1580

loss and began using ligatures for the purpose. However, unlike cauterization, ligatures tended to encourage infection. So some of Paré's colleagues began to combine the two methods.

Medical experts recognized and accepted Paré's abilities and innovations. He helped raise the status of barber-surgeons because their profession gradually merged with that of surgery. His talents also led to his appointment as royal physician to Henry II of France. Working at the royal courts helped

Catches and spring to operate hand

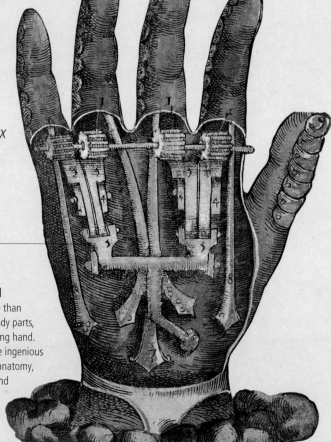

▷ Helping hand
Paré designed more than 50 kinds of false body parts, including this working hand. The mechanics were ingenious and based on true anatomy, but too elaborate and ambitious for routine use.

TIMELINE

- **c.1510** Born in Bourg-Hersent, part of Laval in western France. Paré develops an interest in medicine on account of his older brother being a barber-surgeon, and works as an unofficial apprentice to him.

- **1532** Begins training at the Hôtel-Dieu Hospital in Paris as a barber-surgeon, with hopes of becoming a general physician. He shows early promise and progresses rapidly.

- **1536** Appointed a regimental surgeon in the French army, at a busy time with several battles being fought by France against many enemies, including Spain, Portugal, and the Holy Roman Empire.

- **1537** Runs out of the boiling oil solution used to "detoxify" gunshot wounds and tries a new concoction with considerable success. Paré resolves to become more gentle, more experimental, observe sharply, and follow his instincts.

- **1545** His first major work *La Méthode de Traiter les Plaies Faites par les Arquebuses et Aultres Bastons à feu* (*The Method of Curing Wounds Caused by Arquebus and Firearms*) is published.

- **1552** Joins the House of Valois French royal court as physician to Henry II of France.

- **1559** Henry II dies of septicemia following an eye wound from jousting. Paré is commended for his efforts in trying to save the king and continues as physician to the next three kings.

- **1564** Writes *Dix livres de la chirurgie* (*Treatise on Surgery*), describing the use of ligatures to prevent bleeding after amputation, and other pioneering approaches to treatment.

- **1590** Dies at the age of 79, still holding his position of royal physician.

Repair and Reconstruction

Physical defects and disfigurements can occur for many different reasons, ranging from genetic problems to injuries in warfare. Throughout history, numerous techniques have been developed to repair, reshape, and reconstruct body parts, in order to restore function and create a more natural appearance.

The reasons for malformations and deformities of the face and body have changed through time. In past centuries leading causes were infectious diseases such as smallpox and leprosy, as well as growths and tumors, gangrene, skin ulcers, and radical surgery. Other causes include wounds and trauma, accidental burns, and amputations by machinery. Congenital problems (present at birth), such as a cleft lip and palate, may occur due to inherited or genetic defects or malformation in the developing fetus.

Ancient origins

Reconstructive surgery aims to repair, rebuild, and restore the shape and function of a body part.

It has been used since ancient times in India, Greece, and Rome, along with prostheses (see pp.236–37). One of the first mentions of reconstruction appears in the Indian text *Susruta Samhita* (see pp.30–31), which dates back more than 2,500 years. Conspicuous in this, and other works too, is the nose, partly because in ancient India, nose amputation was a common punishment for crimes such as adultery. *Susruta Samhita* describes the transplantation of patches of skin and even whole noses from one individual to another. The ancient Egyptian Smith papyrus (see pp.20–21), dating back to about the same time, also mentions nose repair. Around 2,000 years ago, Roman writer Aulus Celsus included techniques for the reconstruction of noses and other parts in *De Medicina* (*On Medicine*).

Rebuilding noses

Due to its prominence, the nose is particularly vulnerable to traumatic

◁ **The Indian method**
This 1795 engraving shows an Indian patient about 10 months after undergoing rhinoplasty to repair his nose, which had been cut off during his time as a prisoner of war. His forehead retains a scar from where flesh was sliced and folded over to cover the exposed nasal cavity.

damage. The 16th-century astronomer Tycho Brahe was famed for wearing false noses—allegedly made of silver, gold, copper, brass, or wood—after his was sliced off in a sword duel in 1566. In the same century in Europe, syphilis (see pp.186–87)—a disease newly arrived from the Americas—swept through the population, causing all kinds of terrible symptoms. Among

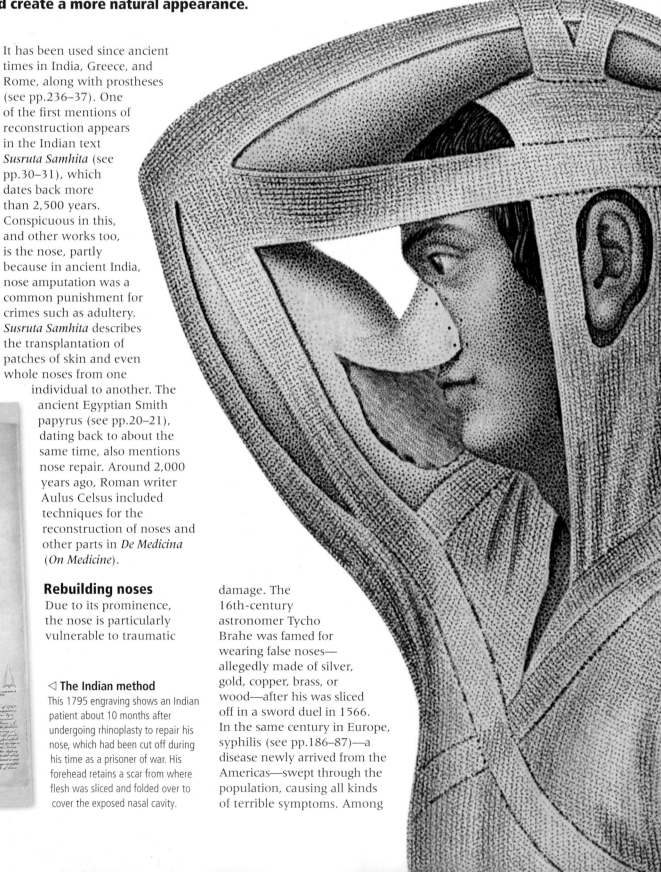

"We… **rebuild** those parts which nature hath given, but… fortune has **taken away.**"

GASPARE TAGLIACOZZI, FROM *DE CURTORUM CHIRURGIA PER INSITIONEM*, 1597

the most visible of these was a collapsed nose, or "saddle nose." As a result, nasal reconstruction, also known as rhinoplasty, became a major medical procedure of the time.

Ancient Indian techniques for rhinoplasty involved slicing a thin flap of skin from the forehead, or perhaps the cheek, angling it around, and applying it to the nose area. The flap was left partially attached by one or more small stalks of skin, called pedicles. The pedicles contained blood vessels and nerves to sustain the transferred skin until it attached naturally to the nasal area. Meanwhile the exposed area on the forehead was reduced and camouflaged by stretching the skin and suturing (stitching) the edges, and by wearing a turban or similar headgear. The Indian method of rhinoplasty was observed by traveling Europeans and also made its way to Europe via Islamic texts.

Refining the art

In 1412 barber-surgeon Gustavo Branca was licensed to practice in Sicily, Italy, where he and his son Antonio soon gained a reputation for reconstructive surgery of the nose and other facial features. In 1456 Italian historian Bartolomeo Facia wrote "Branca was the originator of an admirable and almost incredible procedure. He conceived how to repair and replace noses that had been mutilated or

1 MILLION The number of rhinoplasty procedures conducted in the US every year.

4 THOUSAND The number of rhinoplasty procedures conducted in the UK every year.

cut off and developed his ideas into a marvelous art." Facia reported that Antonio Branca used skin and flesh from the arm rather than the cheek or forehead, binding the patient's arm up against the head for 15–20 days before severing the pedicle. These techniques were refined by Prussian army surgeon Heinrich von Pfolsprundt, who wrote about the procedure in *Buch der Bündth Ertznei* (*Book of Directions for Bandaging*) in 1460.

In 1597 Italian surgeon Gaspare Tagliacozzi published *De Curtorum Chirurgia per Insitionem* (*On the*

◁ **The Italian method**
During the 15th and 16th centuries, a series of Italian surgeons developed a method of using skin from the arm for rhinoplasty. The arm had to be held in place firmly and tightly for weeks, otherwise the skin would easily detach.

THE GUINEA PIG CLUB

Finding volunteers to take part in experimental procedures is a common challenge for medical projects. The UK's Guinea Pig Club was formed in 1941 for World War II military air crew who had suffered disfiguring injuries, especially burns. These were treated by skin grafts and other pioneering reconstructions at Queen Victoria Hospital, East Grinstead, Sussex.

Most members were patients of New Zealand surgeon Archibald McIndoe. While working with the veterans, McIndoe developed new techniques to save lives, restore function, improve appearance, and help rehabilitation. Patients continued to join the club after the war, including servicemen who had suffered injuries during the 1982 Falklands War. The club was officially disbanded in 2007.

Surgery of Mutilation by Grafting). This pioneering account helped establish and advance several kinds of reconstructive surgery, including the Italian method of rhinoplasty based on using skin from the arm, which Branca had developed.

Tagliacozzi reasoned that the option of reconstructive surgery involves weighing the benefits, ranging from the undoubtedly medical to solely cosmetic, against potential disadvantages such as discomfort, pain, infection, and perhaps failure of the procedure. For example, rhinoplasty has several advantages. It conceals the deep nasal cavity visible when the nose is missing, which could be of great psychological benefit to the patient. It also helps keep the mucous membranes lining the cavity moist and free from irritation, directs airflow in the correct way, and restores more normal speech quality and tone. In addition, the nose provides support for eyeglasses, which were rapidly becoming popular during Tagliacozzi's time.

The term plastic surgery, involving reconstructive surgery for medical as well as cosmetic or esthetic reasons, was introduced into medicine in 1818. It was used in German surgeon Karl Ferdinand von Gräfe's report *Rhinoplastik*, which dealt with the procedure of nose reconstruction and improved upon older techniques. The report came 90 years before the invention of synthetic, moldable plastics, and the term "plastic" was used to imply "being shaped or molded."

△ **Prosthetic noses**
Disfigured noses were sometimes covered with prosthetics. Of the examples above, the nose on the left is made of ivory and that on the right of plated metal. They were usually attached by pastes made from natural ingredients, such as plant sap.

Discovering the Circulation

The concept of circulation—blood pumped by the heart and travelling around the body through vessels—seems obvious today, but it was a mystery for millennia. It was not until 1628 that English physician William Harvey gave the first accurate account of this fundamental aspect of physiology.

Early notions of the heart, blood, and vessels were often metaphysical or fantastical. In ancient China, the *Huangdi Neijing* (*Yellow Emperor's Classic of Internal Medicine*, see pp.26–27) described how blood mixed with *qi*, or life energy, and spread around the body. In ancient Greece, Hippocrates (see pp.36–37) believed that the arteries carried air from the lungs and that the heart, thought to have three chambers, was the seat of intelligence, vitality, and warmth. Another Greek physician, Erasistratus, believed that the heart produced a "life vapor," or pneuma, and blood ebbed to and fro in the veins. In ancient Rome, physician Claudius Galen (see pp.40–41) showed that arteries contain bright red blood under high pressure, while the veins contain dark blood under low pressure. He hypothesized a system in which digested food went to the liver, where it was made into new blood, which was then sent via the veins to various body parts, including the heart, where it mixed with air from the lungs. Galen believed that blood emitted from the liver through the veins had a lowly form of "natural spirit." In the heart, blood seeped through tiny pores in the wall, or septum, from the right to left side, and so into the arteries. Here it became charged with a higher form, or

62,000 MILES **The total length of the network of blood vessels in the human body.**

▽ **The dissection of Thomas Parr**
William Harvey carried out many dissections, including the bodies of his father and sister. Here his subject is Thomas Parr, an Englishman who was said to have lived to the age of 152 years.

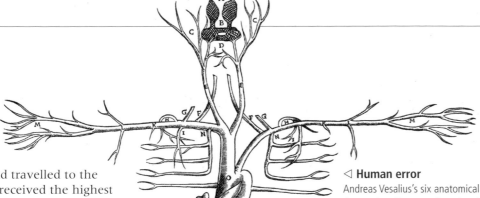

"vital spirit," and travelled to the brain, where it received the highest form, known as "psychic spirit."

Debunking old myths

More than a thousand years passed before anatomists and physicians began to question Galen's theories. Arabic physician Ibn al-Nafis (see p.49) challenged the notion of tiny pores in the heart: "The thick septum of the heart is not perforated and does not have pores... The blood from the right chamber must flow through the vena arteriosa [pulmonary artery] to the lungs, spread through its substances, be mingled there with air, pass through the arteria venosa [pulmonary vein] to reach the left chamber of the heart, and there form the vital spirit." Here was the first account of pulmonary circulation from the heart's right side via the lungs to the left side.

In the early 1500s Italian artist-anatomist Leonardo da Vinci made accurate anatomical drawings of the heart, indicating septal pores in his illustrations, even though he was unable to locate them. Flemish-born anatomist Andreas Vesalius (see p.75) also searched for the pores during studies for his great work *De Humani Corporis Fabrica (On the Fabric of the Human Body)*, and concluded "even a fine bristle cannot be made to penetrate from one ventricle to another."

The gradual debunking of ancient wisdom continued when Spanish physician Andres Laguna affirmed in 1535 that the heart had just two ventricles, rather than three. There was further development in the 1540s when Portuguese-born physician Amato Lusitano showed how valves within blood vessels allowed only a one-way flow of blood, not the two-way ebb and flow of Galen's system.

Double circulation

Al-Nafis's prescient description of a pulmonary circulation was refined by Spanish anatomist and scientist Michael Servetus in his 1553 work *Christianismi Restitutio (The Restoration of Christianity)*. Six years later, Italian anatomy professor Realdo Colombo published *De Re Anatomica (On Things Anatomical)*, which supported the idea of a pulmonary circulation and described how the heart contracted to force blood into the arteries. Italian physician Andrea Cesalpino is credited with founding the concept of general circulation when he concluded in 1569: "the blood is driven to the heart through the veins, where it attains its last perfection, and having acquired this perfection, it is brought by the arteries throughout the body."

The puzzle pieces were finally assembled into the double circulatory system we know today by William Harvey (see pp.84–85)

◁ **Human error**

Andreas Vesalius's six anatomical plates, *Tabulae Anatomicae Sex*, were based on his own dissections. However, he was unwilling to contradict the 1,300 year-old teachings of Galen, and the heart and aorta in this diagram are similar to those of Galen's dissected apes.

in 1628. As chief physician at St. Bartholomew's Hospital, London, and royal physician to James I and his heir Charles I, Harvey had been dissecting animal species and cadavers for almost 20 years. In his seminal book *De Motu Cordis (On the Motion of the Heart and Blood)*, he introduced the idea of a pulmonary circulation pumped from the heart's right side to the left, via the lungs; and a systemic circulation pumped from the heart's left side around the body and back to the right. Harvey had the faith, but no microscope, to identify the connections between the tiny arteries and veins that complete the circulation. Italian scientist Marcello Malpighi revealed these as capillaries in 1661 (see p.96).

▽ **Revolutionary book**

William Harvey's *De Motu Cordis* signaled a new era in medicine. Physicians now understood circulation and why for example, maintaining both arterial and venous blood supply to a body tissue would help avoid gangrene.

" The **concept** of a **circuit** of the **blood** does not **destroy**, but rather **advances** traditional **medicine**."

WILLIAM HARVEY, FROM *EXERCITATIONES DUAE ANATOMICAE DE CIRCULATIONE SANGUINIS*, 1649

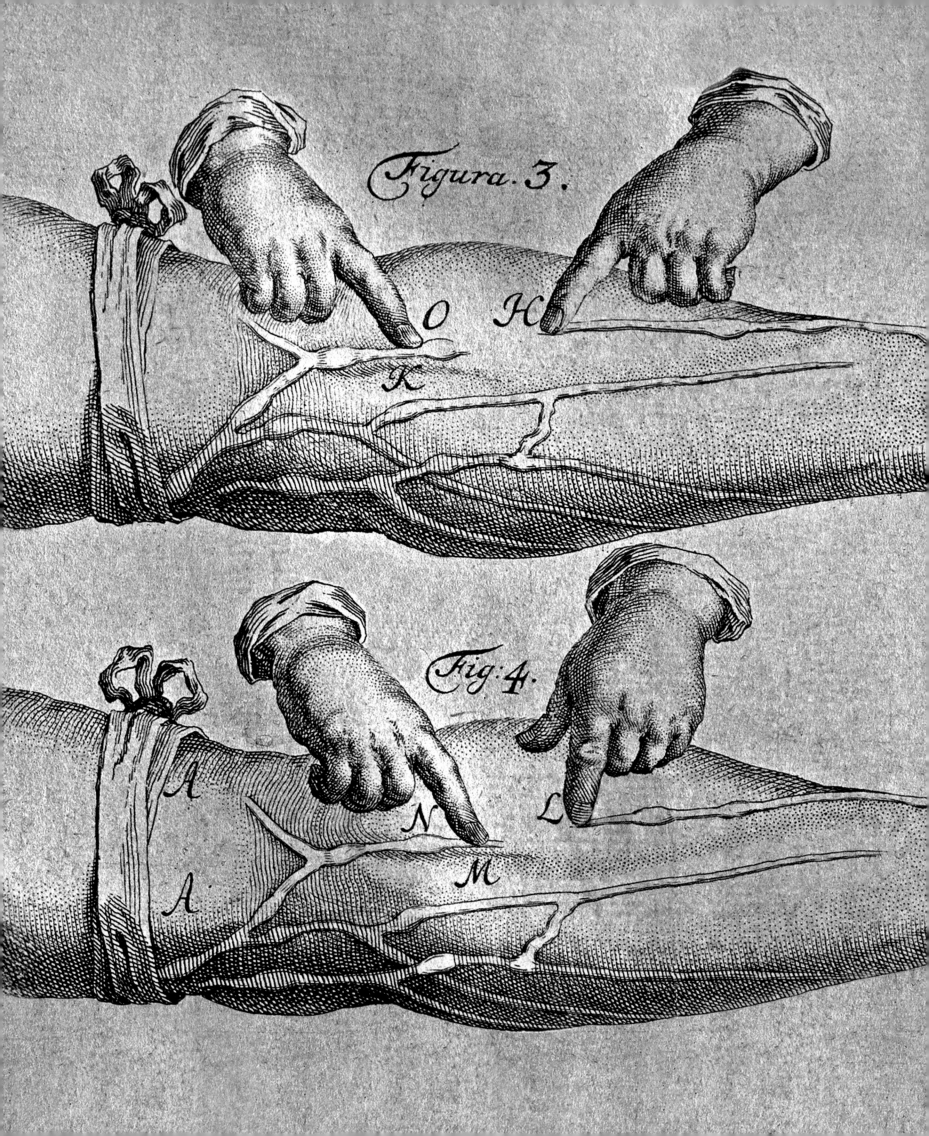

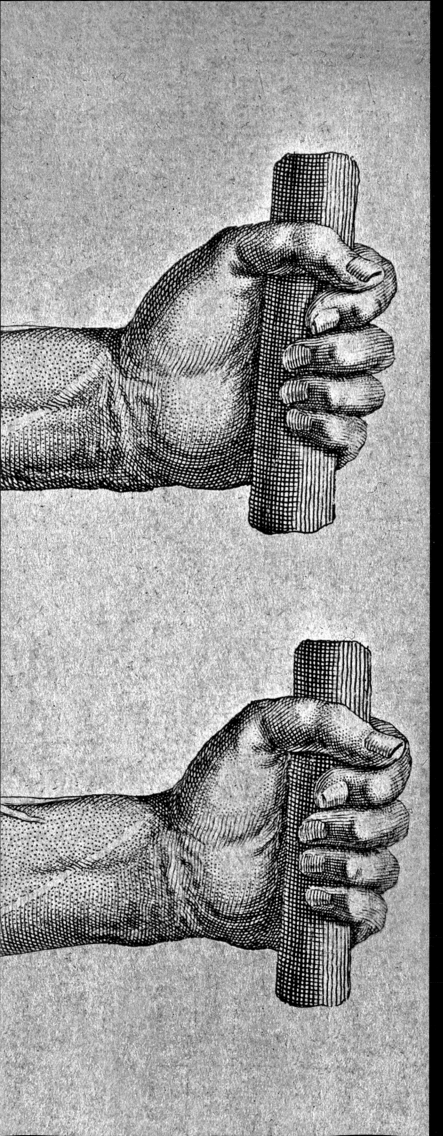

The **Circulation Revolution**

William Harvey's classic work *De Motu Cordis* (1628) was badly printed and relatively short at 72 pages. However, it contained a well-rounded explanation of the circulatory system that revolutionized physiology and medical theory.

In *De Motu Cordis*—short for *Exercitatio Anatomica de Motu Cordis et Sanguinis in Animalibus* (*An Anatomical Exercise on the Motion of the Heart and Blood in Animals*)—British physician William Harvey compiled many concepts to do with the circulatory system, some dating back to ancient Greece and Rome, and integrated them with his own theories and evidence. He carried out various studies, dissections, and experiments on humans and more than 60 animal species over a period of more than 20 years. From this mass of data, he drew a number of sensible conclusions, such as that, "The blood does pass through the lungs and the heart by the pulse of the ventricles, and is... sent into the whole body, and... returns from the little veins to the greater... from whence it comes... into the ear [atrium] of the heart." In particular, Harvey understood that there were two circulations—from the heart via the lungs and back (pulmonary), and from the heart through the body and back (systemic).

De Motu Cordis received a cautious welcome from some but outright hostility from others. Because it denied the teachings of Galen (see pp.40–41) and other revered ancients, critics claimed Harvey was "crackbrained." However, opinions gradually shifted and the science of *De Motu Cordis* prevailed.

> **"The blood is driven into a round by a circular motion... it moves perpetually."**
>
> WILLIAM HARVEY, FROM *DE MOTU CORDIS*, 1628

◁ **Ligature sequence**
This illustration from *De Motu Cordis* shows the valves that prevent the reverse flow of blood in veins. A ligature, or tight band, around the upper arm compresses superficial veins, where blood collects, unable to flow toward the heart. Massaging blood toward the hand has no effect due to the one-way valves, which appear as small lumps.

Cataract Surgery

The world's leading cause of poor vision and blindness is the misting, or clouding, of the eye's lens, known as cataract. While simple treatments began more than 2,000 years ago, major advances made since 1967 now enable sight to be restored to millions each year.

The chief factor in cataract formation is advancing years. Other possible factors are tobacco smoking and prolonged exposure to strong sunlight. As a cataract forms, the clear, flexible lens of the eye—through which light passes after the pupil (hole) and before the retina—gradually develops misty or opaque patches. Eventually in a "ripe" cataract the lens becomes toughened, stiff, and milky, and blocks all vision.

Early removal

Cataracts were mentioned millennia ago, in works such as the Indian *Susruta Samhita* (see pp.30–31). In ancient Rome, Greek philosopher Celsus's *De Medicina* described an already well-established cataract treatment called couching. In this procedure, a sharp-pointed, but not slender, needle was pushed through the eye's surface, its cornea, and the pupil until it met the toughened lens, which was then manipulated downward within the eye. This allowed light to pass to the retina again, although the loss of a focusing lens meant some blurring.

An alternative to needle-couching was to strike the eye with a blunt instrument, so that the tiny ligaments holding the lens in place ruptured and the lens slid away of its own accord. However with both these procedures an "unripe" cataract could rupture and spill lens fragments into the eyeball's jellylike interior, risking inflammation, pain, and further visual problems.

Couching remained the chief cataract treatment for centuries. Progress of a sort occurred in the 10th century with the use of a wider, hollow needle to suck out the whole lens, as described by al-Razi (see pp.48–51) and other Islamic

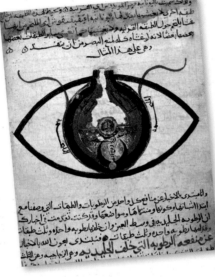

◁ **Anatomy of the eye**
This semi-anatomical illustration is from the 13th-century *Treatise on the Eye* by Arabic physician al-Mutadibid. At that time, treatments for eye conditions varying from bruising to infection were in regular use.

physicians. While the benefit was that the lens could not slip back across the pupil, the risk of it breaking remained and the technique was not widely adopted.

In Paris in 1748, French eye doctor Jacques Daviel pioneered a new technique. He cut a C-shaped slit in the cornea; inserted a narrow spatula to hold the cornea away from the lens; freed the lens from its surrounding capsule with a needle; and manipulated the spatula so that pressure around the lens caused it to pop out of the capsule, and through the incision. Leaving the lens capsule in the eye meant less risk of fragments making their way to the interior. Daviel's method was painful and there were no stitches small enough to

▷ **Surgical detail**
One of the earliest works with pictorial details of cataract excision was *Complete Human Anatomy Treatise Including Surgical Treatments* by French physician Jean-Baptiste Bourgery, completed in 1850. This edition dates from 1866.

suture the incision, so patients were immobilized for days while healing. Topical anesthetics—applied to the surface of a body part to numb it—were developed in the late 1800s, and these, along with smaller, finer sutures, allowed surgeons to experiment with smaller corneal incisions at different sites.

32 MILLION PER YEAR The World Health Organization's (WHO) global target of number of cataract surgeries to be achieved by 2020.

Advances in surgery

In 1967 US ophthalmologist Charles Kelman devised the phacoemulsification, or "lens jellification," technique of cataract removal. This method uses ultrasound vibrations to emulsify the lens, which is then sucked out using a hollow needle. At the same time fluid is washed through the anterior chamber (between the iris and the cornea) to remove any fragments and fill the space in the capsule. With this development, incisions in the cornea shrank to a few millimeters. Cataract removal was transformed from a significant operation to a routine, one-visit procedure that spread worldwide.

Kelman's technique left the rear of the lens capsule in place, which facilitated the next development, a synthetic lens to make vision clearer. This intraocular lens (IOL) was developed in the 1950s by British ophthalmologist Harold Ridley, and after many trials IOLs became routine from the 1970s. An IOL is often inserted into the eye right after cataract removal. The lens is shaped for the individual patient's optical prescription. Newer, flexible materials allow lenses to be folded or rolled so they can be implanted through a small incision, then opened out. Advanced surgery may use accommodative IOLs, which the inner eye muscles can move and alter to focus both far and near, thereby minimizing the need for reading glasses.

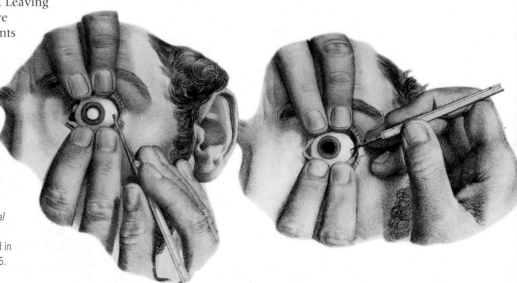

Medieval cataract operation
In 1583 German barber-surgeon Georg Bartisch published the illustrated text *Ophthalmodouleia Das ist Augendienst* (*In the Service of the Eyes*). It described cataract operations, correcting squints, and the removal of growths and foreign bodies.

Exchanging **Epidemics** with the **New World**

When Europeans first came to the Americas in the late 15th century, they triggered one of the greatest series of epidemics in world history. Without natural immunity or appropriate medical care, tens of millions of native Americans succumbed to infectious diseases brought by the newcomers.

△ **Cinchona bag**
Used to treat a variety of maladies, cinchona bark was collected in serons, or rawhide bags, such as this Peruvian example from the 1770s. The bark could be chewed in its natural state or dried, powdered, and added to drinks.

The arrival of Europeans in the Americas is usually dated to explorer Christopher Columbus's voyage there in 1492. At this time, the population of the New World was estimated to be 40–60 million. However, within a century, the number had declined by as much as nine-tenths in some areas, partly due to warfare, but chiefly as the result of huge waves of infectious diseases inadvertently brought by the Europeans.

These imported diseases included diphtheria, measles, bubonic plague (see pp.66–67), smallpox (see pp.100–01), cholera (see pp.122–23), influenza (see pp.196–97), typhus, chickenpox, scarlet fever, yellow fever, pertussis, and malaria (see pp.174–75).

The main reason for the huge death toll was that the native people had no immunity against

5–8 MILLION
The estimated number of Aztecs who died of European diseases around 1519–20.

the new diseases. Through generations of evolution, the human body's immune system has adapted to combat infectious organisms in its environment. People with some degree of natural immunity survive and pass on their resistance to their offspring; those with little resistence do not. The Europeans had lived with most of these diseases for millennia and had inherited resistance to the infections. During this time, they had also developed preventative measures, medical care, and treatments—none of which were available to the native Americans.

Two-way exchange
The Europeans also carried several diseases back home from the Americas. These included syphilis (see pp.186–87); pinta and bejel—skin infections linked to syphilis; and Chagas disease (American trypanosomiasis).

Syphilis arrived in Europe around 1495. In the following decades, the infection had an estimated death rate of more than 75 percent. This rate reduced noticeably within a century as the population built up immunity aided by several factors. One of these was that Europeans had lived closely with domestic animals for thousands of years, and had accumulated some immunity to their diseases—many related to human illnesses such as

5 CENTURIES
The time it took for Central and South America to recover their population numbers after the deaths that occurred following the arrival of the first Europeans.

smallpox and cowpox (see pp.100–01). In contrast, native Americans tended to follow a hunter-gatherer lifestyle, and kept less domestic stock. Also, Europeans lived in towns and cities that were densely populated, and tended to travel extensively for warfare, trade, and other reasons. Native populations in the Americas were less dense and more scattered, and individuals traveled less widely and frequently. So Europeans had a long history of their bodies being challenged by a variety of harmful microbes, which

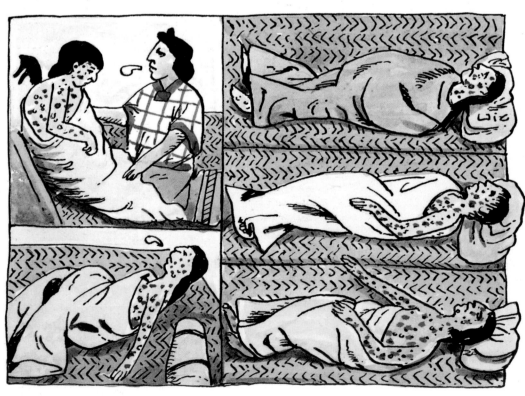

◁ **Decimated empire**
This illustration from the *Florentine Codex* shows Aztecs dying of smallpox, which was allegedly introduced by one African slave in the Spanish army. Almost half of the Aztec population succumbed to the disease, including their ruler Cuitláhuac.

helped build their immunity. Therefore, when new diseases spread from the Americas, resistance to them in Europe's general population developed relatively rapidly—in contrast to the situation in the Americas.

Miracle cure

The exchange of infectious organisms between Europe and the Americas was part of a larger trans-Atlantic phenomenon called the Columbian Exchange, which also involved an interchange of domesticated and wild animals and plants, and human cultures, customs, and technologies.

One of the most significant exchanges of plants was that of the bark of the cinchona tree, native to the Andes in South America. Local people, such as the Quechua of present day Peru and Bolivia, knew that ground preparations of the bark were effective against ailments such as fever, diarrhea, aches, muscle spasms, and fatigue. In the 1620s Jesuit priests in the area discovered that it was especially useful against malaria. In 1630 a cinchona bark preparation produced a malaria cure for Ana de Osorio, the Countess of Chincon and wife of the Spanish Viceroy in Lima, Peru. This encouraged the widespread collection and export of the bark to Europe, where it was heralded as a miracle cure for

△ **Sharing medicine**
Peruvian people offer bark from the cinchona tree to Europeans suffering from malaria. The Europeans learned much from the native Americans about plant treatments, including arrowroot, yerba mate, and tobacco—initially regarded as a cure-all for many illnesses.

malaria and many other diseases. In 1820 the bark's active ingredient was extracted by French chemist Pierre-Joseph Pelletier and his colleagues, allowing the drug to be produced in pure, accurate doses. Named quinine, after the Quechua term for cinchona bark, it has the reputation, after antibiotics, of helping more people than any other medication for infectious diseases.

> " In most provinces more than one half of the population **died... in heaps, like bedbugs.** "

TORIBIO MOTOLINÍA, SPANISH MISSIONARY, ON THE IMPACT OF THE SMALLPOX EPIDEMIC IN MEXICO IN THE 16TH CENTURY

ENGLISH PHYSICIAN   Born 1624   Died 1689

Thomas Sydenham

"You must **go to the bedside.** It is there alone that **you can learn disease.**"

THOMAS SYDENHAM, ADDRESSING A YOUNG PHYSICIAN

One of the most respected names in the history of British medicine, Thomas Sydenham is credited with describing and defining specific diseases, as well as bringing doctors out of the laboratories and into the sick room. His enduring influence led him to be called the "English Hippocrates" after his death.

Sydenham did not devote himself to medical practice until middle age. He had served under Oliver Cromwell as a Puritan in the English Civil War, and only began to practice medicine in about 1656, in London. Here, he made a thorough study of epidemics, inspired by the Great Plague (see pp.66–67), which swept through London in 1665–66. This work led to his first book *Methodus curandi febres* (*The Method of Curing Fevers*) in 1666, which was expanded into *Observationes Medicae* (*Observations of Medicine*) in 1676, a standard medical textbook for over two centuries. His treatise on gout—a condition he suffered from himself—was published in 1683, and is regarded as his masterpiece.

Diagnosis and drugs

A follower of Hippocrates (see pp.36–37), Sydenham shared his belief in the healing powers of nature, and kept

△ **Of the bloody flux**
Sydenham's description of the bloody flux, or dysentery, is part of his collected writings, *The Whole Works of That Excellent Practical Physician Dr. Thomas Sydenham*. Based on personal observation, it is full of vivid descriptions.

an open mind about the medical teaching of the day and his own clinical observations. While the traditional humors (see pp.34–35) provided a foundation for his work, he began to base his clinical practice on what he saw.

Sydenham had little regard for standard professional etiquette or theoretical dogma. A compassionate doctor, he reminded physicians that their main duty was to get to know and care for their patients. He was influential in depicting and classifying identifiable "species" of diseases, which greatly improved medical diagnosis. For example, he described rheumatic fever and Sydenham's chorea, distinguished between scarlatina (scarlet fever) and measles, and made observations

◁ Doctor with compassion
Sydenham did not blindly trust scientific theories when treating patients. Instead, he relied on bedside observations and common sense in his effort to provide effective care and cure.

▽ Laudanum
This drug is made by dissolving opium in alcohol. First discovered in the 16th century by Paracelsus, it was largely unknown until Sydenham popularized it as a treatment for a variety of ailments, particularly pain.

about smallpox and dysentery. Sydenham wanted nature to take its course, and prescribed fresh air, exercise, and drinking beer in moderation.

Sydenham prescribed drugs that were based on herbalism, such as the juice of willow leaves to treat a fever, and he advised restraint, rather than large doses. He believed that a patient's symptoms were not the effect of the disease, but the body's struggle to overcome the disease. The introduction of quinine in Europe to treat malaria in the 1630s was a vindication for Sydenham. It worked, he stated, by stoking up fever and encouraging nature's resistance to disease.

Opium, used to relieve pain, was first mixed to create laudanum by

There, his views were welcomed as a return to encouraging the body's natural defenses, rather than challenging it with harsh, powerful "chemical cures."

Sydenham's contemporaries at home were, however, annoyed by the forceful way in which he expressed his opinions. He was not elected to the College of Physicians, and did not endear himself to that esteemed body by saying "physic is not to be learned by going to the universities; one might as well send a man to Oxford to learn shoemaking as practicing physic."

However, over time, Sydenham became the most respected name in the history of British medicine, for placing great emphasis on

> " … the doctor… should be **diligent and tender** in relieving his suffering patients… "
>
> THOMAS SYDENHAM, FROM *MEDICAL OBSERVATIONS CONCERNING THE HISTORY AND CURE OF ACUTE DISEASES*, 1668

Paracelsus (see pp.70–71), but Sydenham's different mixture— a tincture of opium mixed with wine or water—popularized the medicine. It was so revered it was named *Laudanum Sydenhamii*.

Gaining popularity
It was in mainland Europe that Sydenham made the greatest impact.

clinical observation and accurate descriptions of disease. Sydenham was not concerned with flowery medical theory and derided those who were. He believed that disease "visits" a patient, rather than being an integral and ongoing aspect of the patient—a revolutionary concept and one that changed the way physicians practiced medicine.

TIMELINE

- **1624** Born to wealthy landowners in a small English village in the county of Dorset.

- **1642** Joins Magdalen College at Oxford; his studies are interrupted by the English Civil War, in which he serves as a Puritan.

- **1645** Returns to Oxford and enters Wadham College.

- **1648** Graduates as a Bachelor of Medicine. However, there is conjecture that he was aided by his family's connection with the Parliamentarians. He is elected a fellow of All Souls College at the same time.

- **1665** Leaves London during the Great Plague. While in the countryside, he writes his first book on the subject of fevers. He dedicates the book to his friend, Irish-born chemist Robert Boyle.

- **1666** Popularizes the use of quinine to treat malaria.

- **1676** Includes important studies of the London epidemics of the day in his book *Observationes Medicae* (*Observations of Medicine*) and is the first to attempt to classify diseases (this work is considered the basis of the science of epidemiology). He also graduates as a doctor of medicine from Pembroke Hall, Cambridge, nearly 30 years after graduating from Oxford.

- **1680** Publishes a book on epidemics, *Epistolae responsoriae* (*Letters and Replies*), dedicates it to Regius Professor of Physic at Cambridge, Robert Brady.

EPISTOLAE RESPONSORIAE, 1680

- **1682** Writes about the treatment of smallpox, and hysteria, in his book *Dissertatio epistolaris* (*Dissertation on Letters*).

- **1683** Publishes *Tractatus de Padagra et Hydrope* (*The Treatise on Gout and Dropsy*). It distinguishes gout from rheumatism, and is considered Seydenham's greatest work.

- **1689** Dies in London; is buried in St. James's Church, Piccadilly.

Early Microscopists

Some technological advances, such as X-rays (see pp.172–73), were quickly assimilated into medical practice. The microscope on the other hand, invented in the 1590s, only began to be used for medical research half a century after its invention (see pp.96–97).

Simple magnifiers using a single glass convex lens—bulging in the middle—were in use in ancient Rome some 2,000 years ago. Lens-making improved from the 13th century when the use of eyeglasses began to spread, and magnifiers known as "flea glasses" that could provide magnifications of 10 to 15 times were also invented. In the 1590s the compound microscope, using two or more convex lenses, was invented. Some historians credit the invention of the microscope to the Dutch lens-makers, father-and-son duo Hans and Zacharias Janssen. Others believe the Dutch inventor and eyeglass-maker Hans Lippershey made the first microscope. Italian polymath Galileo Galilei worked on improving microscope lenses in the early 17th century, but early microscopes were mainly curiosities with little scientific use. They suffered from blurring and chromatic aberration—a problem where light waves of different lengths come into focus at different places to produce colored fringes—and their magnification was limited to 15 to 20 times.

Early microscopic studies

One of the first publications to use microscopic studies was the 1625 *Anatomy of the Bee, as Revealed by the Microscope*, by Italian scientist and writer Francesco Stelluti. He achieved clear magnifications of around five to seven times. The device was known in Italy during this time as the "microscopium"; the English term "microscope" came into use in the 1650s. In 1644 Italian astronomer Giovanni Hodierna reported that he had used a telescope modified as a microscope to count 30,000 "little squares" on a fly's eye. In 1655 Peter Borel, physician to King Louis XIV of France, wrote *De Vero Telescopii Inventore* (*The True Inventor of the Telescope*). The telescope was now being improved at a much faster rate and at the end of his text Borel included microscope information and observations, saying: "A microscope, whether it be a flea glass or a fly glass, whereby a flea is enlarged to the size of a camel, and a fly to the size of an elephant, is made out of two glasses enclosed in a small tube: the glass nearest the eye is convex and made out of a small segment of a spherule, whose diameter should be two inches: the other glass is plane [has one flat side]."

Pioneering microscopists

Two people who helped the microscope gain greater fame, and encouraged its use in medicine, were British polymath Robert Hooke and Dutch polymath Antoni

◁ **Janssen's microscope**
Made around 1876, this is a replica of a very early Janssen microscope from the 1590s. It has three tubular sections, which slide in and out to focus, and two lenses. Maximum magnification was about 10 times.

DUTCH SCIENTIST (1632–1723)

ANTONI VAN LEEUWENHOEK

Originally a textile merchant, Antoni van Leeuwenhoek became interested in microscopy while trying to improve magnifier lenses for inspecting cloth threads. He used an unusual single-lens design, through which he achieved magnifications of more than 250 times. As a merchant Leeuwenhoek understood the need for trade secrets and kept his methods to himself—his unique lens-making procedures were not rediscovered until the 1950s. With almost 200 scientific articles published by the Royal Society by the time of his death, Leeuwenhoek can be seen as the first expert microbiologist.

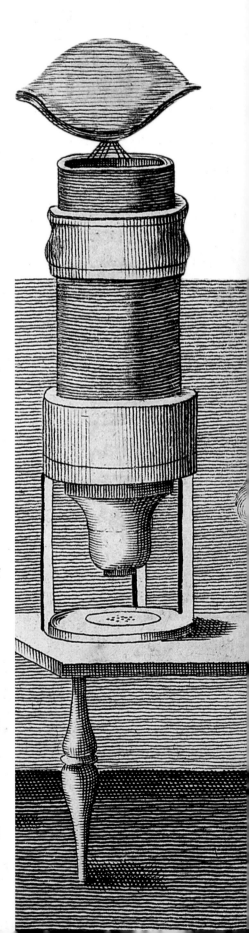

▷ **Campani's microscope**
Dated 1686, this is the first illustration of a microscope in medical use—to examine a patient's leg. The device (enlarged on the left), made by Italian inventor Giuseppe Campani, had a screw thread for focusing. Light concentrated from a candle was used to illuminate the area.

Evolution of Microscopes

The first microscopes were simple devices with two lenses fixed together in a tube. The magnified images they produced revealed a new world for scientists to explore in minute detail. As the quality of lenses improved, so did the images.

1 Small compound microscope This early microscope comprises two lenses. As a result, the image is twice magnified. **2 Hooke's microscope** This replica of British scientist Robert Hooke's compound microscope uses a water-filled glass container to focus light from a lamp onto the specimen being observed. **3 Lyonnet's microscope** Dutch naturalist Pierre Lyonnet designed this simple microscope with a lens mounted on top of a series of ball-and-socket joints attached to a small dissecting table. **4 Culpeper microscope** Built by British instrument-maker Edward Culpeper, this compound microscope had an inflexible, upright style. **5 Simple microscope** This simple aquatic microscope is very similar to the one used by British naturalist Charles Darwin on his exploratory

voyage aboard the *Beagle*. **6 Leeuwenhoek's microscope** Dutch scientist Antonie van Leeuwenhoek built this simple microscope, using a biconvex lens. **7 Polarizing microscope** Designed by British geologist Allen Dick, this device uses polarized light—with light waves undulating in a single plane. **8 Cary-Gould microscope** This Gould-type compound microscope by manufacturer Cary, London, consists of three lenses. **9 Binocular microscope** A complex microscope, this device has a built-in illumination system and its twin eyepieces reduce strain on the eyes when used for longer periods. **10 Electron microscope** This microscope uses an electron beam rather than light to form an image, and allows for increased magnification and improved resolution.

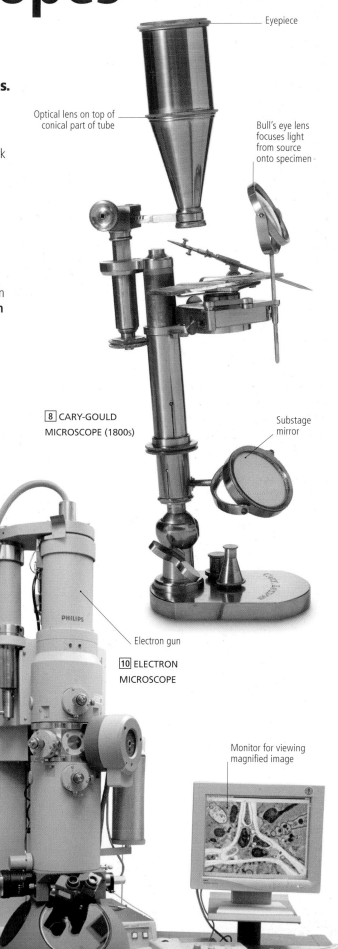

Eyepiece

Optical lens on top of conical part of tube

Bull's eye lens focuses light from source onto specimen

7 POLARIZING MICROSCOPE (c.1890)

8 CARY-GOULD MICROSCOPE (1800s)

Substage mirror

High voltage

Electron gun

Revolving nosepiece holding objective lenses

9 BINOCULAR MICROSCOPE

Coarse focusing

Fine focusing

Illuminator with lens; can be replaced with a mirror

10 ELECTRON MICROSCOPE

Monitor for viewing magnified image

The First Microanatomists

The first microscopists studied tiny objects from the natural world, such as insects. But from the late 17th century, the microscope had become a potent tool for anatomical and medical research and was used to study cells, tissues, and microbial germs.

The invention of the light microscope in the 1590s uncovered a new world of tiny objects and living things. By the late 17th century, several researchers were investigating the previously unknown and unseen world of human tissues and cells, and the harmful microbes, or pathogens, that cause disease.

Microanatomy

In 1653 Peter Borel (see pp.92–93), physician to the French King Louis XIV, provided one of the first accounts of the microscope for medical use. He described how tiny ingrowing eyelashes, which could only be seen using a microscope, produced the irritation and pain of conjunctivitis, and that removing them solved the problem.

Marcello Malpighi (see panel, below) was a principal pioneer in microanatomy and medicine, and studied a huge variety of plant, animal, and human tissues. In about 1661 he identified tiny channels or vessels in frog lungs that had minute bodies moving through them. This was one of the first descriptions of capillaries—the "missing link" between arteries and veins in the circulatory system, as described by William Harvey (see pp.84–85) in 1628. Malpighi also devised new methods to illuminate tiny specimens more brightly, and

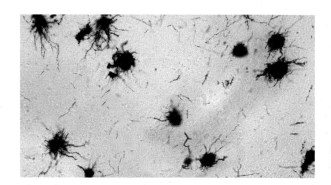

▷ **Seeing neurons**
The long, thin fibres of nerve cells, or neurons, are visualized here using a Golgi stain, containing silver compounds. Golgi discovered this staining technique in 1873 and called it the "black reaction".

also to stain or inject them with substances so they could be better seen under the microscope.

Beginnings of histology

In the late 17th century, Malpighi laid the foundations for histology, a new branch of science. Derived from the Greek word *histos*— meaning web or tissue—histology is the study of tissues, which are a collection of similar cells, such as muscle, bone, nerve, or cartilage. French anatomist Marie-François Bichat further developed the understanding of living tissues in the 1790s.

The quality of microscopes improved with time, and so did the techniques used for examining specimens. One method was to use a very thin slice, or section, of tissue. At first, sections were cut by hand using a razor blade; but in 1770 George Adams invented one of the first automatic cutting machines—the microtome. This device was improved in the late 18th century, by Scottish instrument-maker Alexander Cumming, and then significantly advanced by Swiss anatomist Wilhelm His in the 1860s.

A second area of progress for histology was in the treatment and preservation of tissue samples with chemicals. This made them firm, and therefore, easier to slice. In the 19th century this procedure was improved when the use of salts and acids was replaced by the use of paraffin wax to penetrate and support the sample during sectioning. In the 1890s formalin came into fashion as a preservative-fixative— a compound that hardened fresh tissues, helping retain the minute details of the cells. Another advance in histology was the development of stains, or dyes, to

---

ITALIAN BIOLOGIST AND PHYSICIAN (1628–1694)

MARCELLO MALPIGHI

Born near Bologna in Italy, Malpighi received his doctoral degree in philosophy and medicine from the University of Bologna in 1653. Although he showed some interest in teaching, by 1660 he had become a doctor and researcher in microanatomy, and studied different kinds of plants and animals at his estate near Bologna. He accepted professorships at the universities of Pisa and Messina, in 1656 and 1662, respectively. However, his discoveries challenged the approaches and beliefs current at the time, provoking controversy and making him unpopular among his colleagues.

In 1668 Malpighi became a member of Britain's Royal Society, which reported much of his work. Toward the end of his life in 1691, Malpighi was appointed physician to the Pope in Rome. He died there, probably of a stroke, in 1694. His name is commemorated in many areas of biology and human microanatomy, from Malpighian tubules in the excretory system of insects, to the Malpighian layer of the skin's epidermis, and Malpighian corpuscles— clumps of white blood cells that are found in the spleen.

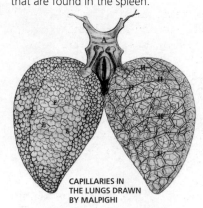

CAPILLARIES IN
THE LUNGS DRAWN
BY MALPIGHI

"Observation by means of the **microscope** will **reveal more wonderful things** than those viewed in regard to mere structure."

MARCELLO MALPIGHI, ON THE DISCOVERY OF CAPILLARIES, *DE PULMONIBUS*, 1661

color certain structures and substances to be viewed under the microscope. One of the first stains, introduced in 1774, was Prussian blue. A version devised in the 1860s to show up iron-containing substances, such as hemoglobin, was known as Perls' blue stain after German pathologist Max Perls. The H&E (hematoxylin and eosin) stain, first described in 1876 by chemist A. Wissowzky, is still the most popular stain used today. Hematoxylin colors the nuclei in cells blue, while eosin stains the cytoplasm or "jelly" pink. Hundreds of stains have since been invented for specialist applications.

Advances in histology

Histology is partnered with histopathology in the study of abnormal tissues and how they lead to diseases. The first work to describe histopathology and its techniques was *On the Nature and Structural Characteristics of Cancer* by German physiologist and scientist Johannes Müller in 1838. During the 19th century, microanatomy, histology, and histopathology were responsible for many momentous medical advances, including germ theory (see pp.146–47), identifying infectious microbes, vaccine development, and unraveling the microstructure of body systems, especially the brain (see pp.160–61) and nerves.

In 1906 the Nobel Prize in Physiology or Medicine was jointly awarded to two histologists—Camillo

△ **Artist at work**

In addition to being a histologist, Cajal was also a talented artist. He produced hundreds of illustrations mapping the nervous system, which are still used as teaching aids today.

Golgi from Italy, and Santiago Ramón y Cajal from Spain. Golgi developed a stain to show the details of nerve cells, while Cajal described the organization of these cells in the brain.

Scurvy

For more than 400 years, scurvy was the bane of sailors. A breakthrough in understanding the disease came in 1747 when Scottish physician James Lind proposed that scurvy was caused by vitamin C deficiency.

Although scurvy had been prevalent since ancient times, the disease did not become problematic until the growth in European exploration and trade saw men set off for increasingly long periods at sea. Crews were forced to eat salted meat and biscuits for long periods, which deprived them of essential vitamins. After about 30 weeks without vitamin C (ascorbic acid) in their diet crews began to show the classic symptoms of scurvy—bleeding gums, blackened skin, rictus of the limbs, and loose teeth.

In the 18th century, James Lind, a physician in a Portsmouth naval hospital, became interested in the disease and carried out a small clinical trial on HMS *Salisbury*. He discovered that scurvy resulted from an inadequate diet and recommended that a ration of fresh fruit be supplied daily to prevent it. In 1753 Lind published his findings in *A Treatise of the Scurvy*. British captain James Cook tried a variety of methods to combat scurvy. In 1768 he carried sauercrat on his three-year circumnavigation of the world aboard HMS *Endeavour* and his crew remained scurvy-free, thus showing the effectiveness of the methods proposed by Lind. But despite a lot of evidence, it took another decade before the navy gave citrus juice to its sailors as standard daily issue.

> **"...**the most sudden and visible **good effects** were perceived from the use of **oranges and lemons."**
>
> JAMES LIND, SCOTTISH PHYSICIAN, FROM *A TREATISE OF THE SCURVY*, 1753

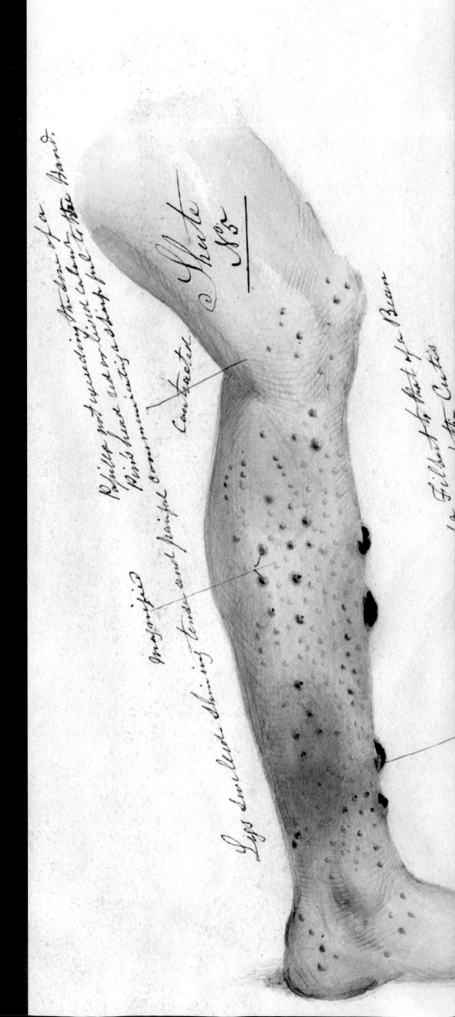

▷ Scurvy

This page from the journal of British naval surgeon Henry Walsh Mahon from his time aboard HM Convict Ship *Barrossa* (1842) shows the effects of scurvy. Here, he describes typical symptoms that develop on a patient's leg, including lesions; open, festering wounds; dark patches; and bleeding.

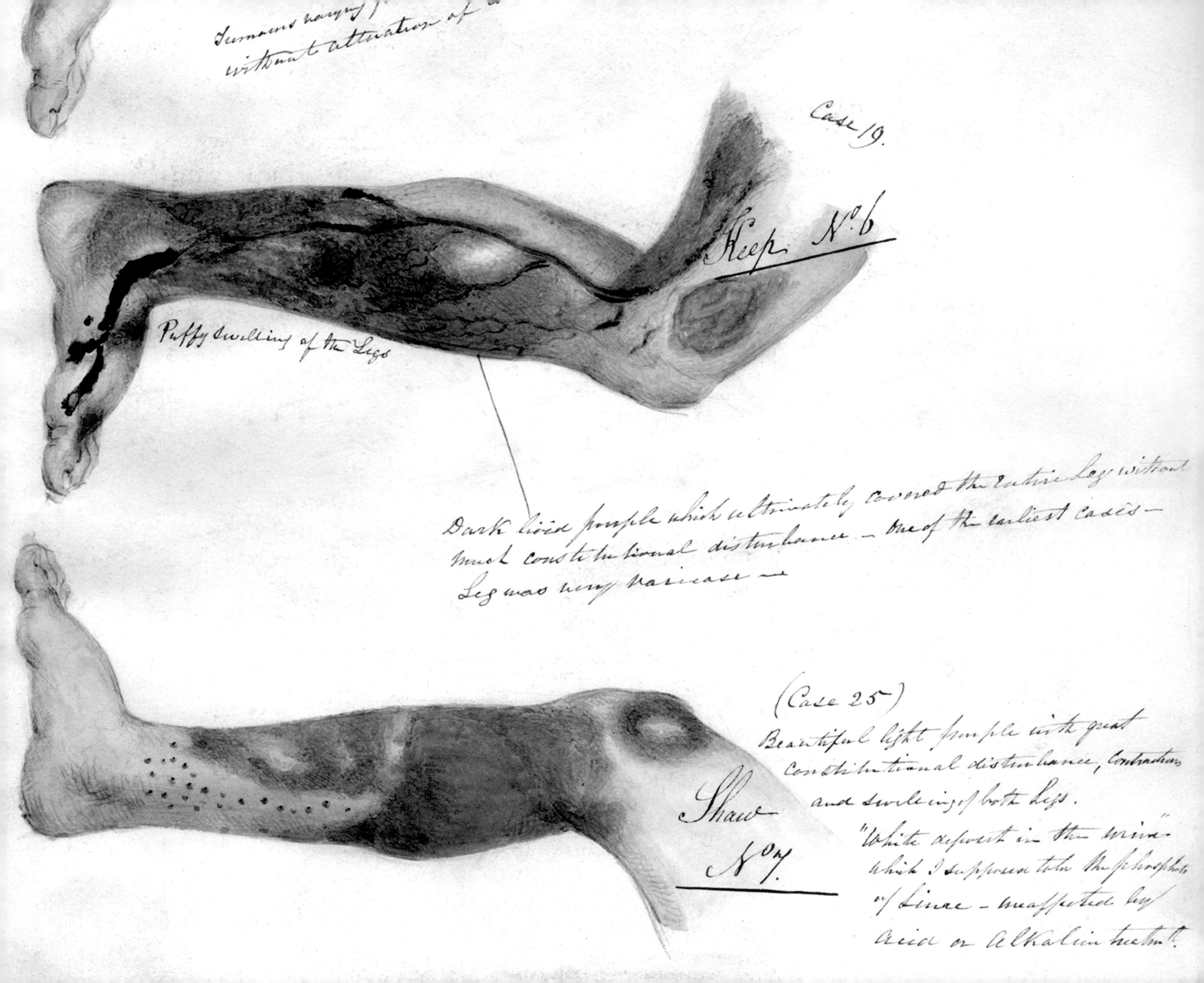

Case 19.

N.b.

Kech.

Common way of
without attention of
without attention —

Dark livid purple which ultimately
... cannot of the entire leg without
... and constitutional disturbance — they the earliest cases —
Leg was very tender —

Puffy swelling of the Legs

(Case 25)

Mr.
No.

Beautiful light purple with great
constitutional disturbance, tenderness
and swelling of both legs.

"White deposit in the urine —
which subsequent to the pimples
"lime — sulphated leg

Leci or Alkaline tracts?

Smallpox: The Red Plague

Of all human diseases, smallpox has perhaps the most claims to fame—or rather, infamy. It has featured in all of recorded history, killed billions, and inflicted lasting suffering on billions more. It was the first infection to be immunized against, as well as the first—and currently only—major global disease to have been eradicated.

Smallpox was caused by several forms of the *Variola* virus. In typical cases it attacked small blood vessels in the skin, mouth, and throat, causing fluid-filled blisters. The most virulent forms killed an estimated one-third of victims. However, during sudden, fast-spreading epidemics the death toll could be as high as 80 percent.

Smallpox spread through the inhalation of airborne droplets from an infected person's mouth, nose, and airways. It also spread by direct contact with bodily fluids or shared objects such as clothing. Survivors of the disease were left with disfiguring scars, physical disabilities that sometimes included blindness, and mental anguish since they were shunned by or even cast out of society.

Smallpox and its virus were part of a group of diseases that included cowpox, horsepox, camelpox, and monkeypox. The term "pox" refers to skin eruptions or sacs that leave pitted pockmarks, and it has been applied to a wide range of diseases from acne to syphilis. The name "small pockes" was introduced in England in the late 15th century to distinguish the viral disease from syphilis, which was then called the "great pockes." Smallpox was also

1 The number of deaths caused per second by smallpox in the 19th century.

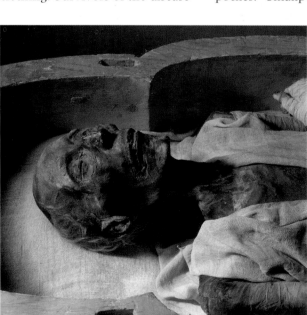

◁ **Poxified mummies**
Several Egyptian mummies have pitted, pockmarked skin indicating smallpox infection. One of the victims was Pharaoh Ramesses II (shown here), who died around 1213 BCE aged 90 years. His mummy was discovered in 1898 and has facial skin lesions. The remains of Ramesses V, who died in about 1145 BCE, show similar evidence.

> ## "For **no one** was ever attacked **a second time,** or not with a **fatal result.**"

THUCYDIDES, GREEK GENERAL AND HISTORIAN, FROM
HISTORY OF THE PELOPONNESIAN WAR, 431 BCE

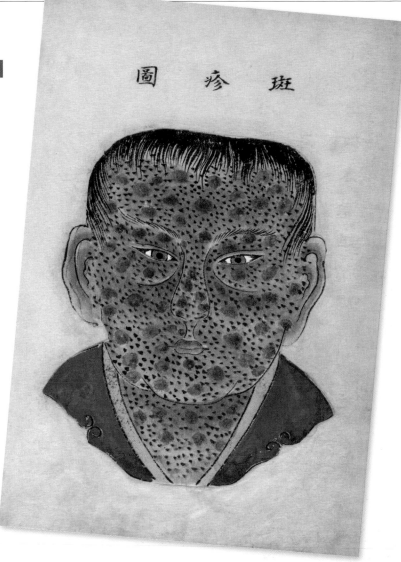

known as the "red plague"—a name derived from the red rash, blisters, and extensive bleeding that characterized the more serious forms of the disease.

Origin of the virus

Studies of the smallpox virus genes suggest that it probably originated in rodents and then transferred to humans about 10,000 to 50,000 years ago. Various forms of smallpox then developed, notably in Africa and Asia.

Because the infection was highly variable—some forms were relatively mild and led to quick recovery, others were severe and fatal—it is difficult to identify the earliest presence of smallpox in history.

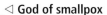

◁ **God of smallpox**
Yu Hoa Long was the Chinese god of smallpox. Many ancient cultures viewed smallpox and similar diseases as punishment from spirits and deities for sins committed in this life or a previous one.

Possible descriptions occur in ancient Chinese and Indian texts dating back more than 3,000 years. Accounts of the Egyptian Hittite Wars also mention a smallpoxlike epidemic in about 1350 BCE, and the ancient Greek historian Thucydides documented a plague in Athens in 430 BCE that killed an estimated 30,000 people and may have been smallpox too.

Gradually, descriptions of the disease became clearer and more accurate. In about 910 one of the greatest Islamic physicians, al-Razi, explained how to distinguish smallpox from other pustule-forming diseases in his *Kitab al-Jadari wa 'l-Hasba (Treatise on Smallpox and Measles).* He also recorded that the disease spread from person to person, and that survivors did not develop it again.

Through the medieval period, new forms of the virus emerged and followed trade, migration, and slave routes in the Old World. When Christopher Columbus and his crew began the European colonization of the Americas in 1492 they brought smallpox to the New World. The indigenous populations had no natural immunity against the disease, and within half a century tens of millions of them succumbed, helping the invading Europeans destroy the Aztec, Inca, and many other civilizations (see pp.88–89).

In the 1790s the disease reached Australia and killed up to half of the Aboriginals in eastern regions.

Conquering the dreaded pox

Back in Europe, Asia, and Africa, smallpox continued to be a major cause of deaths, killing over 500 million during the 18th century. However, in 1798, experimental vaccinations by English physician Edward Jenner, based on the

1978 The year when the last recorded death from smallpox occurred. Janet Parker was infected due to an accidental release of laboratory-kept viruses at her workplace in the University of Birmingham Medical School, England.

procedure known as variolation, provided immunity against the disease (see pp.102–03). Within a decade, immunization programs were being taken up around the world. In Massachusetts, compulsory vaccination was

△ **Symptoms of smallpox**
In 1720 the Japanese doctor Kanda Gensen published *Toshin Seiyo (Essentials of Smallpox),* an illustrated explanatory work that carried numerous coloured illustrations of the different symptoms of smallpox. This illustration shows a face marked with smallpox scars.

introduced in 1809, and the UK followed suit in 1853. By the 1940s research into freeze-dry technology made vaccines less expensive, more stable in storage, and considerably easier to prepare and administer.

In 1967 the World Health Organization set up its Smallpox Eradication Campaign. With a vast effort and much monitoring, case numbers declined. The last case in South America occurred in Brazil in 1971, and in South Asia in 1975, and the last known case of natural smallpox infection was identified in Somalia, Africa, in 1977. The WHO declared the world free of this age-old scourge in 1980, and in 1986 vaccination ceased.

CONCEPT

VARIOLA VIRUS

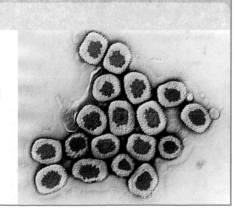

The causative agent of smallpox is the *Variola* virus. It is about 0.3 µm long—placed end to end 3,000 viruses together would stretch one millimeter. In this electron micrograph, the red area shows genetic material— DNA with about 200 genes. The tough outer coat of protein is colored yellow.

The **First Vaccination**

A technique established by Edward Jenner in the 1790s, vaccination greatly reduces the risk of infectious diseases by helping the body develop immunity. Along with antibiotics, it is regarded as one of the foremost advances in medical history.

Immunization is the process of making the body resistant to an infectious disease by working with the body's natural defenses. Natural immunity starts to develop when infecting microbes invade the body and the immune system fights them by releasing antibodies. After infection, the immune system "remembers" those microbes and if it encounters them again it quickly produces antibodies to prevent the body from attack. Vaccination induces immunity artificially by imitating an infection, but without causing illness. An essential part of modern medicine, vaccines have been developed against many dangerous infectious diseases.

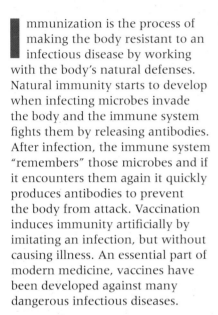

△ **Lady Mary Montagu**
Wife of the British Ambassador to the Ottoman Empire in Constantinople, Lady Montagu started a successful campaign to introduce variolation to Britain. She had suffered from smallpox in her youth and also lost her brother to the disease.

Early variolation

It was widely known in ancient times that the body develops natural resistance to diseases. The earliest attempts to induce immunity artificially may date back more than 2,000 years in India, but the idea of immunization rose to prominence in China in the medieval period, when individuals were inoculated with the smallpox virus (see pp.100–01). Procedures involved taking blister fluids, pus, or scabs from a person infected with a mild case of smallpox and giving them to an uninfected person. This was done by rubbing them into cuts in the skin or blowing ground scabs up the nose. While there was a slight risk of developing severe smallpox, this method had a much greater chance of offering protection, reducing mortality rates caused by smallpox

249 **People given variolation in 1721 by physician Zabdiel Boylston in Boston—the first US inoculations.**

from 30 percent to less than 5 percent. The English later named the process variolation (from the Latin *varius*, meaning speckled).

Increasing popularity

The popularity of variolation in Britain is mostly credited to Lady Mary Montagu who, having seen its success in Constantinople (modern Istanbul, Turkey), was convinced of its worth and had her own son variolated around 1716. She began to gather evidence and campaigned

for its acceptance in Britain. By 1721 with smallpox again on the rise, Lady Montagu persuaded the royal doctor Hans Sloane to try variolation. Informal tests on prisoners were successful and variolation gained popularity, having been accepted by members of royalty.

Variolation became more common throughout the 18th century, but it continued to be unpredictable, with occasional serious cases and even deaths. Another disadvantage was that variolation necessitated the isolation of the recipient for two weeks.

Jenner's breakthrough

Edward Jenner was a successful country physician-surgeon in Berkeley, southwest England, as well as a talented naturalist. He had undergone variolation in his youth, which had made him ill for a time.

> " Future nations will know… smallpox has existed and **by you has been extirpated."**
>
> THOMAS JEFFERSON, IN A LETTER TO EDWARD JENNER, 1806

As a doctor, Jenner was aware of the common belief that catching cowpox somehow gave protection against smallpox—very few milkmaids and cattle herdsmen seemed to suffer from the latter.

▷ **At work**
Jenner's Case 17 in 1796 involved vaccinating eight-year-old James Phipps with cowpox, which caused slight symptoms. Six weeks later, Jenner deliberately infected him with smallpox and recorded: "No disease followed."

Several others had investigated this link, such as the English physician John Fewster, who wrote a paper "Cow Pox and Its Ability to Prevent Smallpox" in 1765, but it was largely ignored. In 1774 Benjamin Jesty, a farmer, reportedly used a darning needle to introduce cowpox sore pus into his family, but he was mocked when his wife became very ill.

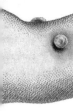

◁ **Cowpox sores**
The sore material used for the vaccination of James Phipps (below) came from the hand of a local dairymaid, Sarah Nelmes. Case 16 in Jenner's report, she had caught cowpox naturally, and did not suffer from the disease.

Jenner knew that for the link to be taken seriously a report on careful medical trials was needed. In 1798 he published *An Enquiry into the Causes and Effects of the Variolae Vaccinae: A Disease Discovered in Some of the Western Counties of England, Particularly Gloucestershire, and Known by the Name of the Cow Pox,* which described his treatment of 23

patients by first vaccinating them with cowpox material and then giving them smallpox. He noted that after the cowpox vaccine his patients did not catch smallpox.

Although the medical community had doubts about the ethics of such an experiment, Jenner's thorough, scientific account of vaccination and its success gained immediate attention. His procedures were later improved by others and rapidly spread around the world.

▷ **Ridiculing the vaccine**
This caricature of Edward Jenner vaccinating his patients against smallpox using the cowpox virus aptly reflects the mindset of the public before the treatment was established. His patients are shown growing cow heads.

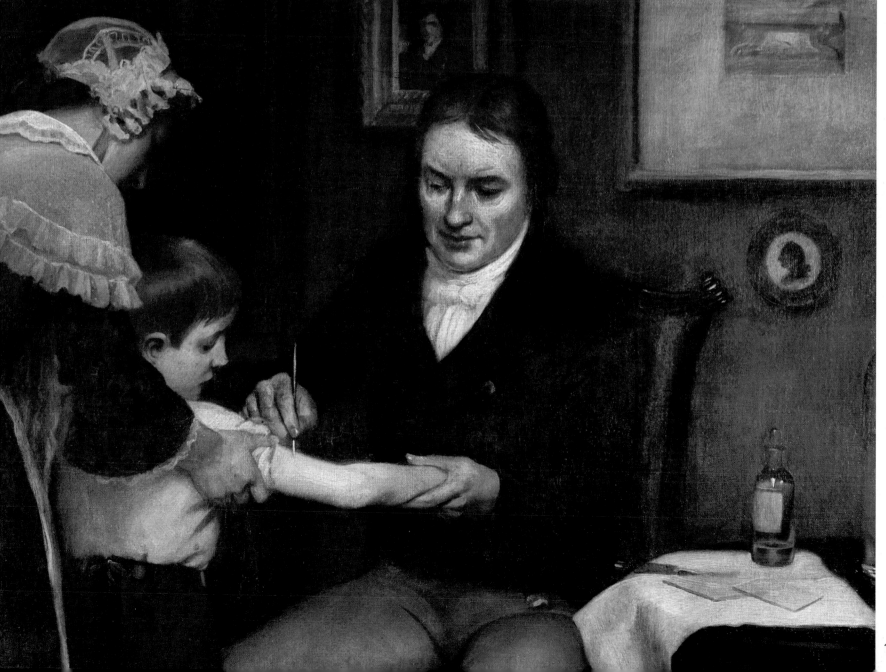

Phrenology

The practice of phrenology—ascertaining a person's character, morality, and intellect by feeling and measuring the contours of the head—is regarded as outdated and unscientific today. Yet this field enjoyed considerable success in the first half of the 19th century, chiefly in Britain, Ireland, parts of mainland Europe, and the US.

Phrenology grew from concepts developed by German physician Franz Gall (1758–1828). At school, he noticed a fellow pupil who had an unusually proportioned head and a great talent for foreign languages. Gall began to investigate links between the shape of the brain and the skull, and traits of the personality. He proposed that the brain was composed of 27 "organs," each the center of a different trait: the larger the "organ," the more it contributed to the character. The skull bone's contours, which could be discerned by observing, feeling, and measuring, followed the location and development of the organs beneath. By 1800 Gall was lecturing and writing articles about his ideas, which were then built upon by his followers.

It is now known that phrenology has no scientific basis, but at the time, it was used by many people as a tool to substantiate controversial causes—for example, to demonstrate the supposed superiority of one ethnic group over others. Although phrenology had waned by the 1850s, some of Gall's ideas are echoed in modern neurology and psychology, such as the belief that certain regions of the brain perform particular mental functions.

> "The **convolutions of the brain...** are the parts in which the **instincts, sentiments, propensities** are exercised..."

FRANZ GALL, FROM *ON THE FUNCTIONS OF THE BRAIN AND OF EACH OF ITS PARTS*, 1796

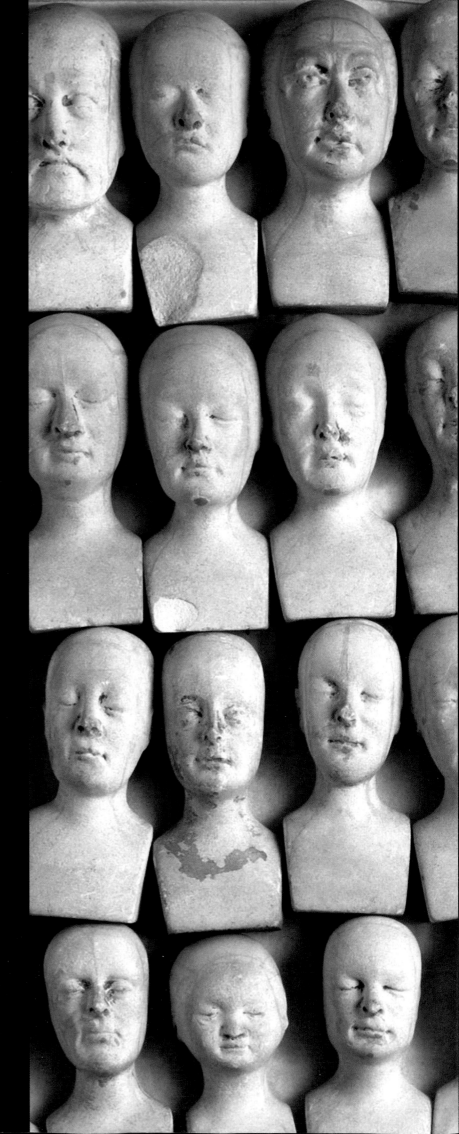

▷ **Head cases**
This collection of around 60 model heads was sculpted by Swiss-born, England-based wax modeler and phrenologist William Bally to explain the principles of phrenology. Sets of plaster casts such as these were sold as teaching aids and displayed at Britain's Great Exhibition in 1851.

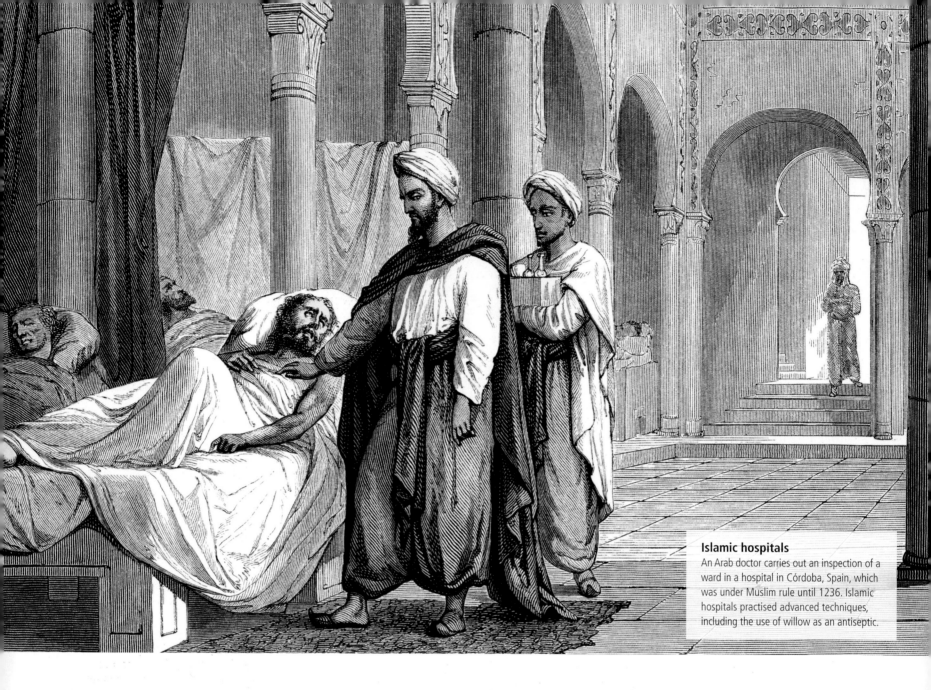

The **Modern Hospital**

The 19th century saw the rapid growth of specialized hospitals, formal medical training schools, and a professional nursing corps. This provided much greater access to hospitals, and far higher levels of care for a greater number of patients from a wider variety of backgrounds.

Although the Roman army had established *valetudinarian*, or hospitals, for wounded or sick soldiers (see pp.38–39), there is no evidence of specialized buildings to provide medical care for civilians before the 4th century CE, when charitable Christian donors began founding establishments to tend to the impoverished sick. Hospitals in medieval times were commonly associated with monasteries, and

most frequently cared for lepers, or, from the 14th century onward, for plague victims, those suffering from other infectious diseases, and the mentally ill.

More formal hospitals did exist in the Islamic world (see pp.48–51), the oldest having originated in Baghdad around 805. Medical training was undertaken in some of them, but they cared mainly for the poor rather than the general populace.

The Dissolution of the Monasteries under Henry VIII, between 1536 and 1540, led to the closure of hundreds of former monastic hospitals in England. Only a few were refounded, so by 1700, London, a city of 500,000 people, had only two substantial medical hospitals—St. Bartholomew's and St. Thomas's. Elsewhere in Europe the situation was slightly better because the Reformation had not led to the wholesale closure

of religiously run institutions. In Vienna, the Allgemeines Krankenhaus (general hospital) was remodeled by Emperor Joseph II in 1784, and included six medical and four surgical wards.

New hospitals
As London's population grew and became more prosperous, there was increased pressure for better medical coverage. Helped by donations from rich merchants,

" A place set up on **purpose for sick children**; where the good doctors… comfort and cure none but children. "

CHARLES DICKENS, DESCRIPTION OF GREAT ORMOND STREET HOSPITAL FOR SICK CHILDREN,
IN *OUR MUTUAL FRIEND*, 1864–65

more hospitals were built: Westminster in 1720, Guy's in 1724, St. George's in 1733, and The London in 1740. Provincial cities acquired their own hospitals—at Bristol in 1737 and at York in 1740—while in Scotland the Edinburgh Royal Infirmary was built in 1745. In the United States, the first general hospital was founded in Philadelphia in 1751, and the New York Hospital followed in 1771.

Specialization begins

For the first time, specialized hospitals began to be established that allowed doctors and surgeons to gain experience in the treatment of a particular ailment. In England, the Moorfields Eye Hospital was the first (in 1804), followed by about 65 others by 1860, including the Royal Hospital for Diseases of the Chest (1814). In the US the earliest specialized hospital was

the Massachusetts Eye and Ear Infirmary, established in 1824. Specialized maternity hospitals appeared for the first time too, beginning with the British Lying-In Hospital in 1749.

The Hospital for Sick Children at Great Ormond Street was founded in 1852, but pediatric hospitals had already been established in Paris (1802), Berlin (1830), and Vienna (1837).

Hospital doctors were now better trained than ever before. In 1750 the Edinburgh Royal Infirmary established a special clinical ward, where medical students were taught with direct reference to the patients, and by the 1770s the concept of clinical lecturing on wards had spread to Vienna. The formalization of medical education took a step further in 1834 when University College London established its own hospital dedicated to instructing medical students.

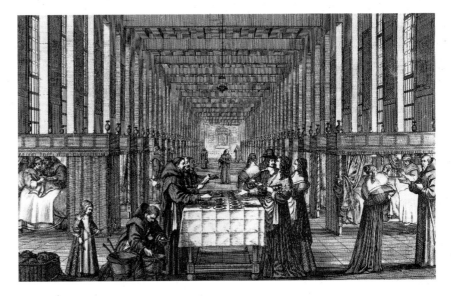

Training nurses

In the 19th century nursing also became a formal profession. Theodore Fliedner, a Lutheran pastor, established the Deaconess Institutions at Kaiserwerth near Düsseldorf, Germany, in 1836, to provide training for women to become "nurse-deaconesses" in religious orders. It became a magnet for nursing reformers from other European countries. Florence Nightingale (see pp.142–43) spent three months at the institute in 1851 before practicing what she had learned in field hospitals for

△ **Charity hospital in Paris**
In France, a series of *charités*, or hospitals for the poor, were founded in the early 17th century. Like most hospitals at the time, The Hôpital de la Charité, established in 1602, was staffed by a religious order—the Brothers of Charity.

the British troops during the Crimean War (1853–56). On her return, a public subscription of more than £44,000 was raised, which enabled her to found a

1900 PERCENT
The increase in the number of outpatients attended to at St. Thomas' Hospital, London, between 1800 and 1890.

nursing school in Britain. From 1860 Nightingale's institute provided trained nurses to the new English hospitals.

As the medical services offered by hospitals increased, there was a danger that poor patients would be squeezed out. Hospitals started charging patients a small fee, and middle-class patients began to pay more for access to private rooms. To counter this trend, new "dispensaries" appeared, which provided medical care to the poor for free. These institutions, such as the New York Dispensary (1790), the Public Dispensary of Edinburgh (1776), and the Finsbury Dispensary (1780), were the true descendants of their medieval forerunners.

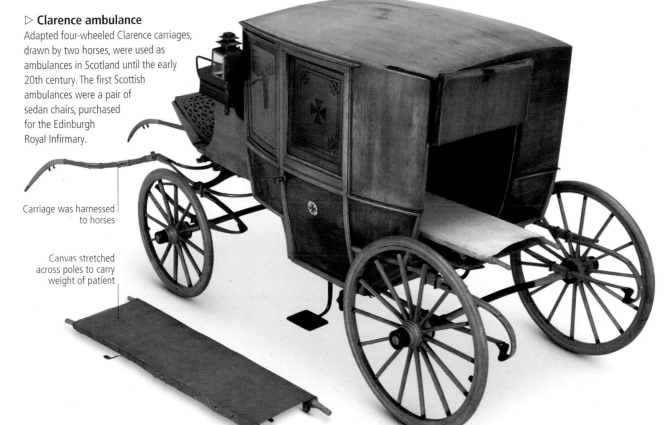

▷ **Clarence ambulance**
Adapted four-wheeled Clarence carriages, drawn by two horses, were used as ambulances in Scotland until the early 20th century. The first Scottish ambulances were a pair of sedan chairs, purchased for the Edinburgh Royal Infirmary.

Carriage was harnessed to horses

Canvas stretched across poles to carry weight of patient

Homeopathy

A healing system developed in Germany in the 19th century, homeopathy is based on the principle that "like cures like" or the "law of similars." It is one of several therapies that takes a different approach from that of conventional Western medicine.

The basis of homeopathy—that a substance that causes certain symptoms in a healthy body can, in lesser quantities, be used to treat an illness with the same symptoms—was first recognized in ancient Greece and later developed by the Romans. In the 4th century BCE, the Greek physician Hippocrates was making homeopathic remedies, and homeopathic medicine was described by Greek-born Roman apothecary Dioscorides in his *De Materia Medica* (see pp.38–39).

In the 1790s German physician Samuel Hahnemann began to develop a set of therapies based on this theory, which became known as homeopathy. Prior to this, Greek-Swiss physician Paracelsus (pp.70–71) and Austrian physician Anton von Storck, among others, had suggested that materials that

were poisonous in quantity could be beneficial in smaller doses. Von Storck reported on experiments using some of the most feared herbs, such as hemlock. However the technology to extract active ingredients in their pure form was not available at the time, so von Storck's results were inconclusive. Hahnemann began to investigate these claims, often using himself as an experimental subject. He tested plant materials such as cinchona bark— later found to be a source of the antimalarial compound quinine (see pp.174–75)— and the leaves and berries of belladonna (known as deadly nightshade).

Diluted remedies

Hahnemann suspected that if smaller doses of a substance could treat a symptom, even smaller ones would have a greater effect while reducing any unwanted side effects. He developed the technique of diluting his extracts in water or alcohol many times shaking the container at each stage of dilution (known as "succussion"). He also devised the centesimal scale, or "C-scale," in order to measure the potency of the solution. A 1C dilution consisted of

▷ **Homeopathic medicine chest**
This early 19th-century medicine chest contains 69 small glass vials and six large bottles. Professional homeopaths prepared and prescribed dozens of remedies, and dispensed them according to various lists and guides compiled by Hahnemann and his followers.

△ **The cinchona bark experiment**
Hahnemann consumed an extract of the cinchona plant—traditionally used as a cure for malaria— to show that in a healthy person it led to symptoms similar to malaria.

"That which can produce... symptoms in a healthy individual, can treat a sick individual who manifests similar... symptoms."

MOTTO OF SAMUEL HAHNEMANN, c.1800

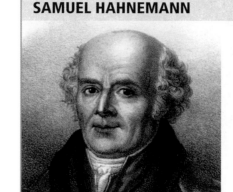

GERMAN PHYSICIAN (1755–1843)

SAMUEL HAHNEMANN

Born in Meissen, near Dresden, Germany, Hahnemann started off as a country physician in Saxony, but was quickly disillusioned by the crude methods and unproven treatments that were prevalent at the time, especially in rural areas. He gave up conventional medicine by 1785 and turned his attention to chemistry and writing. Having a great flair for foreign languages (he spoke a total of 10), Hahnemann made his living as a translator while traveling widely and developing his "art" of homeopathy. He died in Paris in 1843.

one part remedy in 99 parts water, 2C referred to a 1C solution diluted in another hundred-parts liquid, and so on. This process of dilution is called "potentization" because, paradoxically, the more dilute the remedy is, the higher the potency; some remedies are so dilute they no longer contain any molecules of the original substance.

Growing popularity

Hahnemann set forth his findings in *The Organon of the Healing Art* (1810). He proposed that diseases were caused by underlying weaknesses ("miasms") and that homeopathy could gently coax these out of the body. His publications were circulated widely and homeopathic practitioners, journals, and organizations began to emerge in Europe and North America. The German Central Association of Homeopathic Doctors was founded in 1829, and many other similar groups followed, such as the American

Institute of Homeopathy in 1844. This popularity was probably led by the fact that homeopathy was gentler than some of the other brutal treatments of the day. Another advantage was that patients could be treated at home rather than in a hospital, where they sometimes caught additional infections or faced conventional treatments that often did more harm than good. Another wave of popularity came in the 1960s

and 1970s, alongside other aspects of counterculture or "alternative" lifestyles, literature, and music.

The placebo effect

Despite millions vouching for the effectiveness of homeopathy, many studies claim it is in fact the "placebo effect" at work—that is, if someone believes that they will get better, they have an increased chance of improvement. This is especially true if a patient takes a substance that

he or she considers to be helpful, not knowing that it is a placebo (ineffective preparation). Even if there is no discernible improvement in objective terms, the patient may perceive one. Modern medicine is still investigating the mechanism of the placebo effect, which is often observed but is difficult to explain. Some studies tie it to active substances found naturally in the brain, such as endorphins, which cause an improvement in health.

▷ **The need for an alternative**
This 1857 painting by Alexander Beydeman shows the figure "Homeopathy" horrified at the practices of conventional treatments. Hahnemann was driven to a new approach in medicine after experiencing the harm done by common medical treatments in the 18th century, such as blood-letting.

3

SCIENCE TAKES CHARGE

1800–1900

SCIENCE TAKES CHARGE
1800–1900

1800	1840	1860

1842
William Clarke anesthetizes a patient for tooth extraction; Crawford Long uses anesthesia to remove a patient's neck cysts.

1847
Ignaz Semmelweis speculates that "cadaverous material" is responsible for many cases of childbed fever in Vienna; his hand-washing regime drastically reduces the death toll but his work is not recognized for some time.

1860
The first modern nursing school opens at St. Thomas' Hospital, London.

1862
Louis Pasteur carries out his "swan-neck" glass flask experiments, proving that if contaminating microbes are kept away from a nutrient liquid, germs do not grow.

1873
Camillo Golgi introduces a silver-staining technique to show details of nerve tissue under the microscope.

1802
Europe's first pediatric hospital, Hôpital des Enfants Malades, opens in Paris.

1808
Johann Christian Reil introduces the term psychiatry, proposing it should become a recognized medical speciality.

⌃ The Burke and Hare murders

1828
Burke and Hare not only rob graves in Edinburgh, Scotland, but also murder people, to sell their corpses to doctors for anatomical study.

1845
Dentist Horace Wells conducts a demonstration to show the effects of ether as an anesthetic but the patient cries out in pain.

1849
Elizabeth Blackwell is the first woman to receive a medical degree in the US.

1816
René Laënnec invents a simple but hugely significant diagnostic instrument—the stethoscope.

1830s
Pediatrics becomes more established as specialized wards and hospitals open in Berlin, St. Petersburg, Vienna, and Wroclaw.

⌃ Morton ether inhaler

1854
Florence Nightingale arrives at Scutari Barracks to care for soldiers wounded during the Crimean War.

⌃ Lister's carbolic spray to deliver antiseptic

1867
Joseph Lister publishes a report—*Antiseptic Principle of the Practice of Surgery*.

1876
Robert Koch shows that a bacterium, now known as *Bacillus anthracis*, causes anthrax—dealing a death blow to miasma theory.

⌄ Early stethoscope in use

1846
William Morton successfully demonstrates ether anesthesia at Massachusetts General Hospital.

1838
Ueber den feineren Bau und die Formen der krankhaften Geschwülste (On the Nature and Structural Characteristics of Cancer) by Johannes Muller lays foundations for the field of histopathology.

1847
James Simpson begins the use of chloroform for pain relief during childbirth.

1858
Anatomy: Descriptive and Surgical is published. Written by Henry Gray, later editions will come to be known as *Gray's Anatomy*.

⌄ Illustrations from *Gray's Anatomy*

1868
Jean-Martin Charcot, a principal founder of neurology, begins his studies of Parkinson's disease.

1876
The hematoxylin and eosin, or H&E, stain is first described and becomes one of the most useful techniques for visualizing cells and tissues in histology (the study of cells and tissue).

1828
James Blundell revives the idea of human-to-human blood transfusions to treat mothers suffering from excessive blood loss after childbirth.

1839
Publication of the first dental journal, *The American Journal of Dental Science*.

1872
Elizabeth Garrett Anderson founds the New Hospital for Women and Children, London (later renamed the Elizabeth Garrett Anderson Hospital).

1879
Pasteur makes his first vaccine discovery, for chicken cholera, and extends his research into human diseases.

Some of medicine's greatest achievements came during the 19th century, including anesthesia, antiseptic procedures, and rapid advances in vaccination. Louis Pasteur and Robert Koch led the way in replacing age-old ideas of spontaneous generation and miasma with germ theory, while microscopic studies revealed the bacteria responsible for mass killers such as cholera, tuberculosis, and tetanus. Microscopes also encouraged great progress in histology and pathology. Women began to qualify as doctors, and nursing became a recognized profession. In the last decade, X-rays opened up a new world of noninvasive medical imaging.

1880

1895

1881
The first professional midwives organization, Matrons' Aid Society, is founded in Britain, and soon changes its name to the Midwives Institute.

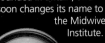

« Sphygmomanometer

1881
Samuel von Basch invents the first sphygmomanometer, a device to measure blood pressure.

1882
Robert Koch identifies the cause of tuberculosis: *Mycobacterium tuberculosis.*

1884
Robert Koch isolates the causative germ for cholera and describes how it is spread, and prevention and control measures.

1885
Louis Pasteur carries out the first successful rabies vaccination, on a young boy.

❯ Administering rabies vaccination

1890
Giovanni Grassi and Raimondo Filetti discover there are several kinds of malarial parasites; Ronald Ross demonstrates that mosquitoes transfer these parasites between humans.

1893
William Einthoven introduces the term "electrocardiogram" and publishes *New Methods for Clinical Investigation* concerning the heart's electrical activity and its relevance to disease and diagnosis.

1894
Kitasato Shibasaburo and Alexandre Yersin independently identify the microbe of bubonic plague, which is named *Pasteurella pestis,* and later renamed *Yersinia pestis.*

⌃ Early X-ray examination

1895
Wilhelm Röntgen discovers X-rays and their ability to "see" bones and hard tissues inside the body.

1895
In Vienna, Karl Landsteiner begins his studies of immunity, antibodies, and blood, especially how and why it clots.

1895
Sigmund Freud and Josef Breuer coauthor *Studies on Hysteria,* the first main work in psychoanalysis.

1896
Almroth Edward Wright develops and introduces the first effective typhoid vaccine.

1896
John Hall-Edwards uses X-ray imaging for the first time during a surgical operation. The same year the first reports of harm caused by X-rays, including hair loss, blisters, burns, and swelling, appear.

1896
The sphygmomanometer is improved by Scipione Riva-Rocci, who adds a cuff around the arm to apply even pressure to the limb.

1897
A vaccine for plague is developed, but limited effectiveness and the infection's complex nature mean it does not become widely used.

1897
Chemists at Bayer in Germany, including Felix Hoffman and Heinrich Dreser, produce a synthetically modified version of salicylic acid that is better tolerated by the body; it is named Aspirin.

1899
Aspirin goes on sale worldwide and becomes one of the most successful and adaptable medical drugs of all time.

« Asprin carton

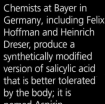

1899
Sigmund Freud publishes *The Interpretation of Dreams* setting out various psychological theories, including a model of mental structure based on the unconscious, preconscious, and conscious.

1899
Santiago Ramón y Cajal publishes *Comparative Study of the Sensory Areas of the Human Cortex,* greatly advancing the neurosciences.

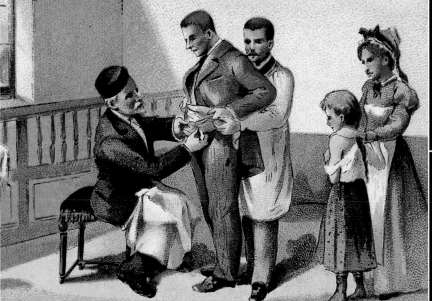

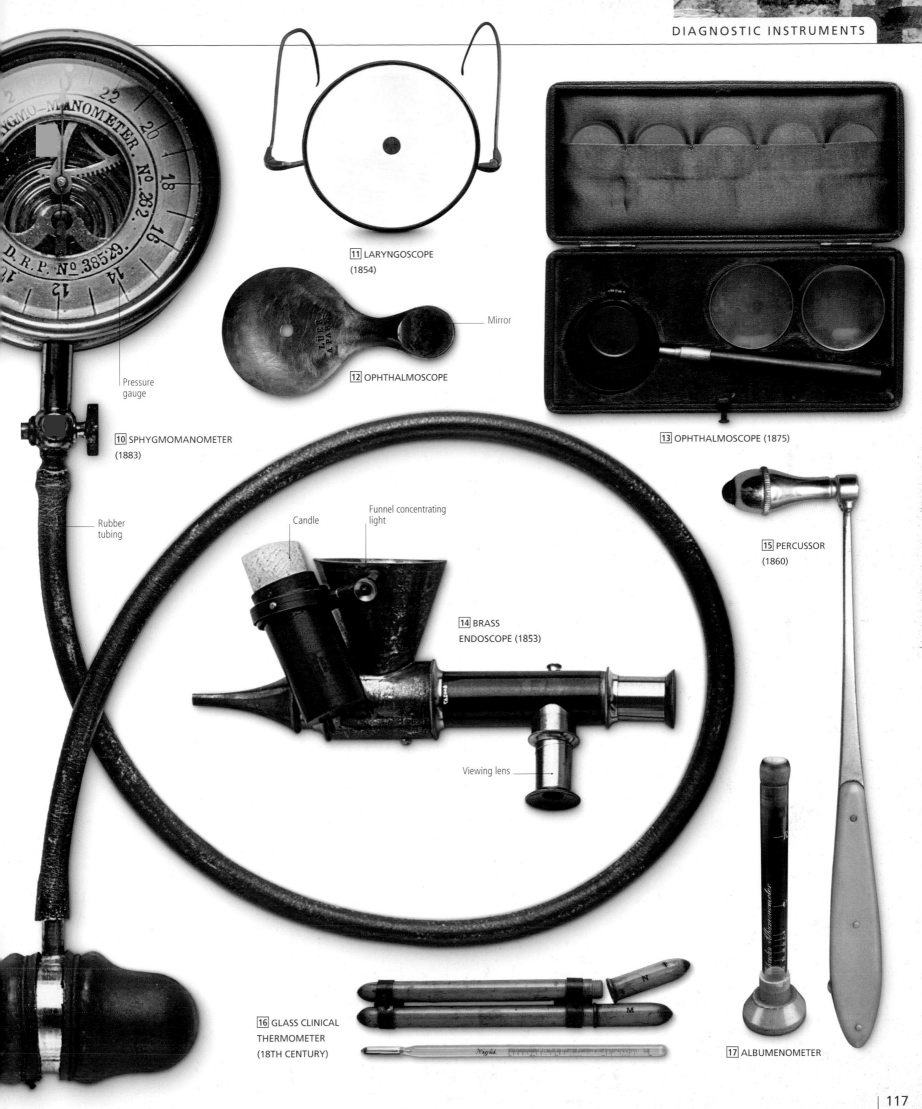

11 LARYNGOSCOPE
(1854)

Mirror

12 OPHTHALMOSCOPE

13 OPHTHALMOSCOPE (1875)

Pressure
gauge

10 SPHYGMOMANOMETER
(1883)

Rubber
tubing

Candle

Funnel concentrating
light

15 PERCUSSOR
(1860)

14 BRASS
ENDOSCOPE (1853)

Viewing lens

16 GLASS CLINICAL
THERMOMETER
(18TH CENTURY)

17 ALBUMENOMETER

Resurrection Men

In the 18th and 19th centuries the insufficient supply of corpses to medical schools in Britain for the purpose of dissection gave rise to the "resurrection men." Often operating in gangs, resurrectionists disinterred fresh corpses and supplied them to anatomists. Outrage at such activity led to a change in the law so that medical schools could acquire cadavers legally.

Advances in anatomical science since the medieval period had come about as a result of the dissection of human corpses. Although Pope Benedict VIII had forbidden the practice in 1300 on pain of excommunication, the authorities in most European countries took a more pragmatic approach, allowing the unclaimed corpses of the poor to be supplied to anatomical schools. The detail and accuracy of Flemish physician Vesalius's (see pp.72–75) drawings of many anatomical features in his 1543 work *De Humani Corporis Fabrica* (*On the Fabric of the Human Body*) was only made possible through human dissections.

In Britain, the law was much stricter. In 1540 Henry VIII gave the Company of Barber Surgeons the right to take four corpses of executed felons each year to use for dissections. Even after the Murder Act 1752 decreed that criminals could be dissected by anatomists after execution, the supply was wholly inadequate to meet the needs of medical schools. For this reason, surgeons turned to the service of resurrection men, who would disinter freshly buried corpses and sell them for dissection for a fee ranging from 2 to 20 guineas (the latter being well over 20 times the average weekly wage of a surgeon at the time).

This sinister but highly lucrative practice became so widespread that the Edinburgh College of Surgeons introduced a clause in their contracts in 1721 forbidding trainees from dealing with the resurrectionists. However, the restriction was largely ignored, since anyone with aspirations to be a surgeon needed to witness

▽ **The Burke and Hare murders**
Burke and Hare's first corpse was a pensioner, who died of natural causes at their lodgings. After that they lured potential victims—mainly vulnerable women—with the promise of alcohol, then got them drunk and smothered them.

or to perform dissections, and so the resurrection men found a ready supply of keen customers for their wares.

The resurrectionist gangs

Professional resurrection men often operated in gangs, supplying dozens of dead bodies to medical schools each year (one gang that was uncovered in Lambeth in 1795 had 15 members). For corpses of well-known people or "freaks"— such as the highwayman Dick Turpin, whose grave was robbed in 1739, or the "Irish Giant" O'Brien, who was over 7 ft (2.1 m) tall, respectively—the fee could rise to as much as £500.

The activities of the resurrection men became so widespread that at times confrontations broke out in cemeteries as mourners realized that the men with shovels and

349

The number of corpses supplied by resurrection men in 1809–10, from evidence given to the House of Commons Select Committee in 1828.

These murders and those carried out in London in 1831–32 by the "Burkers"—those who modeled themselves on Burke and Hare— led to a call for reform. Medical self-interest played a part; when a Liverpool surgeon William Gill was convicted of receiving a corpse in 1828, doctors realized that they, too, were now liable to prosecution for the activities of the resurrection men. The same year, a House of Commons Select Committee was set up and issued a report about the need for anatomical science and dissection, but it faced initial opposition from those who were against a relaxation of the law.

Finally, in 1832 an Anatomy Act was passed that allowed licensed anatomy lecturers to use unclaimed dead bodies from workhouses, hospitals, and prisons. As medical schools no longer needed illegally acquired corpses, demand dropped and the price that the resurrection men could charge for their services plummeted; within a few years, they had disappeared entirely.

picks lurking in the shadows were not in fact gravediggers, but "body snatchers." In desperation, some local communities funded graveyard patrols, and wealthy families paid for security measures, such as the mortsafes (iron cages) or the "Patent Coffin," invented in 1818, with its metal-sprung catches that were designed to thwart tools used by the resurrectionists to open coffins.

The resurrectionists were never popular—a riot in Greenwich in 1832 against the activities of the West Kent gang involved several

thousand people. However, as long as the authorities turned a blind eye, little was done about them. In England, the removal of a corpse was not officially an offense until 1788, when courts ruled that "common decency" required the practice to stop; even then, there was no specific statute against it.

Motive for murder

Such was the demand for corpses, some resurrectionists took things even further. Between 1827 and 1828 Irish immigrants William

Burke and William Hare sold 16 corpses to the Edinburgh-based physician Robert Knox. It turned out, after the body of a dead woman was found under a bed at their address, that they had never dug up any corpses. Instead, they had murdered their victims and sold the fresh corpses to Dr. Knox. After a notorious trial, Burke was hanged on January 28, 1829, and his body was publicly dissected the following day. Hare escaped by giving evidence against his former partner.

△ **Caged graves**
In Scotland, the graves of well-to-do residents were often protected by sturdy iron cages ("mortsafes"), to foil the efforts of grave robbers and "body snatchers." These cages encased buried coffins or were set in a concrete foundation and covered the whole grave.

> **"**The coffin was forced… and the melancholy relics, clad in sack-cloth **after being rattled for hours** on moonless by–ways, were at length **exposed to uttermost indignities** before a class of gaping boys.**"**
>
> ROBERT LOUIS STEVENSON, SCOTTISH AUTHOR, FROM *THE BODY SNATCHER*, 1884

Miasma Theory

Bad smells, which are associated with rot and decay, have long been linked with illness. The miasma theory, believed since ancient times, held that diseases were caused and spread by a mix of foul-smelling vapors, gases, and possibly tiny particles present within them.

The notion that poisonous air was the cause of illness grew from observations that disease was more common in crowded areas and in places where unsanitary conditions such as rot, mold, dirty water, excrement, and putrid odors abounded. In the medieval period, as towns and cities grew, outbreaks of diseases such as plague, tuberculosis, cholera, and malaria (from the Italian for bad air, "mala aria") increased with them.

By the 18th century, with the discovery of many previously unseen microscopic menaces, the miasma theory was redefined. It was believed that poisonous vapors and tiny particles from decomposing matter, too small for microscopes but identified by their offensive smell, were released into the air, and made their way into the body and caused disease. Although the work of John Snow during the London cholera outbreaks (see pp.122–23) pointed to contaminated water as the disease's source, rather than bad air, his findings were dismissed at the time as the ideas of the miasma theory prevailed. It was not until the 1870s and the work of Robert Koch and others, that miasma theory was finally replaced with the germ theory of disease (see pp.146–47). Yet, despite their inaccurate basis, anti-miasma public health measures, such as clean drinking water and sanitation, had been beneficial as they had helped not only remove the smells but also the germ-causing microbes.

> **"** First rule of nursing, to **keep the air** within as **pure** as the air without.**"**
>
> FLORENCE NIGHTINGALE, ENGLISH NURSE, FROM *NOTES ON NURSING: WHAT IT IS, WHAT IT IS NOT,* 1898

◁ Poisonous air
This mid-19th century cartoon by British illustrator Robert Seymour, titled *Cholera Tramples the Victor and the Vanquished Both,* shows a ghostlike figure spreading cholera across the battlefield.

Cholera

One of the most virulent diseases ever known, cholera has killed millions of people and had a huge social impact around the world. The study of microbiology during the 19th century contributed to the understanding and control of this disease but, without safe water available to all, outbreaks continue to occur.

Cholera has affected people for many centuries. Records from India, dated to around 1000 CE, describe a disease thought to be cholera that induced severe diarrhea and vomiting, leading to dehydration, and often death. However, cholera did not spread beyond the subcontinent until 1817, when infected travelers carried it out of India along trade routes. By the 1830s it had reached as far as the US.

Before the study of bacteria gained importance, and prior to the linking of germs, or microorganisms, to infectious diseases, it was believed that cholera was caused by excessive production of bile—the term "cholera" is derived from the Greek word *khole*, meaning "illness from bile." It was hard to discriminate cholera from other diseases associated with diarrhea and vomiting, but the enormity of human suffering in the 19th century led to intensive research and lengthy debates about the nature and causes of the disease. The scientific world became embroiled in discussing the merits of germ theory (see pp.146–47) over miasma theory (see pp.120–21).

One of the key people involved was English physician John Snow (see pp.124–25), who believed that the disease was not airborne.

Instead, he suggested that excrement contained infectious material that could infect the populace if it found its way into the water supply. During a cholera outbreak in 1854, he noted that a number of cases were clustered around one hand pump on Broad Street in Soho, London; when he removed the handle of the pump, the cholera stopped spreading. Snow was a pioneer but, despite his attempts, he could not identify the pathogen that caused cholera.

Identifying the cause

In the middle of the 19th century, when cholera reached Florence, Italian scientist Filippo Pacini—an expert microscopist—was determined to study the onset of the disease and find out how it was transmitted. He performed autopsies on victims and studied their intestines. His tests resulted in the isolation of a comma-shaped

△ **Water-testing kit**
Frederick Danchell, a civil engineer, introduced this simple water testing kit in the 1860s to test for organic matter and chemical pollutants, after John Snow argued that cholera was a waterborne disease.

▷ **Treatment centre**
Cholera emerged in Haiti following the 2010 earthquake, which resulted in water supplies being infected and rapidly transmitting the disease to thousands. In this treatment center, patients lie on a "Watten bed," or cholera bed, with holes cut out to catch the watery diarrhea common to cholera patients.

> " Death from **sickness** at a **level** not seen since the Black Death."
>
> MARTIN DAUNTON, PROFESSOR AT THE UNIVERSITY OF CAMBRIDGE, FROM *LONDON'S "GREAT STINK": THE SOUR SMELL OF SUCCESS*, 2004

bacterium, belonging to the bacillus group, which he called *Vibrio*. However, his findings did not become well known until 1965—more than 80 years after his death.

In 1883 thirty years on from Pacini's studies, German physician Robert Koch began researching the cholera-causing microorganism. He traveled to Egypt, where cholera was widespread, and studied the intestines of deceased victims. Like Pacini, he also found *Bacillus* in their intestinal mucosa. He moved to India to further his research. There he was able to grow the bacterium in a pure culture, and noticed the distinctive commalike shape of the *Bacillus*; in 1965 it was officially named *Vibrio cholerae*.

Koch observed that the bacteria flourished in moist places, such as wet linen. His scientific peers accepted his findings and acknowledged him as the discoverer of the cholera-causing bacterium.

Koch's discovery had important social consequences—people became aware that exposure to contaminated water caused disease, and that bacteria could return to the water supply through sewage.

The introduction of filtered water pipes led to a dramatic fall in the incidence of the disease. However, the knowledge that cholera was caused by contaminated water was not enough to cure people or save their lives—clean drinking water was a luxury many in the developing world could not enjoy.

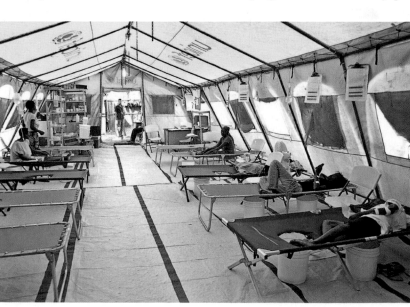

Developing treatment

The recurring epidemics of the 19th century made the need for effective treatment increasingly pressing. In the 1830s physicians began to realize that dehydration was the real cause of death in cholera patients. This led to new experiments with fluid replacement therapies, involving intravenous injections of water and salt. Improvements in the salt concentrations, the amount of fluid given, and the rate of delivery gradually reduced fatalities, but it took until the mid-20th century for major advances to be made.

In 1958 US Navy medical researcher Raymond Watten invented a cot with a hole in the middle, allowing for an accurate measurement of excrement, so that rehydration fluids could be given in the right amount, and of the same chemical composition as was being lost. The "Watten cot"

3–5 MILLION Number of recorded cholera cases every year, killing more than 100,000, according to the World Health Organization (WHO).

is still used routinely in treatment centers. Even more significant was the discovery in the 1960s that glucose helps the gut absorb salt, enabling the creation of the first oral rehydration therapies. Effective, easy to administer, and relatively cheap, this treatment (and appropriate antibiotics) has become the most widely used means of managing cholera and other diarrhea-related diseases.

Since the first recorded pandemic in 1817, there have been seven further outbreaks of cholera. The need for improved vaccination, and effective approaches to both prevention and control remains.

▷ **Cholera defeats the Turkish Army**
The triumphant grim reaper, Death, is shown on the cover of this Paris newspaper in 1912. The Turkish army is defeated not by the enemy, but by cholera. The disease swept through the camps, killing 100 men a day during the First Balkan War (1912–13).

Le Petit Journal

ADMINISTRATION
61, RUE LAFAYETTE, 61
Les manuscrits ne sont pas rendus
On s'abonne sans frais
dans tous les bureaux de poste

5 CENT. SUPPLÉMENT ILLUSTRÉ **5 CENT.**
23me Année — ** — Numéro 1.150
DIMANCHE 1er DÉCEMBRE 1912

ABONNEMENTS

	SIX MOIS	UN AN
SEINE et SEINE-ET-OISE..	2 fr.	3 fr. 50
DÉPARTEMENTS............	2 fr.	4 fr. »
ÉTRANGER	2 50	5 fr. »

LE CHOLÉRA

ENGLISH PHYSICIAN Born 1813 Died 1858

John Snow

> " The **cholera** extended to… houses in which the **water** was thus **tainted…** "

JOHN SNOW, FROM *MODE OF COMMUNICATION OF CHOLERA*, 1855

Water bath inside the bottle helps ether to vaporize

▷ **Ether inhaler**
This device was invented by Snow in 1847, just one year after the first demonstrations of ether in the US. The temperature of the water bath, top right, could be altered to adjust the dose.

Mouthpiece

A quiet, modest, hard-working English physician, John Snow brought about vast changes in our understanding of how infectious diseases spread, the need for public health and sanitation, and the significance of epidemiology (see pp.126–27) as a specialist area of study. However, Snow's reasoning, which ultimately led to these advances, was rejected at the time and he did not live to see his work accepted, dying at the early age of 45 years old.

After an education in which he showed an aptitude for mathematics and statistics, Snow gained early medical experience in Newcastle upon Tyne.

In 1836 he moved to London, where he acquired membership of prestigious medical colleges, became president of the Westminster Medical Society, and in 1849 became a founder member of the Epidemiological Society of London, which aimed to examine the origin, propagation, mitigation, and prevention of epidemic diseases.

Ether and anesthetics
During the 1840s Snow developed an interest in anesthesia (see pp.128–31). The medical use of chemicals to dull sensation and pain and to induce unconsciousness was a popular area of research at the time. In 1846 there was news from Boston, Massachusetts, that ether could be safely used as an anesthetic in dentistry and general

◁ **Death's dispensary**
A cartoon from 1866 shows how Snow's deductions about the spread of cholera by water were accepted a decade later.

surgery. Snow read avidly on the subject of anesthetics and began to devise his own equipment. He tested new gases—especially chloroform—on animals and, to his detriment, himself (modern-day scientists speculate that his self-experimentation may have exacerbated preexisting health problems and led to his early death). He wrote articles on the subject and also created the profession of "specialist anesthetist." The Royal Medical and Chirurgical Society (a forerunner of the Royal Society of Medicine) described him as "more extensively conversant with its operation, and more successful in administering it, than any living person." Snow gained much recognition and was instrumental in making anesthetics safer, more effective, and more widely accepted.

Studying cholera
Snow's first encounter with the bacterial infection of cholera (see pp.122–23) was in 1831–32

in Killingworth colliery in northern England. In 1849 he witnessed more cases and began to investigate the cause and spread of the disease. Since its major early symptoms were vomiting and diarrhea, he suspected that it was a digestive problem and probably transmitted by eating or drinking contaminated matter. However, the miasma theory (see pp.120–21) was also prevalent at the time, and many experts regarded cholera as a blood-based sickness. In the first edition of his pamphlet, *On the Mode of Communication of Cholera* (1849), Snow wrote: "It is quite true that a great deal of argument has been employed on the opposite side, and that many eminent men hold an opposite opinion."

In 1854 Snow applied an epidemiological approach when he studied a cholera epidemic centered on Broad Street in Soho, London. He visited houses, interviewed residents, and delved into plans of the area's water supplies and sewage disposal.

> " This journal… **failed to recognize** Dr. Snow's… **visionary work** in deducing the mode of **cholera transmission.** "

APOLOGY FOR OMISSIONS IN SNOW'S OBITUARY, FROM THE MEDICAL JOURNAL *THE LANCET*, 1958

He recorded straightforward information but his skill lay in his analysis of the data. He made maps showing that cases were clustered around a public water pump in Broad Street—an innovative idea at the time. In light of his suspicions, and with the help of the parish authorities, Snow arranged for the handle of the public water pump to be removed so that local people had to obtain their supplies elsewhere. The outbreak was already subsiding, but Snow believed that disabling the pump would speed its end.

The next year, Snow published the updated edition of *On the Mode of Communication of Cholera*. Although his evidence was convincing, it was passed over for various reasons, including the high costs of public works to provide clean water supplies and hygienic sewage disposal, and rival theories, such as that of Bristol-based physician William Budd, who blamed the cholera outbreak on a fungus spread through drinking water. Snow was rebuffed and disappointed and, when he died

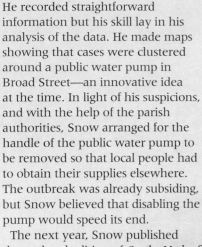

▷ **A simple man**
Snow was far from flamboyant and fame-seeking. A close friend remarked that he "clothed plainly, kept no company, and found every amusement in his science books, his experiments, and simple exercise."

three years later, he had not witnessed the rewards of his work. However, the next decade saw further cholera outbreaks with more detailed studies, and the establishment of germ theory (see pp.146–47), all of which vindicated Snow's conclusions and sealed his place in medical history.

TIMELINE

- **1813** Born in York, England, the eldest son of a farm and general laborer. He attends a local private school.

- **1827** Becomes apprentice to Newcastle surgeon William Hardcastle, and works as a colliery physician during the cholera epidemic of 1831–32.

- **1836** Enrolls as a student at the Hunterian School of Medicine, London; later works at Westminster Hospital.

- **1838** Becomes a member of the Royal College of Surgeons and, a few months later, of the Society of Apothecaries.

- **1846** Becomes interested in the properties of the anesthetizing agent ether, and works to make improvements in its administration, along with testing other agents.

- **1847** Publishes *On the Inhalation of the Vapour of Ether*.

- **1849** *On the Mode of Communication of Cholera*, his first report on the transmission of cholera through contaminated water supplies, wins an award from the Institut de France.

- **1850** Joins the Royal College of Physicians.

- **1853** Administers chloroform to Queen Victoria during the birth of Prince Leopold. He will do the same again in 1857 for the birth of Princess Beatrice.

- **1855** Publishes an updated edition of *On the Mode of Communication of Cholera* that includes the Soho, London, outbreak of 1854.

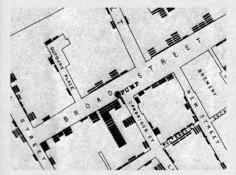

JOHN SNOW'S MAP OF SOHO, ILLUSTRATING INCIDENTS OF DEATH BY CHOLERA

- **1858** *On Chloroform and Other Anaesthetics* is published. Snow dies from a combination of stroke, and kidney failure brought on by experimenting with anesthetic gases. He is buried in Brompton Cemetery, London.

Epidemiology and Public Health

Until the 19th century, little progress was made in containing epidemics in the rapidly growing cities. However, the breakthrough came when medical scientists began to discover the causative agents of diseases, leading to effective control and prevention strategies and considerable advancements in public health.

In the 4th century BCE, Greek physician Hippocrates tried to explain diseases in terms of external and environmental factors rather than divine displeasure, as had always been the case. However, doctors were unable to understand, let alone control, the spread of infectious diseases. Nonetheless, during the Black Death in Italy in the 14th century, the introduction of quarantine and isolation hospitals (see pp.68–69) showed an awareness that reducing contact with infected persons was the most obvious way to contain the disease.

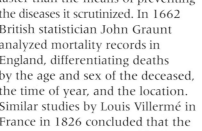

30.6 PER THOUSAND The death rate in Paris's poorest districts in 1826.

19.1 PER THOUSAND The death rate in Paris's richest districts the same year.

Supporting sanitation

The science of epidemiology—the study of disease patterns, causes, and epidemics—at first progressed faster than the means of preventing the diseases it scrutinized. In 1662 British statistician John Graunt analyzed mortality records in England, differentiating deaths by the age and sex of the deceased, the time of year, and the location. Similar studies by Louis Villermé in France in 1826 concluded that the poor had higher rates of mortality than their middle- and upper-class counterparts. The "miasma theory" (see pp.120–21), which was popular in the 19th century, maintained that bad vapors in the air caused by filth were the primary agents of disease, and efforts were made to clean up cities that were growing uncontrollably as a result of the Industrial Revolution, which drew workers from rural to urban areas.

The cholera epidemic that struck London in 1831–32 led to calls for reform. In 1842 British lawyer Edwin Chadwick compiled a report on sanitary conditions in cities. This prompted the establishment of a Royal Commission on the Health of Towns and also local boards of health, which were responsible for enforcing sanitary and hygiene regulations in their districts. Public Health Acts gave these bodies greater powers, starting in 1848, when they were given a remit to inspect lodging houses and provide sewers. Clean water had become a concern after British physician John Snow's discovery of the waterborne nature of cholera (see p.122). In 1858 the British parliament gave £3 million to the Metropolitan Board of Works to build new sewers for London; when completed in 1870, these finally put an end to the cholera epidemics of the previous four decades.

Mass vaccination programs

The discovery that diseases were transmitted by bacteria and viruses (see pp.166–67) meant that, in the late 19th century, public health

△ "Typhoid Mary"

In the early 20th century, it became clear that people without symptoms could still carry the typhoid pathogen and transmit the disease. Mary Mallon, a cook, infected more than 50 people in several households where she worked.

efforts turned to the use of vaccines (or in some cases drug treatments) for fatal diseases. Britain began the first mass vaccination program—for smallpox (see pp.100–01)—in 1853, which extended worldwide over following decades, eventually leading to the disease's global eradication in 1977. Other similar programs for polio, typhoid, mumps, and measles gradually led to these once-common, often fatal, infections becoming rarities.

Noncommunicable diseases

As epidemics of infectious diseases became rarer in industrialized countries after World War II, global public health efforts turned to noncommunicable diseases—for example, cancer and diabetes—and to those, like malaria, whose main impact was felt by poorer countries. Studies in the early 1950s linked

BRITISH PHYSICIAN 1877–1967

JANET LANE-CLAYPON

The first woman to receive a research scholarship from the British Medical Council, physician Janet Lane-Claypon pioneered two research methods that are key to the field of epidemiology. She used cohort studies to compare weight gain between one group of children who were breast-fed and another group who were bottle-fed milk. In 1923 she used a case-control study to conclude that women who married earlier, had more children, and breast-fed them were less likely to develop breast cancer.

" The primary and most important measures… are drainage, the **removal of all refuse** from habitations, streets and roads, and the **improvement of the supplies of water.** "

EDWIN CHADWICK, BRITISH LAWYER, FROM *THE SANITARY CONDITION OF THE LABOURING POPULATION OF GREAT BRITAIN*, 1850

▷ **Crimean War deaths**
Produced by Florence Nightingale during the Crimean War, this chart illustrates that more soldiers died as a direct result of infectious diseases than battlefield wounds. Nightingale used it in her hard-fought campaign to improve the standards of hygiene in field hospitals.

smoking and lung cancer for the first time, eventually leading to attempts to curb tobacco usage through taxation, public health campaigns, and, in some countries, banning smoking in public areas.

Post-war health organizations
At a national level, epidemiology and public health campaigns came to be managed by bodies such as the US Communicable Disease Center, known today as Centers for Disease Control and Prevention (CDC), established in 1946, and Britain's National Health Service (NHS), founded in 1948. At a global level, the World Health Organization (WHO), set up in 1948, coordinates international responses to crises—for example, the Ebola epidemic in West Africa in 2014–15—as well as setting up longer-term global eradication programs (pp.266–67).

△ **Smoking and lung cancer**
Before medical studies emerged making the connection between smoking and lung cancer, some advertisements actually promoted the habit as having health benefits. In 1960 over one-third of American doctors still did not believe that smoking and cancer were linked.

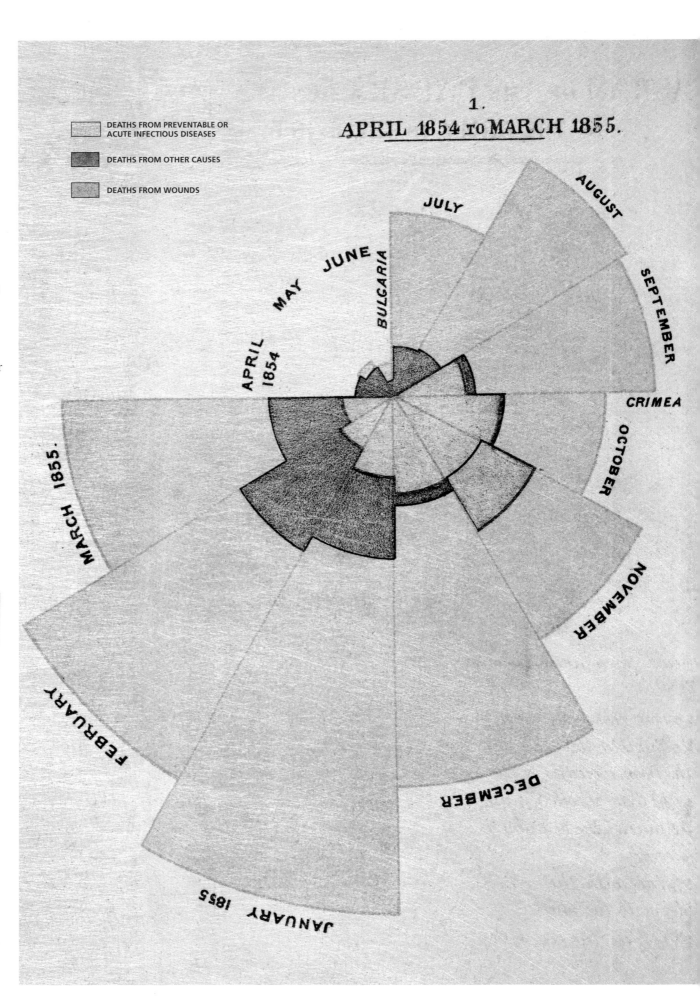

Anesthetics

Since antiquity surgeons have looked for ways to dull the pain experienced by their patients during an operation. In 1846 American dentist William Morton finally came up with an effective solution by using gas to anesthetize a patient, and the era of modern pain-free surgery began.

In ancient times surgery was not only dangerous but also very painful, although surgeons tried many methods of pain relief. Hemp was used as an anesthetic in China in the 2nd century CE, while in the medieval period Arab doctors soaked "sleep sponges" in aromatics and soporifics, such as mandragora and opium. Extreme compression of the nerves near the part of the body being operated on using screw-clamps was tried in the 18th century, but this often caused the patient as much pain as the operation itself. More effectively, in the 1770s German physician Anton Mesmer pioneered mesmerism (see p.160)—a form of hypnosis that could induce a trance in patients and reduce their sensitivity to pain.

80 PER CENT The approximate rate of death after operations before the 19th century.

Nitrous oxide

A more promising avenue for pain relief during surgery proved to be the inhalation of gases and vapours. In 1799 British chemist Humphry Davy observed the intoxicating effect of nitrous oxide and suggested that "it may be used with advantage during surgical operations". He did not pursue this idea, however, and nitrous oxide, often called "laughing gas", was for decades taken mainly at parties.

The real advances came from dentists in the US. In the 1840s dentist Horace Wells experimented with administering nitrous oxide through a wooden tube attached to an animal bladder. He even had one of his own teeth extracted under the influence of nitrous oxide to prove that the procedure

▷ **Mandrake**
The root of the mandrake plant contains hallucinogenic and narcotic compounds and was used in the medieval period as an anesthetic, sometimes mixed with opium. In too large a dose, it could cause delirium and even death.

was pain free. However, in 1845 a demonstration by Wells in Boston failed, since the patient experienced pain. This operation was performed on William Morton—a former dental partner of Wells—and Morton resolved to try a different approach.

"This Yankee dodge, gentlemen, beats mesmerism hollow."

ROBERT LISTON, SCOTTISH SURGEON, AFTER PERFORMING THE FIRST AMPUTATION USING GAS ANESTHESIA IN BRITAIN, DECEMBER 21 ,1846

Ether and chloroform

The properties of diethyl ether (commonly known as ether) had been known since the 16th century and had been used as a general anesthetic in 1842 by Crawford Long, a general practitioner from Georgia. However, Long did not publicize his findings, and it was Morton who was credited with the first successful series of operations under anesthetics.

Having first tried ether on himself, a dog, and several assistants, on September 30,1846, Morton carried out a tooth extraction on a patient, Eben Frost, using ether saturated in

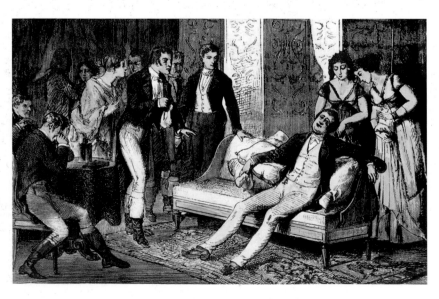

◁ **Laughing gas**
After Humphry Davy's discovery of the exhilarating effects of nitrous oxide, it became popular at parties. By the 1840s it was substituted by ether, which could be more easily transported, giving such events the nickname "ether frolics."

a handkerchief. The patient felt no pain and word of Morton's success spread. He was invited a few days later to conduct an operation to remove a benign tumor from a patient's neck at Massachusetts General Hospital. By this time he had refined his method of ether delivery to incorporate a double-necked glass globe, with air entering one section that then passed via an ether-soaked sponge to be inhaled by the patient. The operation was carried out in front of a crowd of medical professionals and was again a success.

By November that year, surgeons felt confident enough in Morton's methods to perform an amputation on a seven-year-old girl, who was suffering from tuberculosis of the knee, under the influence of ether. Use of the technique spread rapidly, and on December 19, 1846, the first anesthetic operation in Britain—the extraction of a molar—was carried out, and the second—an amputation—was performed just two days later. The amputation was so successful that the patient asked when the operation was going to begin, after his leg had been sawn off.

By January 1847 anesthesia reached France, and six months later an operation was carried out in Australia. However, ether fell out of fashion because it was slow to take effect and often induced vomiting in the patient. A new

gas—chloroform—was pioneered by obstetrician James Young Simpson, Professor of Midwifery in Edinburgh, Scotland, who first used it in 1847. It was faster-acting and gentler than ether, and in the 1850s became a popular method of

easing pain during childbirth after pioneer anesthetist John Snow gave Queen Victoria chloroform for her last two births (pp.124–25).

The road ahead

Within a year of Morton's first effective anesthetic operation, surgery had been revolutionized. Operations could be longer, and surgeons could work more slowly and carefully without fear of their patients dying from shock. Through the second half of the 19th century anesthesia continued to undergo many refinements. As the gases improved, better masks and pumps were devised to administer them more effectively. Local anesthetics appeared in 1884—the first one to be used was cocaine, used as eye drops in optical surgery. Intravenous anesthetics, which acted far more swiftly than those administered through inhalation, were first used in 1874, and spinal anesthesia was introduced in the 1890s.

The remarkable developments that had occurred in anesthesia in the 19th century transformed surgery, and they paved the way for more complex operations in the 20th and 21st centuries, most notably those on internal organs.

Hanaoka Seishu
Japanese physician Hanaoka Seishu devised an anesthetic drink made from a variety of herbs, including angelica. In 1804 he used it as a general anesthetic during a mastectomy operation.

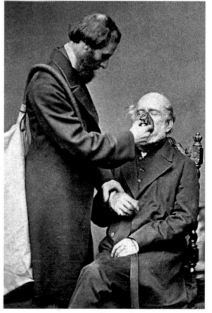

△ **Chloroform apparatus**
In 1862 English doctor Joseph Thomas Clover devised an apparatus, shown here, to deliver chloroform in accurate and measured doses, overcoming the earlier problem of patients dying from an overdose.

Seed capsule

1 POPPY SEEDS AND CAPSULE

4 19TH-CENTURY
REPLICA OF MORTON
ETHER INHALER

2 CHLOROFORM INHALER (1848)

3 HEWITT DROP BOTTLE (1886)

5 MINNITT GAS-AIR ANALGESIA APPARATUS (1950)

6 ANESTHETIC
FACE MASK (19TH
CENTURY)

Gauze
mask
cover

Early Anesthetics

The administration of anesthesia (see pp.128–29) at first required complex apparatus to create, mix, store, and deliver the gas. Over time instruments became more compact and manageable.

7 COMBINED-GAS
APPARATUS

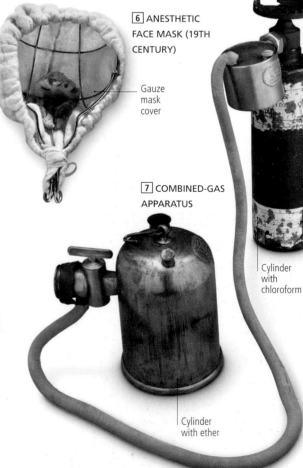

Cylinder
with
chloroform

Cylinder
with ether

1 **Poppy seed capsule** The seeds of the opium poppy have a sedative effect and were used in ancient times to provide pain relief. 2 **Chloroform inhaler** Invented by John Snow, this inhaler had two tubes. Chloroform was pumped in through one tube and breathed out of the other. 3 **Hewitt drop bottle** This bottle was used to administer drops of chloroform or ether at a controlled rate. 4 **Morton ether inhaler** This inhaler was first used by William Morton in 1846. Ether was passed through the tap, soaked by the sponge, and released through a rubber tube and mask. 5 **Minnitt gas-air analgesia apparatus** This gas-air machine was designed to produce a mixture of nitrous oxide and air to provide pain relief for women in labor. 6 **Anesthetic face mask** This 19th-century face mask consists of a gauze cloth stretched

over a wire frame, with a sponge soaked in ether that sat over the patient's nose and mouth. 7 **Combined-gas apparatus** This had a large cylinder from which chloroform or ether was passed to the smaller portable brass cylinder. A tube connected the smaller cylinder to the patient's face mask. 8 **Hypodermic syringe** This allowed the easy intravenous injection of drugs. 9 **Boyle's apparatus** The Boyle bottle allowed anesthetists to control the vaporization of gas from a liquid to create a safe mixture of gases. 10 **Clayfield's mercurial holder** This device measured the amount of nitrous oxide inhaled by a patient. 11 **Basket Boyle anesthesia machine** This machine allowed a continuous flow of anesthetic gases. 12 **Nitrous oxide cylinders** These were commonly used in dentistry from the 1850s.

8 HYPODERMIC SYRINGE (20TH CENTURY)

9 BOYLE'S APPARATUS (1930)

10 CLAYFIELD'S MERCURIAL HOLDER (20TH-CENTURY REPLICA)

Gas tubing

Ether vaporizer

Patient circuit through which gas is administered

Movement of weights indicate level of gas remaining in the jar above the mercury

11 BASKET BOYLE ANESTHESIA MACHINE (1950)

12 NITROUS OXIDE CYLINDERS (20TH CENTURY)

Connector for mask

Dentistry

Advances in dental technology have dramatically improved oral health. Where once complete extraction was the only solution for widespread tooth decay, now dental patients are far more likely to retain most, if not all, of their own teeth.

Contrary to the popular perception that people in the medieval period suffered from rotting and missing teeth, most understood the importance of dental health and cleaned their teeth regularly. Although not a fully fledged profession at the time, dentistry was practiced among the wealthy and included extractions, fillings, and the fitting of false teeth, but tools and techniques were basic, and procedures painful.

Historians estimate that 20 percent of the medieval European population suffered from tooth decay. With the widespread intake of sugar, this figure had risen to 90 percent by the 19th century. As the demand for treatment increased, dentistry was transformed. Many advances occurred in the 1800s, such as the reclining dental chair, amalgam fillings, and the use of anesthesia.

In the late 19th century, drills replaced files and chisels, to remove decay and prepare cavities, and the filling of teeth became a viable alternative to extraction. The Harrington windup dental drill of 1864 was followed by the foot-operated drill invented by American dentist James Morrison in 1872. Just a few years later, in 1875, the invention of the first electric drill heralded the dawn of modern dentistry. In 1957 the advent of the air turbine drill ushered in the era of high-speed dentistry and joined other innovations of the latter half of the 20th century such as fluoride toothpaste, lasers, resin filling materials, ceramic polymer implants, and "invisible" braces, which all helped bring dentistry into the modern era.

"**Dentistry** is the practice of a special branch of **medicine.**"

CHARLES MAYO, AMERICAN MEDICAL PRACTIONER, IN AN ADDRESS TO THE AMERICAN MEDICAL ASSOCIATION, 1928

▷ Elecro-anesthesia
From the 1840s electricity was tried as a dental anesthetic, especially in France, as shown in this 1870s dental school in Paris. Results were disappointing and injectable anesthesia eventually took over.

Pregnancy and Childbirth

In many cultures, care of women and their babies during pregnancy, birth, and infancy was often separate from mainstream medicine. The role of the midwife only became formally recognized from about 100 years ago.

Modern medical specialties relating to women's health, childbirth, and children include gynecology, for dealing with female reproductive health; midwifery, for health care during uncomplicated pregnancy and birth; obstetrics, for more medically involved pregnancy and birth; and pediatrics, for infants and children through to puberty. However, such specialties have not always been in existence.

Ancient wisdom

For millennia, pregnancy and birth were private matters involving female family members and close friends, who were usually non-medical. In ancient Mesopotamia and Egypt, female birth attendants helped the mother give birth, and specialists—the midwives of their day—were described in the Ebers papyrus (see pp.20–21).

One of the first texts on women's health and childbirth was *Gynaikeia* (*Gynecology*), written by the 1st-century CE Greek physician Soranus of Ephesus. The first major Chinese work on obstetrics and gynaecology was *Jing Xiao Chan Bao* (*Treasured Knowledge of Obstetrics*) published c.850 by the Chinese physician Zan Yin. It covers treatments from traditional Chinese medicine (see pp.26–27), and herbal remedies for pregnancy-related conditions from morning sickness to miscarriage.

Cesarean section—the delivery of a baby through an incision—is one of the oldest known surgical procedures, with descriptions of this surgery dating back 3,000 years in China and 2,200 years in India. The term is said to be derived from the name of Roman emperor Julius Caesar, allegedly born by this method in 100 BCE, but the more likely origin is *caedare*, Latin for "to cut."

In 1598 French royal surgeon Jacques Guillemeau introduced the term "section" rather than operation in his book on midwifery. German gynecologist Ferdinand Kehrer is credited with successfully performing the first modern cesarean section in Meckesheim village, Germany, in 1881. This involved making an incision across the lower part of the mother's uterus to deliver the baby, while minimizing blood loss.

Men in women's health

In 16th-century Europe medical men such as French military barber-surgeon Ambroise Paré (see pp.78–79) discovered that their general medical knowledge could be applied to a field that was dominated by female birth attendants who were not medically trained. German

physician and apothecary Eucharius Rösslin helped disseminate medical knowledge with his 1513 publication *Der Schwangeren Frauen und Hebammen Rosengarten* (*The Rose Garden for Pregnant Women and Midwives*).

In 1609 the practical and progressive midwife to French royalty, Louyse Bourgeois, became the first woman to write a medical treatise on obstetrics, *Observations diverses sur la stérilité, perte de fruits, fécondité, accouchements et maladies des femmes et enfants nouveaux-nés* (*Various Observations on the Sterility, Fruit loss, Fertility, Childbirth and Diseases of Women and Newborn Infants*). But the male takeover of the traditionally female practice of midwifery continued, giving rise to the often derogatory term *accoucheur*, or man-midwife.

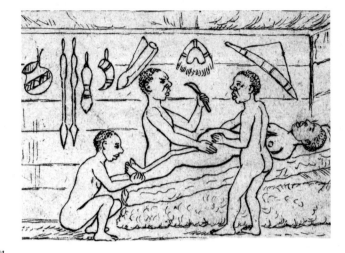

△ **Cesarean operation in Uganda**
The original purpose of a cesarean section was to save a baby when the mother would probably die. However, medical advances in the 19th century, such as anesthesia and antiseptics, improved the mothers' chances of survival too.

△ **A Man-Midwife**
This 1793 cartoon satirizes the movement into traditional female midwifery of men. Often eminent surgeons, these men were seen as keen to extend their own fame and influence, rather than to do the best for mothers and babies.

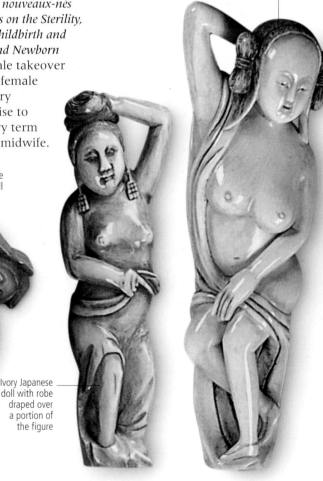

Japanese ivory doll

Japanese ivory doll

Ivory Japanese doll with robe draped over a portion of the figure

In the first half of the 1700s obstetric forceps were introduced by Scottish obstetrician William Smellie, who also published *A Treatise on the Theory and Practice of Midwifery* in the 1750s. The vaginal speculum, known since Roman times (see pp.42–43), also came into wider use. By this time, more births were happening in hospitals than homes, reinforcing the power of obstetricians over midwives. This trend had its own problems, such as childbed fever (see pp.138–39)—caused by lack of hygiene, leading to infections on the wards—and the new male "experts" often lacked the empathy experience, and traditional knowledge of female midwives.

Midwives recognized

Following the work of Florence Nightingale and other pioneers in nurs ing (see pp.142–43), midwives, too, began receiving recognition. Gradually, midwifery

was acknowledged and formalized the world over. In 1861 the Professional Midwifery Education Foundation was set up in the Netherlands. In Britain, meanwhile, women's rights campaigner Louisa Hubbard founded what became the Midwives' Institute in 1881. In 1902 the Midwives Act in England and Wales established midwifery as a specialized profession with training and certification. The UK Midwives' Institute became the Royal College of Midwives in 1947, the French College of Midwives was set up in 1949, and the American College of Nurse-Midwifery in 1955. By the mid-20th century many other nations had established similar recognitions and qualifications.

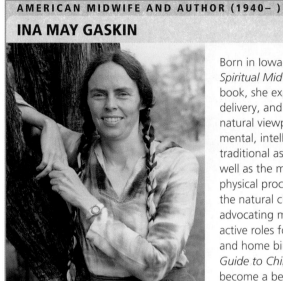

AMERICAN MIDWIFE AND AUTHOR (1940–)

INA MAY GASKIN

Born in Iowa, Gaskin published *Spiritual Midwifery* in 1977. In this book, she explained pregnancy, delivery, and infant feeding from a natural viewpoint, emphasizing the mental, intellectual, emotional, and traditional aspects of childbirth, as well as the medically mediated physical processes. She supported the natural childbirth movement, advocating minimal intervention, active roles for family and friends, and home births as the norm. Her *Guide to Childbirth* (2003) has become a bestseller.

> " **Our bodies must work pretty well,** or there wouldn't be so many humans on the planet. "

INA MAY GASKIN, AMERICAN MIDWIFE AND WRITER, *INA MAY'S GUIDE TO CHILDBIRTH*, 2003

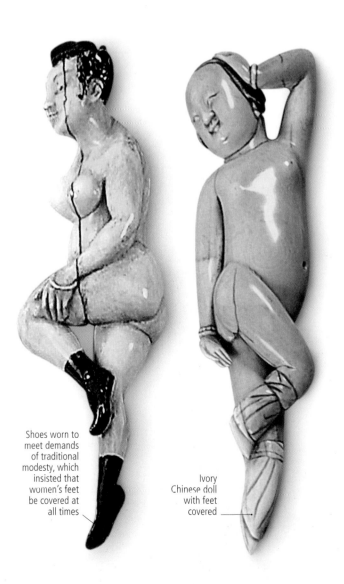

Bun indicative of an adult woman

Ivory Chinese doll wearing bangles

Shoes worn to meet demands of traditional modesty, which insisted that women's feet be covered at all times

Ivory Chinese doll with feet covered

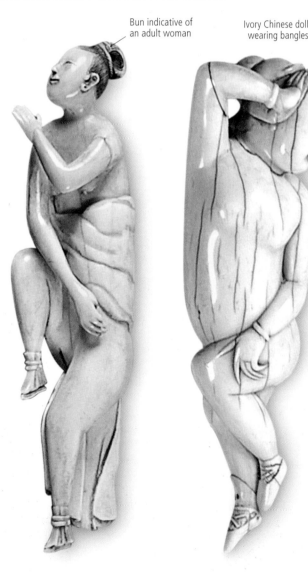

▽ **Diagnostic dolls**
Cultural taboos, or perhaps simple modesty, often prevented male physicians from examining a woman's genital area, so the female patient would explain her predicament using a diagnostic doll. These examples of such dolls are from China and Japan in the 18th and 19th centuries.

Midwives

Sculptures and papyri from ancient Egypt record specially trained women attending mothers during pregnancy and birth, and in Islamic medicine the midwife was a highly regarded specialist. However, this status did not last, and not until the 19th century did female midwives regain their standing within the medical community.

During the medieval period, especially in Europe, the profession lost some of its importance, and the role of the midwife was usually given to an older woman of the community. She was often illiterate and had no formal training, but she did have experience and knowledge of traditional techniques and folk remedies. Most countries continue to have this kind of "lay midwife," or Traditional Birth Attendant (TBA). In the 1400s the midwife's role became recognized again, albeit informally. However, its practitioners still had low status in what was a male-dominated medical system. In Britain, a 1512 Church Act brought in some regulation and necessitated that midwives swear an oath concerning their training and duties.

From the 1600s male physicians and surgeons began incorporating midwifery into their practices. The next century, especially in Britain, was the era of the "man-midwife," and various advances were made, such as improved obstetric forceps by Scottish obstetrician William Smellie in the 1750s. The 19th century saw a swing toward female midwives with recognized qualifications, and the establishment of professional bodies such as the UK Matron's Aid (1881). Midwifery joined mainstream medical specialities in many countries, with the International Confederation of Midwives established in 1919.

> **" A world where every childbearing woman has access to a midwife's care."**
>
> THE VISION OF THE INTERNATIONAL CONFEDERATION OF MIDWIVES

▷ **School for midwives**
The Maternité de Paris in Port-Royal, France, was a "lying-in" hospital for poor women as well as a school for midwives. In this illustration, midwives of the Maternité de Paris are seen attending to infants in the first incubators, introduced there in the 1880s.

△ **Handwashing in maternity wards**
In 1847 Ignaz Semmelweis noted that after he advocated regular handwashing, death rates at the First Clinic in the Vienna General Hospital fell from 12–13 percent to 1–2 percent.

Childbed Fever

In the 1840s simple observations and actions by Ignaz Semmelweis dramatically reduced occurrences of childbed (or puerperal) fever. However, his work was initially ridiculed and its importance was only recognized years later, once the germ theory was widely accepted.

Childbed fever has long been a dreaded infection for new mothers and infants, but the first major reduction in the death rate did not come until Ignaz Semmelweis implemented changes on a maternity ward in Vienna, Austria.

After completing his medical training in Vienna, Semmelweis was appointed assistant to the professor at the maternity unit in the Vienna General Hospital. At the time, new mothers were dying from childbed fever in epidemic proportions—but only in one of its two maternity clinics. Semmelweis was puzzled by the difference in infection and death rates between the First and Second Clinic: it was well known that there were many more maternal fatalities in the First Clinic, but no one knew why. Methodically, Semmelweis eliminated possible factors, such as food and drink, temperature, humidity, and other environmental conditions; he noted the age of patients, their backgrounds, and even religion.

The only significant difference that he discovered was the visiting staff: the First Clinic was a training center for apprentice physicians, while the Second Clinic was for the teaching of student midwives only.

Deadly particles

In March 1847 Semmelweis was saddened by the untimely death of a colleague and professor of forensic medicine, Jakob Kolletschka. The postmortem showed that he had suffered an accidental knife wound during an autopsy demonstration, and that the course of his infection was very similar to that of childbed fever. From this, Semmelweis inferred that Kolletschka had died from the same disease and it was likely that the wound made by the contaminated knife had caused his colleague's death.

While there seemed to be a link, the nature of the contamination remained a mystery, because the existence of germs was not yet proven. Semmelweis suggested that some kind of infective matter, which he named "cadaverous particles,"

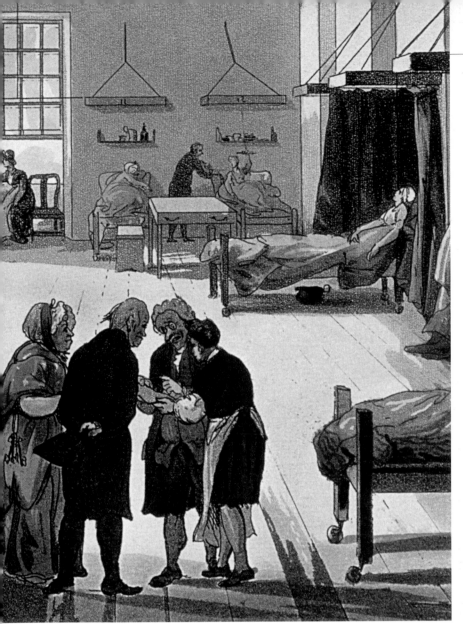

HUNGARIAN PHYSICIAN (1818–1865)

IGNAZ SEMMELWEIS

Born in Budapest, Hungary, Semmelweis received his doctoral degree in medicine from the University of Vienna, Austria, in 1844. He was then appointed to the Vienna General Hospital's obstetrics clinic, where he became involved with the problem of childbed fever. After being passed over for promotion, in 1850 he returned to Budapest and joined the Szent Rokus Hospital as Head of Obstetrics, where he introduced the same hand-washing routine that he had introduced in Vienna. In 1855 he was appointed professor at the University of Pest, Hungary, and he published his principal work on childbed fever in 1861 but generally it was not well received. Semmelweis's behaviour became increasingly erratic after he developed a kind of dementia, and he died only two weeks after being admitted to an asylum in Vienna.

was to blame for both Kolletschka's death and childbed fever. He argued that surgeons and medical students often came from autopsies and corpse dissections directly to the maternity clinics (in the case of Vienna, the First Clinic), and that they carried the particles on their hands and equipment, which then infected the mothers.

Handwashing routine

Semmelweis was convinced that the solution to the problem of cross-contamination was thorough handwashing. Believing that soap was not sufficently powerful, he introduced a routine of regular handwashing using *chlorina liquida* (calcium hypochlorite) for all his staff. The results were sudden and very startling. Death rates from childbed fever fell drastically in the First Clinic to about the same level as those in the Second Clinic, and they continued to fall through the following year.

Semmelweis regarded his views as proven and vitally important. Yet he was met with enormous criticism and inaction by the medical establishment who, typically, were sceptical of the new and untested ideas. Semmelweis could not prove that the "cadaverous particles" existed and his theory did not fit in with long-established beliefs, such as the concept of the four humors (see pp.34–35) or the miasma theory (see pp.120–21). Also, the surgeons he accused of carrying the contamination were important men who refused to accept that they were to blame. In addition, political and religious factors came into play, given that Semmelweis was a Jewish Hungarian living in Austria.

In 1861 Semmelweis published a book about his findings titled *Die Ätiologie, der Begriff und die Prophylaxis des Kindbettfiebers* (*The Etiology, Concept, and Prophylaxis of Childbed Fever*), but his work was generally rejected. Semmelweis died in obscurity in Vienna in 1865. The same year, pioneering British surgeon Joseph Lister began using phenol antiseptics (see pp.154–55) after reading Louis Pasteur's theory—partly derived from an interest in childbed fever—that unseen germs cause disease (see pp.146–47). It was only after this advance that the work of Semmelweis came to be fully appreciated. Today, he is praised for his great work on childbed fever and improving hygiene in hospitals, as well as his research into antiseptics, how contagious diseases spread, and how microbial germs cause disease.

> # "Cleanliness was out of place. It was considered finicking and affected."

SIR FREDERICK TRAVERS, BRITISH ROYAL SURGEON TO KING EDWARD VII, 1853–1923

◁ **Infection carriers**
Childbirth forceps came into general use from the early 1700s. However, without an understanding of hygiene, they were a reservoir of infection that repeatedly spread childbed fever.

Women in Medicine

Over the millennia and around the world, the medical profession has often reflected wider society. As a result, medical practitioners, especially at senior levels, have been overwhelmingly male. A degree of equality has only been achieved during the past century, but not yet in all nations.

Women have always played important roles as caregivers, nurses (see pp.142–43), and midwives (see pp.136–37), but until the 19th century, only a few rose to higher ranks in the medical profession.

One of the earliest known female physicians was ancient Egypt's Merit-Ptah, around 4,700 years ago. Not much is known about her, except her tomb inscription, which reads "chief physician." At Heliopolis in Egypt, female students attended medical school around 3,500 years ago, but little detail is known. Women's involvement in medicine in ancient Greece was also limited. Greek physician Metrodora is recognized as the first female writer on medicine. She wrote *On the Diseases and Cures of Women* around 2,300 years ago, however nothing more is known of her life. Another ancient Greek woman, Agnodice, is said to have practiced medicine while disguised as a man.

Early influencers
There are records of female healers in the medieval Islamic world from the 8th century, although, in common with many other cultures through history, they only treated other women. Female surgeons are depicted in the illustrated manual *Cerrahiyyetu'l-Haniyye* (*Imperial Surgery*) by male Turkish surgeon Sabuncuoglu Serefeddin. Christian Europe was far less enlightened and only a few female physicians are known from the period. Hildegard of Bingen (see pp.56–57) was a prominent abbess, poet, musician, and physician. Her works from the 1150s include *Liber Simplicis Medicinae* (*Book of Simple Medicine*), later called *Physica*, which describes hundreds of treatments made from minerals, herbs, and animal parts. Trotula de Ruggiero, a more shadowy personality who, if real, lived during the latter half of the 11th century, is associated with several medical publications. "The Trotula" became the collective name for several works, including *Diseases of Women*, *Treatments for Women*, and *Women's Cosmetics*. The writings were refreshingly practical, and covered a wide range of

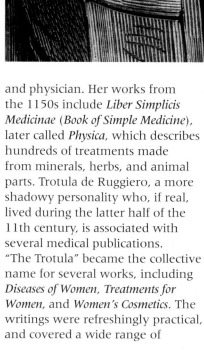

◁ **Hildegard of Bingen**
This altarpiece depicts the arrival of Hildegard with her family at the Benedictine Abbey of Disibodenberg, in about 1112. Hildegard wrote a number of scientific and medical works, and founded several monasteries. In 2012 she was named "Doctor of the Church" by the Pope.

△ **Interview panel**
Despite gaining a medical license in England, Elizabeth Garrett Anderson was forbidden from working in hospitals. She traveled to Paris to gain a French medical degree and worked there. This illustration shows her being interviewed by the Faculty of Medicine of the Sorbonne, Paris.

subjects such as feminine hygiene, fertility, conception, pregnancy, and childbirth.

Female medical pioneers
The acceptance of women into the medical profession began to happen in the 18th century. In 1732 Italian Laura Bassi was named professor of anatomy at the University of Bologna, before continuing her career in physics. In Prussia, Dorothea Erxleben, with special permission from King Frederick the Great, graduated in medicine from the University of Halle in 1754. But these were still isolated cases.

BRITISH–AMERICAN PHYSICIAN (1821–1910)

ELIZABETH BLACKWELL

In 1847 Elizabeth Blackwell enrolled at Geneva Medical College, New York State. She graduated in 1849, becoming the first woman to receive a Doctor of Medicine degree from a US medical school. She faced prejudice when trying to enter the profession, so in 1851 she set up her own medical practice and dispensary in New York for disadvantaged women, followed by the New York Infirmary for Indigent Women and Children in 1857.

"It is not easy to be a pioneer— but oh, **it is fascinating!**"

ELIZABETH BLACKWELL, BRITISH–AMERICAN PHYSICIAN

In 1849 Elizabeth Blackwell became the first American medical graduate, and went on to have a long and distinguished career, pioneering women's roles in medicine (see panel, right). In England, she helped establish the London School of Medicine for Women in 1874 with British physicians Sophia Jex-Blake and

▷ **Agnodice**
Around the 4th century BCE, in ancient Greece, Agnodice disguised herself as a man to help women during pregnancy and childbirth. At this time, women were banned from working as doctors and could be executed.

Elizabeth Garrett Anderson. Jex-Blake was one of the first female doctors in Britain, and she went on to found the Edinburgh School of Medicine for Women in 1886.

In 1859 Garrett Anderson had met and been inspired by Blackwell. She became a nurse at London's Middlesex Hospital, and in 1862 joined the Society of Apothecaries to gain a license for medical practice—a first for a British woman. She opened a private practice, then St. Mary's Dispensary for Women and Children, and in 1872 the New Hospital for Women (later renamed the Elizabeth Garrett Anderson Hospital). Continuing her pioneering work, she became the first female member of the British Medical Association, and carried

on campaigning for women's rights for the rest of her life. In 1876 British law changed to allow women full access to the medical profession, although an underlying prejudice still remained for many decades.

Women were also gaining access to the medical profession in other nations, especially in Europe. Madeleine Brès was the first Frenchwoman to receive a medical license in 1875. The trend spread and, in Japan, physician's and women's rights campaigner Yoshioka Yayoi founded the Tokyo Women's Medical University in 1900. By this time the women's rights and suffragette movements were also gaining momentum, and from about 1914 feminist campaigner Margaret Sanger (see pp.226–27) also fought for women as patients and health services users.

149 PERCENT **of all general practitioners in the UK in 2015 were women.**

Nursing

Although nursing is one of the oldest medical occupations, it has not always had a good reputation. It took the influence of one extraordinary woman—Florence Nightingale—to transform nurses from uneducated "ward maids" to the academically qualified, skilled professionals that we know today.

In Europe during the medieval period hospitals were usually attached to religious institutions, such as monasteries and convents, with patients nursed by monks and nuns. However, in the 16th century many hospitals were shut down as a result of Protestant reformations. With the growth of industrialization in the 18th century, new secular hospitals were founded. During this period, sometimes termed the "Dark Ages of nursing," the quality of care was frequently dire—nurses tended to be recovering patients, or hired men and women who could not read or write and often drawn from the poorhouses. Nurses gained a reputation for ignorance, drunkenness, and promiscuity.

The push for nursing reform in Europe began in the 19th century, largely instigated by the Christian community. Many visitors to Germany were impressed by the work of pastor Theodor Fliedner, who opened a hospital on the Rhine in 1836 (see pp.106–07). Nurses were given simple clinical instruction and studied pharmacy—the practice of preparing and dispensing drugs. The nursing course was quite advanced for its time, and Fliedner's most famous student—Florence Nightingale— spent three months at his hospital in 1851. By the mid-19th century, the concept of women being trained to nurse was well established.

Nurses go to war

The advent of the Crimean War (1853–56) transformed nursing. Cholera spread rapidly in the British army camp, and surgeons had to perform major operations and amputations without light, anesthetics, or even bandages. When the British press reported that the wounded and the sick were not being properly cared for, the government responded by sending female nurses abroad to tend to the casualties. Florence Nightingale was appointed as the "Superintendent of the Female Nursing Establishment of the English General Hospitals in Turkey"—a powerful position that gathered huge attention.

Nightingale enforced a strict code of discipline, discouraging nurses from fraternizing with the patients and doctors, as well as promoting hygiene, sobriety at all times, and good manners. Nightingale and her small band of nurses were a great inspiration to women, showing that war was no longer a male preserve. When the American Civil War broke out in 1861, the Sanitary Commission—a forerunner to the Red Cross—was founded. Armed with the knowledge of good hygiene practices from the Crimean War, it recruited a large number of nurses.

Nursing was on the threshold of reform. In 1860 Nightingale realized her dream of establishing a training school for nurses at St. Thomas' Hospital in London; it became a blueprint that was copied throughout the British Empire and the US. Nursing associations were established across the world, which brought in standardization of training and finally recognized nursing as a profession. In 1863 the International Red Cross (see pp.266–67) was set up to offer neutrality and protection to those wounded in armed conflict, and it endorsed the training of nursing.

A modern profession

Until World War I the Nightingale legacy prevailed. Nurses were seen as the guardians of hygiene, the dispensers of compassion, and the center of calm amid the chaos of the hospital. However, the nurses' actual duties were rather vaguely described. During World War I the

90,000 The number of volunteers with the Red Cross's Voluntary Aid Detachments during World War I.

boundaries between medicine and nursing broke down. As doctors struggled to cope with emergency surgery, trained nursing staff took on duties that would not normally

△ **Wartime nursing recruitment poster**
Thousands of nurses flocked to the Western Front in the early months of World War I as a result of recruitment posters such as this. The first trained nurses reached France just eight days after the war on the Western Front began.

BRITISH NURSE (1820–1910)

FLORENCE NIGHTINGALE

Born into a wealthy English family, Florence Nightingale reformed the profession of nursing. A woman of very strong will, her tireless work caring for soldiers during the Crimean War established her as "The Lady with the Lamp." Her reforms led to a dramatic reduction in deaths. She founded a training school for nurses at St. Thomas' Hospital, London, in 1860, and helped promote nursing as a respectable career for women.

" … the very first requirement … do the sick no harm."

FLORENCE NIGHTINGALE, FROM *NOTES ON HOSPITALS*, 1859

fall to them, including triage (see p.256), the administration of saline drips and intravenous injections, and the dispensing of narcotic drugs. The nursing staff were also responsible for implementing many of the new developments aimed at combating infection and passing on their knowledge to volunteers from the Red Cross's Voluntary Aid Detachments (VADs), which were set up to provide supplementary first aid and nursing to the medical service in wartime. In addition, nurses had to cope with the effects of new wartime technology—for example, learning how to use oxygen cylinders for soldiers with lungs filled with mustard gas, and applying sodium bicarbonate to their blinded eyes.

World Wars I and II emphasized the growing need for fully trained, well-educated nurses, and today many countries demand that nurses have a university degree. From an occupation of the poor and illiterate, nursing has evolved to become one of the most important professions within the healthcare industry.

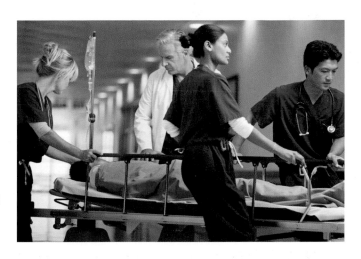

▷ **Modern nurses**
The role of nurses has developed to occupy an ever-wider range of healthcare duties. Modern nurses are not merely caregivers— they have to display a high level of technical competence and may also act as clinicians, diagnosing illness and making decisions about suitable treatments.

Night After the Battle
This painting by Robert Neal and D.J. Pound shows Florence Nightingale tending to a wounded soldier amid the carnage of the battlefield during the Battle of the Alma (1854) in Turkey.

Fig. 161.—Bones of the Left Hand. Dorsal Surface. Fig. 162.

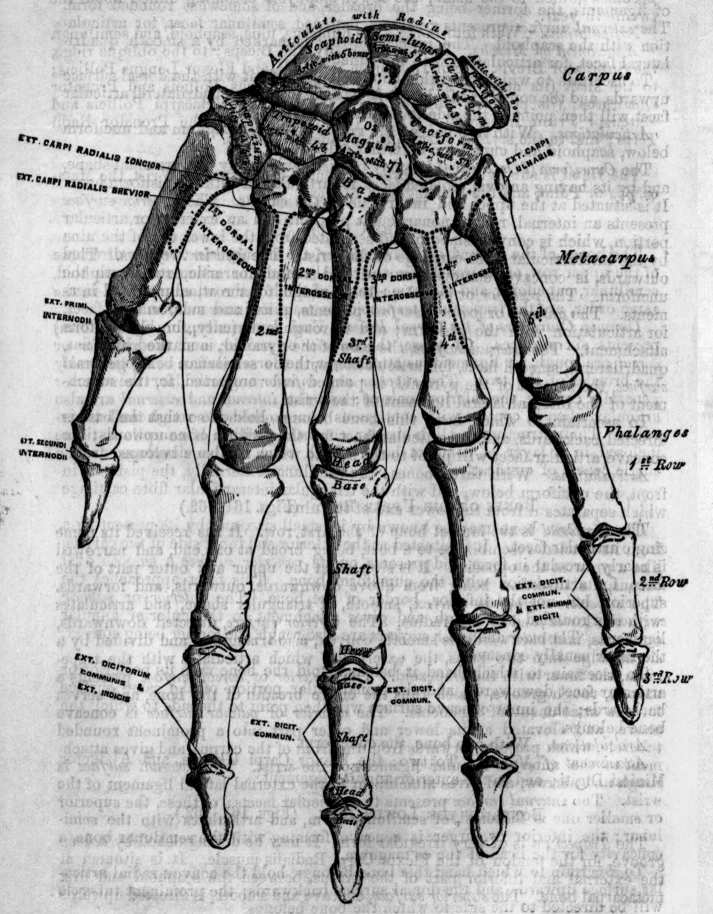

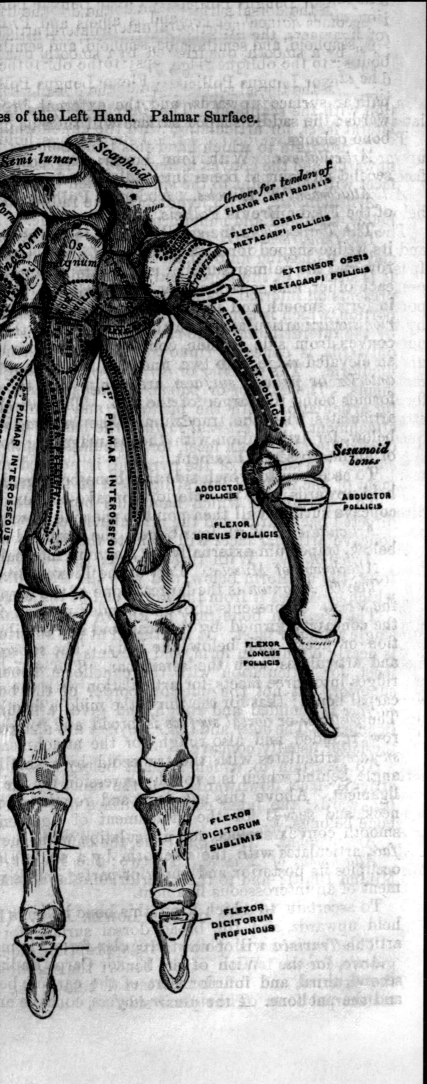

es of the Left Hand. Palmar Surface.

Medical Publishing

In 1858 English anatomist Henry Gray wrote *Anatomy: Descriptive and Surgical*, illustrated by his colleague Henry Vandyke Carter. Gray died just three years later, at the age of 34 years, but his name lives on in the best-known educational and reference work in all of medicine.

In 1853 Gray became an anatomy lecturer at St. George's Hospital Medical School, London. His aim was to write a compact illustrated textbook for students that was low-cost yet accurate and authoritative. He enlisted the artistic skills of Carter, who was studying at St. George's for his medical qualifications. The two men dissected the bodies of deceased people who had no family or friends and wrote and illustrated their findings. The work rapidly expanded and the first edition ran to 750 pages with more than 360 pictures.

Gray died of smallpox soon after preparing the second edition in 1860, while Carter moved to the Indian medical service in 1858. Their book was retitled *Gray's Anatomy* and there followed regular updated, enlarged editions with distinguished editorial panels. The scope widened to include new material such as microscopy, X-rays, scans, and physiology diagrams; by the time the 38th edition was printed in 1995, it had more than 2,000 pages. The work entered a new era in 2004 with a newly organized, slimmed-down 39th edition of 1,600 pages, with almost 2,000 illustrations—400 of them new—and digital and online versions. *Gray's Anatomy* has remained a vital teaching and reference work for generations of medical students, surgeons, and all other health practitioners.

> **"Every** living **physician** today has been **exposed** to **Gray's Anatomy."**
>
> JOHN CROCCO, ASSISTANT PROFESSOR OF CLINICAL MEDICINE, INTRODUCTION TO THE COLLECTOR'S EDITION OF *GRAY'S ANATOMY*, 1977

◁ Bones of the hand
Basic human anatomy remains much the same, but updates to Gray's original work regularly provide additional details. After Carter's departure, John Westmacott made the illustrations for the second and later versions, such as this one from the 20th edition.

Microbiology and Germ Theory

Less than 200 years ago, the existence of the germs that are now known to cause infections was unsuspected. The gradual discovery of these harmful microbes and methods to combat them were among the greatest advances in all of medicine.

Nature can generate life almost anywhere, minute plants sprouting and infinitesimal animals appearing as if out of the air. With no evidence to the contrary, people assumed that living things could arise from nonliving matter—a concept known as spontaneous generation. Another popular notion was miasma theory (see pp.120–21), which stated that noxious vapors and gases somehow penetrated the body to produce diseases. After the invention of the microscope (see pp.94–95) in around 1600, these perceptions gradually began to change. This novel instrument showed for the very first time that there were minute animals, or "animalcules," everywhere, and

scientists and physicians began to infer that these could be responsible for the transmission of diseases.

In 1668 Italian naturalist-physician Francesco Redi began investigating the supposedly spontaneous appearance of maggots on dead meat. He carried out experiments with old meat in jars—some open to the air, some covered with cloth, and some stoppered. Redi noted that maggots would develop only if flies could land on the meat.

A century later, Italian priest Lazzaro Spallanzani boiled meat broth and sealed some samples in glass vessels while leaving others open. The sealed samples stayed uncontaminated, but the others soon began to deteriorate.

The 19th century saw a steady stream of discoveries. In 1835, while studying a silkworm disease, Italian entomologist Agostino Bassi deduced that the condition occurred due to some kind of "contagion" or "transmissible particle" spread by contact or close proximity. In 1840 German anatomist and histologist Jakob

Henle proposed: "The material of contagions is not only an organic but a living one." In 1847 Hungarian obstetrician Ignaz Semmelweis reasoned that "cadaverous particles" caused childbed fever (see pp.138–39). In 1854 British physician John Snow suspected contagion during a cholera outbreak (see pp.122–23). The contagion, or germ, theory of disease—according to which transmissible living particles are responsible for human diseases—was gaining ground, although miasma theory still prevailed.

Isolating harmful microbes

In 1862 French biologist Louis Pasteur (see pp.148–49) performed pivotal experiments with boiled meat broth and glass "swan-neck" flasks. He concluded that some kind of contagion led to the development of molds in broths open to the air, but not in those broths that were protected from contamination. Despite protests from supporters of spontaneous generation, Pasteur's evidence boosted the notion of germ theory—and that transmissible living particles might cause human diseases.

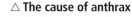

△ **The cause of anthrax**
Koch cultured and tested 20 generations of the rod-shaped anthrax bacterium (*Bacillus anthracis*) to prove that it caused the disease. He also noted that the bacteria could survive tough environments by transforming into dormant spores, which would reactivate when conditions improved.

Initially a colleague of Pasteur but later a bitter rival, German physician Robert Koch qualified with distinction in medicine at Göttingen University in 1866 and was inspired by his professor Jakob Henle to pursue microbiology. He set up a home laboratory in Wollstein (now Wolsztyn, Poland), where he began a series of studies with far-reaching effects. Koch's first subject was anthrax—a highly infectious disease of herbivores. He inoculated some mice with samples from healthy, and some with samples from diseased, farm animals. The former did not develop the disease, but the latter did. He then set about purifying anthrax bacteria, growing them in a laboratory culture medium,

▽ **Refuting spontaneous generation**
Francesco Redi's 1668 work *Experiments on the Generation of Insects* showed that maggots hatched in old meat not through spontaneous generation but from eggs laid by visiting flies. However, the theory of spontaneous generation persisted for another two centuries.

MOSCH DE BACHI DEL SAMBVCO

"The earth... has never produced any kinds of **plants or animals...** everything we know… [comes] from the **true seeds** of the plants and animals themselves."

FRANCESCO REDI, FROM *EXPERIMENTS ON THE GENERATION OF INSECTS*, 1668

"The **pure culture** is the foundation for all research on **infectious disease.**"

ROBERT KOCH, FROM *STUDIES OF PATHOLOGICAL ORGANISMS*, 1881

and studying them under the microscope. He published his findings in 1876, establishing for the first time the connection between a specific disease and a microorganism. In 1880 Koch

▽ Attempting to cure tuberculosis

A patient is given Koch's treatment for tuberculosis at the Royal Hospital, Berlin, in around 1890. Koch's remedy for tuberculosis, named tuberculin, failed amid great controversy, as did a revised version in 1897. Tuberculin was later used to develop a test to diagnose tuberculosis infection.

devised a set of criteria linking particular microbes to specific diseases. These came to be known as Koch's postulates and are still in use today.

Next, he studied tuberculosis and discovered its causative agent—Koch's bacillus, or *Mycobacterium tuberculosis*—in 1882. After that he turned his attention to cholera (see pp.122–23), traveling to Egypt and India as part of his research. In 1884 he isolated the causative germ, since named *Vibrio cholerae*, defined how it spread via

▷ Bacteria cultures

This illustration shows laboratory test tubes containing bacteria cultures of tuberculosis (left) and cholera (right), both discovered by Koch. His team devised many techniques for growing, staining, observing, identifying, and photographing microbes that benefited medical research.

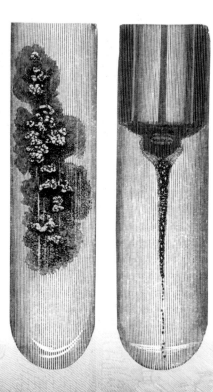

contaminated water and food, and suggested prevention and control measures.

Koch's contributions to medical research were recognized in 1905 when he received the Nobel Prize in Physiology or Medicine for his "investigations and discoveries in relation to tuberculosis." The award gave credence to the work of Koch and others, who replaced ideas of miasma and spontaneous generation with the germ theory of disease.

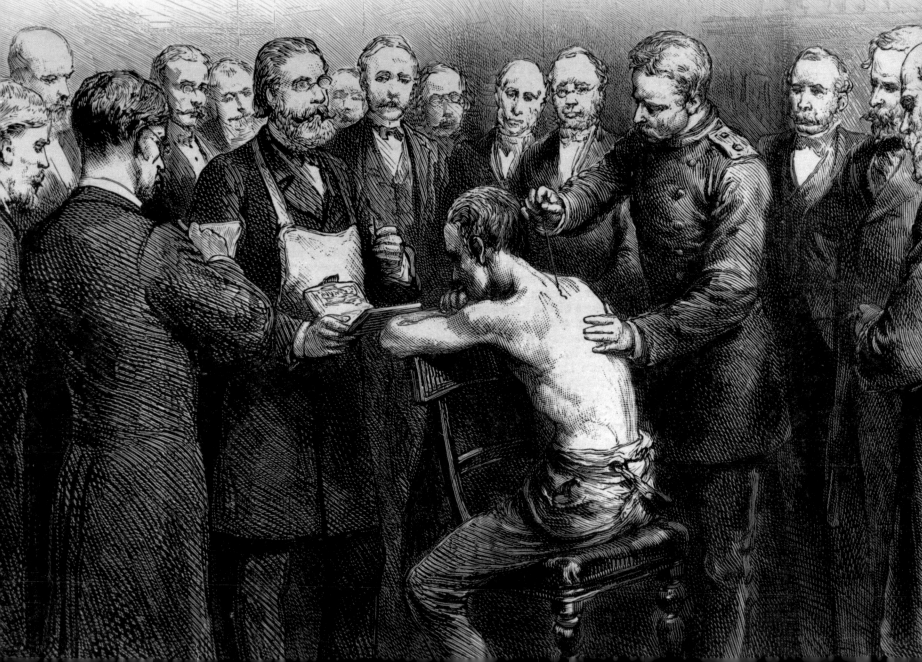

FRENCH CHEMIST Born 1822 Died 1895

Louis Pasteur

"In the field of observation, chance favors only the prepared minds."

LOUIS PASTEUR, ON APPOINTMENT AS DEAN OF SCIENCE FACULTY, LILLE UNIVERSITY, 1845

◁ **Founder of microbiology**
Along with Robert Koch—at first his colleague but then great rival—Pasteur placed the study of microbial life onto a scientific footing and moved it into mainstream medical research.

One of France's greatest scientists, Louis Pasteur made significant contributions to almost every field he ventured into. He developed the process of killing germs using heat, now called pasteurization; helped replace the theory of spontaneous generation with germ theory (see pp.146–47); and aided the silk industry by identifying a disease of silkworm caterpillars. From the 1870s he developed vaccinations for chicken cholera, forms of animal anthrax, and rabies in animals and humans (see pp.168–69).

Breakthrough research

Pasteur's first major contribution to life sciences was to investigate why alcoholic drinks sometimes "spoil"(go bad or sour)—a costly problem for the French beer and wine industries. After exhaustive microscopic studies he drew two conclusions. First, fermentation was not a simple chemical change, as believed, but a living process carried out by yeast microbes. Second, souring was caused by contamination with bacterial microbes. The remedy he devised in 1864 was to heat the drinks briefly to 122–140°F (50–60°C), to kill off disease-causing bacteria without altering the beverage's aging process, taste, or appearance. In the 1880s the process became known as pasteurization—in his honor. Medically it helped save many lives, for example preventing diseases such as tuberculosis, which spread through contaminated milk. Pasteur also questioned the theory of spontaneous generation—that living things could arise from nonliving matter. In 1862 he conducted experiments using glass flasks with S-shaped necks (see below). He proved that if contaminating microbes were kept away from a nutrient liquid, germs did not grow even if the liquid was in contact with air. Pasteur's experiments were powerful evidence against spontaneous generation and led to its rapid demise in the following few decades, it was replaced by germ theory, according to which microbes cause infections and contamination. In 1865 Pasteur's research revealed

1881
The year that Pasteur coined the term vaccination, from Latin *vacca*, meaning cow.

that harmful microbes were responsible for a ruinous disease affecting silkworms. He was also able to isolate infected silkworms from healthy ones and prevent further contamination.

Vaccination

The same year France suffered a cholera epidemic and Pasteur began studying it, as well as anthrax and other human and animal diseases. He made little progress until 1879, when he started culturing, or

▷ Making milk safe

Contaminated milk causes many diseases. The heat treatment devised by Pasteur for alcoholic drinks was applied to mass milk production from the 1880s. This milk pasteurization equipment is from French science magazine *La Science Illustree*, 1898.

Milk container

growing, fresh cholera microbes. However, his research was interrupted by a vacation. On his return, he gave the month-old cultures to chickens, who did not die of infection. Pasteur suspected that the germs were weakened and gave immunity to the chickens (see pp.158–59). This finding subsequently resulted in his developing a vaccine using a weakened form of the disease-bearing organism.

Animal anthrax was another disease causing great damage to French farming. In 1881 Pasteur gave a group of cows, sheep, and goats a vaccine of weakened anthrax

germs; a similar group was not vaccinated. When both groups later received full-strength anthrax, the treated animals survived while the untreated ones died.

In 1885 he carried out the first successful rabies vaccination on a young boy. This was among his last research projects, although he continued to lecture, fundraise, accept awards and medals, and set up the prestigious Institut Pasteur in Paris. His death in 1895 was deeply mourned across the world. Although he never actually qualified as a medical doctor, Pasteur's work helped save countless human—and animal—lives.

"There are **no such things** as applied sciences, **only applications** of science."

LOUIS PASTEUR, FROM *REVUE SCIENTIFIQUE*, 1871

▽ The swan-neck experiment

Pasteur did many experiments with S-shaped, or swan-neck, flasks. Once the flasks were sterilized by heat treatment, nutrient broth did not spoil if dust, microbes, and other particles were prevented from falling into it by the long, bent tube—even if it was open to the air.

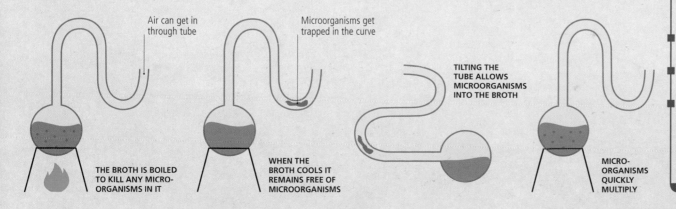

Air can get in through tube

Microorganisms get trapped in the curve

TILTING THE TUBE ALLOWS MICROORGANISMS INTO THE BROTH

THE BROTH IS BOILED TO KILL ANY MICRO-ORGANISMS IN IT

WHEN THE BROTH COOLS IT REMAINS FREE OF MICROORGANISMS

MICRO-ORGANISMS QUICKLY MULTIPLY

TIMELINE

1822 Born in Dole, eastern France; grows up in Arbois.

1840 Gains his Bachelor of Arts degree, followed by a science degree at the Royal College of Besançon, France.

1847 Receives a doctorate from the École Normale Supérieure, Paris.

1848 After researching and teaching at various locations, Pasteur is appointed chemistry professor at the University of Strasbourg, France. He marries Marie Laurent. They have five children, although three die young from infections, which inspires Pasteur's later work. Studying a chemical called tartaric acid, Pasteur discovers that molecules can exist as mirror-image left- and right-handed versions. This fundamental discovery leads to the field called stereochemistry.

1854 Appointed professor of chemistry and dean of science at Lille University, France, and begins work on "souring" alcoholic drinks.

1857 Becomes director of Sscience at the École Normale Supérieure.

1865 Shows that microbes attack silkworm eggs causing disease, and that this can be prevented. His advice is quickly adopted by silk producers around the world.

REPLICA OF THE FLASK USED BY PASTEUR TO SHOW THAT GERMS CAUSE DISEASE

1868 Partially paralyzed by a stroke, but continues to work.

1879 Develops his first vaccine for chicken cholera; extends this research to human diseases.

1882 Already an associate member of the Académie de Médecine since 1873, Pasteur is accepted into the Académie Française.

1885 Vaccinates Joseph Meister, a boy bitten by a rabid dog.

1888 Sets up the Institut Pasteur in Paris, France, for the study of microbiology.

1895 Dies and is buried in Notre Dame Cathedral. The next year, his remains are moved to a special crypt at the Institut Pasteur.

Cell Theory

Before the 19th century no one had a clear notion of the basic building blocks of life. The use of microscopes (see pp.92–93) allowed the development of cell theory—that living organisms are composed of cells—which had enormous effects in many fields of medicine.

The invention of the microscope in the 1590s made it possible for the first time to observe animal and plant matter at a level that had previously been invisible to the naked eye. Plant cells were first described by British polymath Robert Hooke in 1665. He coined the term "cell" because the angular structures reminded him of the cells, or living quarters, of monks. In 1682 Dutch polymath Antoni van Leeuwenhoek observed the nucleus of a cell in the red blood corpuscles (RBCs) of a salmon. Some hundred years later, in 1800, French anatomist and physiologist Marie-François Bichat took advantage of improvements in the magnification levels of microscopes to catalog the structure of human skin, which he compared to a woven fabric.

However, it was not yet understood that all life forms are composed of these small structures, or that all cells are derived from other cells by cell division or cell-based reproduction. Indeed, in the early 19th century it was thought that cells could spontaneously generate from nonorganic matter, or from decomposing living material.

Recognizing building blocks

In 1838 Matthias Schleiden, professor of botany at the University of Jena, Germany, wrote the article "Contributions to Plant Phytogenesis," in which he used previous scientific observations and his own observations through a microscope to deduce that all parts of plants are composed of cells. He explained his theory to his friend German physiologist Theodor Schwann, who had seen similar cell-like structures in the internal skeletal rod (notochord) of primitive fish. Schwann took Schleiden's theory a step further, applying it to animals as well as plants, and defining the three structural parts of a cell—the wall, nucleus, and cellulose, or fluid content. In 1839 Schwann published the paper *Microscopic Investigations on the Accordance in the Structure and Growth of Animals and Plants*, in which he famously observed that, "all living things are composed of cells and cell products."

However, how cells are created and how growth occurs were not yet understood. Schleiden believed, and Schwann accepted, that new cells were crystallized from the fluid that lay between previously existing cells. This focus on the material outside the cell held back cell biology for some years.

Finally in 1851 German botanist Hugo von Mohl proposed that new cells are formed by the division of existing ones—a process he had observed in algae.

As early as 1842 Swiss botanist Karl von Nägeli had identified small structures in the nucleus, which came to be called chromosomes, and which contain the cell's genetic material. By the 1850s microscopes had become powerful enough to allow scientists to see cell division taking place, and in 1879 German military physician Walther Flemming observed chromosomes separating as the cell divided, in a process he named "mitosis." Other components of cells were also identified. These included, in 1890, mitochondria—the cell's "powerhouses," which play a role in processing sugars and oxygen to produce energy, decribed by German pathologist Richard Altmann.

A basis for new advances

The development of cell theory gave scientists a firm basis for the understanding of heredity. In 1869 Swiss biochemist Friedrich

> **37 TRILLION** A 2013 estimate of the number of cells in the human body.

GERMAN BOTANIST (1804–1881)

MATTHIAS SCHLEIDEN

Educated in law at Heidelberg University in Germany, Schleiden found legal practice distasteful and became a botanist instead. He had already rejected the contemporary botanical preoccupation with classification—which he mockingly described as recognizing plants "with the least possible bother"— in favor of examining samples with the microscope. His observations led him to conclude that all plants are composed of cells, which formed the basis of cell theory. After a brief stint as a lecturer in Russian-ruled Dorpat (in Estonia), he returned to Germany, where he was a private teacher.

▷ **Schwann's drawings**
Schwann's 1839 publication included drawings of different types of animal cells. Although they varied widely in form, the presence in all of them of a nucleus and an enclosing membrane, or cell wall, convinced him that they were all versions of the same basic cellular building block.

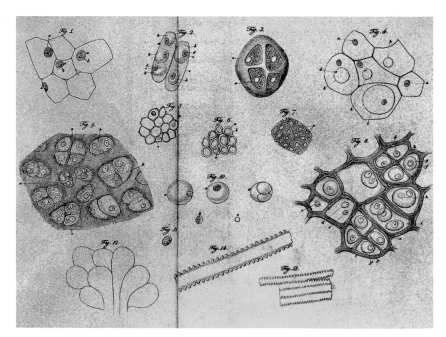

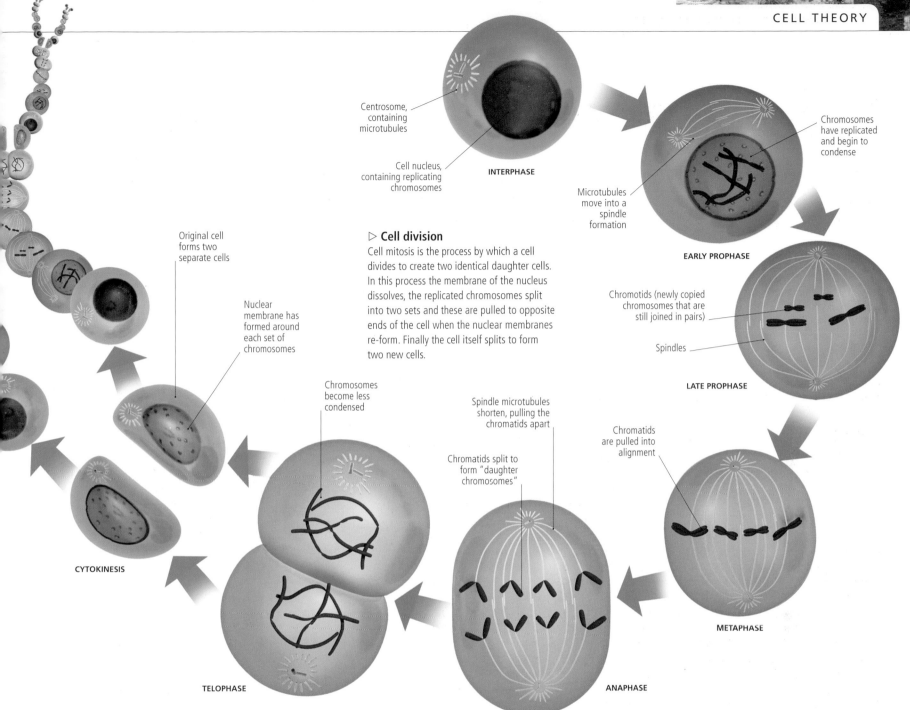

Cell division

Cell mitosis is the process by which a cell divides to create two identical daughter cells. In this process the membrane of the nucleus dissolves, the replicated chromosomes split into two sets and these are pulled to opposite ends of the cell when the nuclear membranes re-form. Finally the cell itself splits to form two new cells.

INTERPHASE

Centrosome, containing microtubules

Cell nucleus, containing replicating chromosomes

EARLY PROPHASE

Chromosomes have replicated and begin to condense

Microtubules move into a spindle formation

LATE PROPHASE

Chromotids (newly copied chromosomes that are still joined in pairs)

Spindles

METAPHASE

Chromatids are pulled into alignment

ANAPHASE

Spindle microtubules shorten, pulling the chromatids apart

Chromatids split to form "daughter chromosomes"

TELOPHASE

Chromosomes become less condensed

CYTOKINESIS

Nuclear membrane has formed around each set of chromosomes

Original cell forms two separate cells

Miescher identified nucleic acid, which in the form of DNA (deoxyribonucleic acid) is the building block of genes and chromosomes. In 1905 English biologists John Farmer and John Moore coined the term "meiosis" to describe a consecutive, double division of cells that halves the amount of chromosomes passed to spermatazoa or ova in sexually reproducing organisms.

Cell theory also contributed to the understanding of cellular pathology and disease. In 1863 the Prussian anatomist Rudolf Virchow advanced the idea that cancer occurs at sites of chronic inflammation in the body, and that this can cause cells to proliferate unnaturally, causing tumors.

"Omnis cellula e cellula (all cells come from cells)."

RUDOLF VIRCHOW, PRUSSIAN ANATOMIST, 1855

None of these advances would have been possible without the work of Schleiden and Schwann in establishing the universal nature of cells. Building on their discoveries, cell theory continues to inform our understanding of the structure and mechanics of the body, as well as underpinning modern research in reproductive medicine, genetics, pathology, and pharmacology.

▷ Single-celled parasite

Even tiny unicellular organisms such as the *Trypanosoma brucei* protozoa, which cause African sleeping sickness, have a nucleus. Bacteria, however, are even simpler in structure, and lack nuclei.

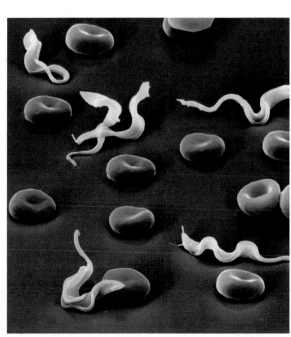

Pathology and Medical Autopsy

Much medical knowledge arose from examining dead bodies—autopsy. At first this was with the naked eye, but microscopes allowed massive advances in understanding disease and the birth of cellular pathology.

The period from the late 1700s to the early 1800s proved to be a watershed for the science of pathology—the branch of medicine that focuses on the examination of organs, tissues, and bodily fluids in order to diagnose disease. This emerging field thrived on autopsies (postmortems).

Research through autopsy

The study of disease through autopsy was not new. Dissections had been used to further scientific discovery since ancient times, although human dissection was against Roman law. Autopsy was legalized in several European countries from the 13th century onward, and during the 17th century it became the practice of a number of leading physicians,

including Italian anatomist Marco Aurelio and the Dutch surgeon Nicolaes Tulp. Some physicians even published "autopsy reports," the most important of which was Italian anatomist Giovanni Batista Morgagni's *De Sedibus et Causis Morborum per Anatomen Indagatis*

1832 The year that the Anatomy Act was introduced in England, allowing licensed anatomists to dissect unclaimed bodies.

(*On the Seats and Causes of Diseases as Investigated by Anatomy*) in 1761, which described his observation of more than 640 autopsies. Modern pathology emerged from these precise accounts. Disease was now linked to body organs rather than

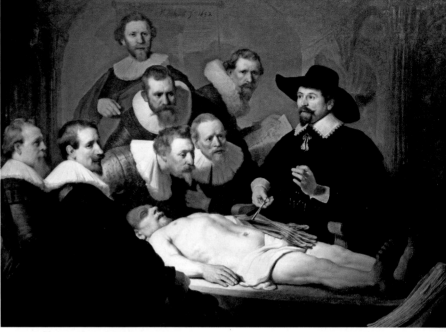

△ **Anatomy lesson**
In this 1632 painting by Rembrandt, Nicolaes Tulp explains the musculature of the arm to an engrossed audience. People paid to attend such autopsies, which could only be performed on male criminals.

to an imbalance of body fluids (see pp.34–35), an idea that had been a part of mainstream medical belief for almost two thousand years.

From the mid-19th century onward, a more scientific approach to the study of disease, led by two brilliant figures, Karl Rokitansky and Rudolph Virchow, drove pathology into a new era. Austrian physician Rokitansky radically

altered how the autopsy was performed. He insisted on a routine for dissection that was thorough and systematic, and was concluded with an accurate documentation of the findings in a report. However, Rokitansky was reluctant to use a microscope, and some of his theories about diseases proved to be incorrect.

Shift to cellular pathology

Unlike Rokitansky, Rudolph Virchow (see panel, left) was an advocate of the use of microscopes, urging his students to "think microscopically."

In 1858 Virchow published *Die cellularpathologie* (*Cellular Pathology*), in which he asserted that the cause of disease should always be looked for in the cell. He argued that diseases arise from abnormal changes within cells, and from those altered cells then multiplying through the process of cell division.

GERMAN PATHOLOGIST (1821–1902)

RUDOLF VIRCHOW

Regarded as the most important figure in the history of modern pathology, Virchow studied medicine at the Kaiser Wilhelm Institute in Berlin, Germany. He showed that blood clots were caused by changes in the walls of blood vessels, in blood's flow, and its composition. This became known as "Virchow's Triad." He was the first to use microscopes extensively in tissue analysis. In 1855 he published his seminal work popularizing the idea of *Omnis cellula e cellula*, "Every cell stems from another cell"; so launching the field of cellular pathology. He explained how tumors grow and, for the first time, gave hope that malignancies—cancers—could be treated. He coined the term *leukemia* in 1847 to describe blood cancers, after noting that they cause an excess of white blood cells.

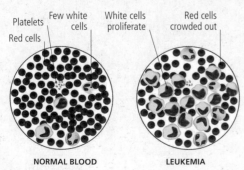

Platelets
Red cells
Few white cells
White cells proliferate
Red cells crowded out

NORMAL BLOOD   LEUKEMIA

▷ **Postmortem instruments**
A more rigorous scientific approach to postmortems in the 19th century required dedicated instruments. This box includes a head clamp, bone saw, chisels, scissors, and a mallet.

This shift from the idea of organ-based disease to cell-based disease was an important step in the "new pathology."

Later in the 19th century German Friedrich von Recklinghausen rose to prominence. A pupil of Virchow, he published important studies on thrombosis (blood clots); embolism (blockage in a blood vessel); and infarction (tissue death due to lack of oxygen), among many other pathology-based conditions.

Links and methods

Another student of Virchow, German-Swiss pathologist Edwin Klebs, made connections between bacteriology and infectious disease; he is chiefly credited with identifying the bacteria that cause diphtheria in 1883.

Another German pathologist, Julius Cohnheim, devised a method for freezing tissue before slicing it into thin sections for microscopic examination that is still a standard procedure today. His pupil, Carl Weigert, went on to describe the mechanisms of degeneration and necrosis—the death of cells and living tissue as a result of disease or injury.

By the 20th century pathology was well established and the pace of development accelerated. Today's advances in technology, especially in microscopy and computer-aided image processing, enable more precise diagnoses than ever before.

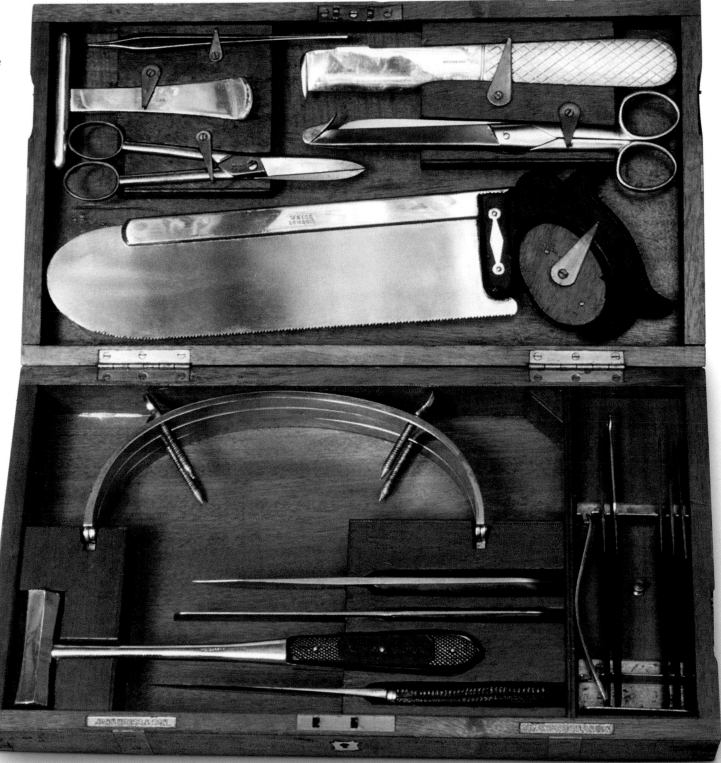

" Those **who have dissected** or inspected many [bodies] **have at least learned to doubt;** while others who are ignorant of anatomy… are in no doubt at all."

GIOVANNI BATTISTA MORGAGNI, ITALIAN ANATOMIST, FROM *DE SEDIBUS ET CAUSIS MORBORUM PER ANATOMEN INDAGATIS*, 1761

The First Antiseptics

The tendency of wounds—in particular incisions in the flesh made during surgery—to become infected led to significant loss of life in premodern times. It was not until the mid-19th century when Joseph Lister devised a solution—the use of antiseptics—that the number of postoperative deaths was radically reduced.

The fact that pus forms in wounds when they become infected and fester was well known to physicians. This "sepsis," or putrefaction of flesh, was so difficult to treat that many doctors even came to regard it as a natural part of the healing process, in spite of the fact that so many of their patients died. Attempts were made to combat the problem: in the 4th century BCE, Hippocrates (see pp.36–37) recommended the use of wine and vinegar in wound dressings, as both are mild antiseptics that he thought should prevent sepsis. Although it was successful in some instances, it never worked with compound fractures. These injuries are especially susceptible to infection since many shattered bone fragments are exposed to air, allowing germs access to the body.

Progress did come in 1812 when French chemist Bernard Courtois discovered iodine—a more potent antiseptic agent—while searching for a substitute for the saltpeter used in making gunpowder. It was not widely adopted at the time as its use was not backed by research.

Banishing filth

The belief that "miasmas" (bad vapors in the air, see pp.120–21) caused infections was common in the 19th century, and led to an emphasis on cleanliness, which did yield results. For example, in 1847 Hungarian physician Ignaz Semmelweiz, who worked in Vienna, ordered a regime of rigorous hand-washing in chlorinated water and the cleansing of surgical instruments and dressings (see pp.138–39), which reduced infection rates.

However, the real cause of such infections was not fully understood until the 1850s when Louis Pasteur (see pp.148–49) showed that the culprits were microorganisms entering wounds, not just bad vapors. Joseph Lister, a young Edinburgh doctor, hypothesized that finding a way of preventing the microorganisms from entering a wound might solve the problem. He experimented with a variety of substances, including zinc chloride, but nothing seemed to work with compound fractures. He then heard about the use of carbolic acid in the treatment of sewage in Carlisle, UK, and asked for samples of the acid. In August 1865 he applied it to the wound of an 11-year-old boy with a compound fracture of the leg during surgery. Although the acid caused mild flesh burns, the boy's leg did not become infected. Over the next year he used the acid on nine patients, seven of whom came through surgery without infection.

Spraying clean

So effective was Lister's "antiseptic" acid that its use soon became routine at his Glasgow hospital, and death rates from infection during amputations fell. In 1869 Lister devised an antiseptic spray that combined a local anesthetic with carbolic acid. Where before, surgeons had been reluctant to make incisions in skin, for fear of infection, more complex operations now became possible.

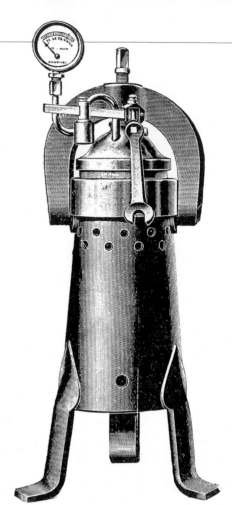

▷ **Sterilizing instruments**
The autoclave, a closed pressure chamber for the sterilization of surgical instruments using high-pressure steam was invented by French microbiologist Charles Chamberland in 1879. It represented a major advance in aseptic surgery.

BRITISH SURGEON (1827–1912)

JOSEPH LISTER

Lister inherited his scientific curiosity from his wine-merchant father, an amateur physicist with an interest in microscopes. Lister studied medicine at University College London, where he wrote a paper on inflammation. He transferred to Edinburgh University in 1853, and then to Glasgow in 1860 as Regius Professor of Surgery. It was here that he carried out his work on antiseptics. He returned to England in 1877, and had to overcome initial strong resistance to his ideas on antiseptics. In 1897 he was the first surgeon to be given a British peerage.

46 PERCENT The proportion of amputation patients at Glasgow Royal Infirmary who died of infection before antiseptics.

15 PERCENT The proportion of amputation patients who died of infection after the introduction of antiseptics.

> **"**It occurred to me that decomposition of the injured part **might be avoided...** by applying as a dressing some material capable of **destroying the life of the floating particles."**

JOSEPH LISTER, DELIVERING THE HUXLEY LECTURE, CHARING CROSS HOSPITAL, 1900

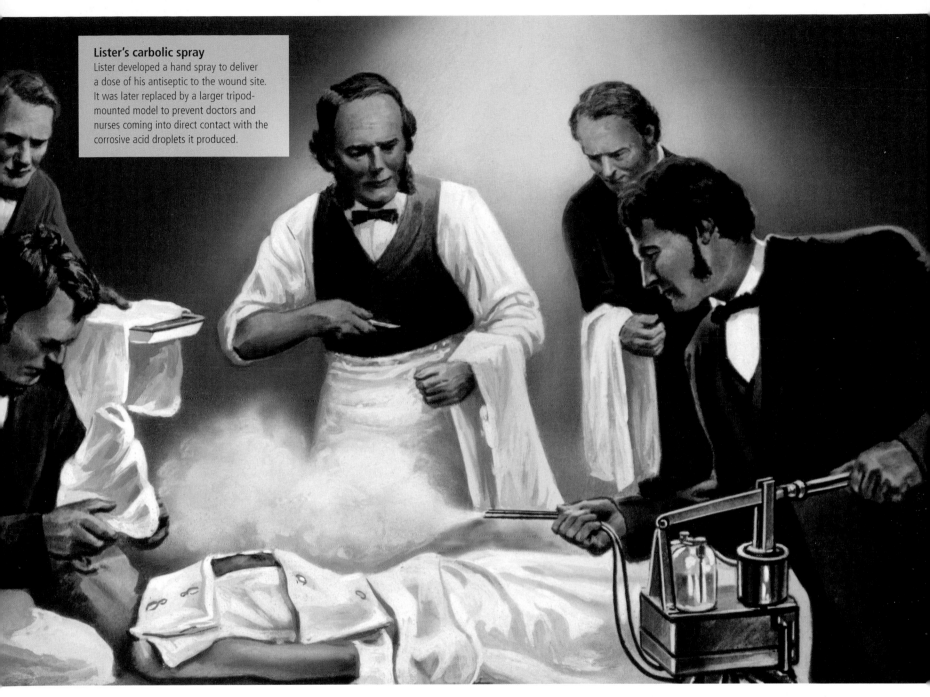

Lister's carbolic spray
Lister developed a hand spray to deliver a dose of his antiseptic to the wound site. It was later replaced by a larger tripod-mounted model to prevent doctors and nurses coming into direct contact with the corrosive acid droplets it produced.

By the 1870s, however, the use of carbolic sprays began falling out of fashion as attention shifted from the risk of infection from airborne pathogens to the greater risk posed by poorly cleaned instruments and unwashed hands. Scottish surgeon William Macewen pioneered the use of steam to cleanse surgical instruments and masks. He also devised a set of all-steel surgical instruments that could be sterilized at high temperatures. The use of rubber gloves that could be boiled (their first recorded use was in 1897 in Estonia) further reduced the number of cases of infection—until then surgeons had worked with bare hands. "Asepsis" (the absence of microorganisms) combined with "antisepsis" (killing microorganisms that had become present) led to a new era in surgery. The risk of infection, although not eliminated, was dramatically reduced.

▷ **Iodine tincture**
Iodine was found useful as an antiseptic in its diluted form. Its main medical use, however, was as treatment for goiter, an enlargement of the thyroid gland causing swelling in the neck.

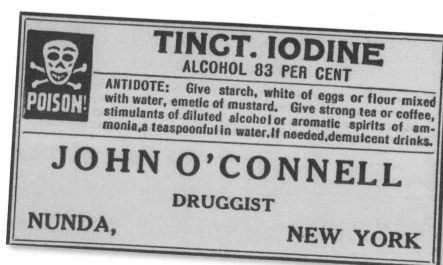

TINCT. IODINE
ALCOHOL 83 PER CENT

POISON!

ANTIDOTE: Give starch, white of eggs or flour mixed with water, emetic of mustard. Give strong tea or coffee, stimulants of diluted alcohol or aromatic spirits of ammonia, a teaspoonful in water. If needed, demulcent drinks.

JOHN O'CONNELL
DRUGGIST

NUNDA, NEW YORK

Tuberculosis

Also known as TB, phthisis, consumption, and white plague, tuberculosis is one of the world's longest-known, most widespread, and deadliest diseases. Even today, it affects 8–10 million people every year.

Stone Age remains suggest that tuberculosis was present more than 15,000 years ago, and evidence of it from recorded history goes back 7,000 years. Hippocrates (see pp.36–37) claimed that it was the most widespread disease of his time, and that it was hereditary. The expanding cities of Renaissance Europe saw numerous outbreaks, and a number of theories for its cause emerged. It is now known that tuberculosis is a bacterial disease that mainly affects the lungs, and that it spreads through the air. Yet symptoms can be so varied that it was only recognized and named as a single disease in the 1830s. The first sanatoria for TB patients opened soon after. They were mostly situated in upland locations where patients could rest, breathe pure air, and eat well, in the hope that this would help them recover.

In 1882 Robert Koch (see pp.146–47) identified the microbe that causes TB—*Mycobacterium tuberculosis*. However, it was not until 1947, when medical trials showed the curative effects of the recently discovered antibiotic streptomycin, that the disease came under partial control. Tuberculosis remains common in developing regions, chiefly Africa, and South, East, and Southeast Asia. One of the World Health Organization's major goals is ending the TB epidemic by 2030.

> **"… the overcrowded dwellings of the poor… [are] the real breeding places of consumption."**
>
> ROBERT KOCH, GERMAN PHYSICIAN, FROM THE ADDRESS TO THE BRITISH CONGRESS ON TUBERCULOSIS, 1901

▷ *Fading away*
During the 18th to early 20th centuries, tuberculosis was "romanticized" by writers, poets, playwrights, and artists as a disease of the able, intelligent, and creative. This serene scene is part of a five-image montage of a young, dying woman that was composed by English photographer Henry Peach Robinson.

Vaccines Come of Age

In the late 19th century, discoveries concerning the mechanisms of disease transmission and immunity revolutionized medicine. They led to the development of new vaccines to protect against infectious diseases that had hitherto killed tens of thousands of people each year.

Research by French microbiologist Louis Pasteur (see pp.148–49) on microorganisms and the germ theory (see pp.146–47) gave hope that agents might be developed to combat a range of diseases, and that more vaccines might be created in addition to the one Edward Jenner had devised against smallpox (see pp.102–03).

A major breakthrough came in 1879, when Pasteur produced the first laboratory-developed vaccine. He was researching chicken cholera by injecting birds with live bacteria and then observing the progression of the fatal disease. One day, he asked his assistant to inject the birds with a fresh bacterial culture, but the assistant

neglected to do so. A month later, the chickens were injected with the "old" batch, still sitting on the shelf. The birds, while showing mild signs of the disease, survived. Pasteur then injected them with fresh bacteria and they did not become ill. He had discovered the principle of attenuation, by which a weakened form of a disease-bearing organism bestows immunity to the full form of the disease if administered to a patient. In 1882 Pasteur extended the principle to anthrax and, in 1885, to rabies.

◁ **Symptoms of diphtheria**
This disease is characterized by a fever, severe cough, and a gray coating over the infected areas, particularly the throat and tonsils. If left untreated, diphtheria is fatal in young children in about 20 percent of cases.

GERMAN PHYSIOLOGIST (1854–1917)
EMIL VON BEHRING

Born into a poor family, Emil von Behring could not afford to go to university and so he undertook his medical studies in the German army. In the early 1880s he showed that while the compound iodoform did not kill microbes it seemed to neutralize the bacterial toxins they produced, rendering them harmless.

In 1888 he began work at the Institute of Hygiene in Berlin. Here he discovered that cultures in which diphtheria microbes had been killed leaving their toxin could provoke immunity in animals injected with them. Von Behring was awarded the first ever Nobel Prize for Medicine in 1901 in recognition of his work.

New vaccines are discovered

By the late 1880s researchers had realized that toxins in blood serum, released by certain bacteria such as diphtheria and tetanus, were responsible for the diseases' symptoms. German physiologist Emil von Behring discovered that by injecting nonlethal doses of diphtheria into guinea pigs, and later horses, he was able to extract a serum from the test subjects that then conferred immunity on other animals that were injected with it. The discovery of the tetanus toxin in 1889 enabled von Behring and his colleague Shibasaburo Kitasato to develop an antitoxin against the disease the following year. A diphtheria vaccination became commercially available in 1892, and the death rate declined dramatically. In 1921 there were 206,000 cases of diphtheria in the US; now, there are less than five cases annually.

Working on viruses

Most of the early advances in the development of vaccinations involved diseases transmitted by

△ **Vaccine ampoules**
The ampoules shown here, from 1915, contain serums that were used to vaccinate against typhoid and paratyphoid. These vaccines were especially important in wartime, as more soldiers tended to die from typhoid than battle injuries.

Noguchi found that he could grow the *Vaccinia* virus (related to the *Variola* virus that caused smallpox) inside the testes of live rabbits. By the 1930s viruses were being cultured inside chicken eggs,

> " The **immunity… against tetanus** consists in the power of the **cell-free blood fluid** to **render innocuous** the **toxic substance** that the tetanus bacilli produce. "
>
> EMIL VON BEHRING AND SHIBASABURO KITASATO, IN A PAPER IN
> *DEUTSCHE MEDIZINISCHE WOCHENSCHRIFT (GERMAN MEDICINE WEEKLY)*, 1891

bacteria rather than viruses. Although viral agents had been discovered in the 1890s, they proved more difficult to cultivate than bacteria. However, in 1915 Japanese physician Hideyo

which enabled the widespread production of vaccinations against typhus (which had first been tested in 1898) and the creation of an effective vaccination against polio (which was first trialed in 1954).

△ **Plague inoculation**
The 1906 outbreak of plague in Burma (now Myanmar) led to a widespread program of inoculation using a vaccine devised in 1897. Unfortunately, this vaccine was limited in its effectiveness and 6,000 people still died.

Promoting vaccination

As vaccines became more widely available, many nations introduced public health programs to promote their take-up or even to make them compulsory. One of the earliest such programs was established in the UK, where the Vaccination Act was passed in 1853 ordering the mandatory vaccination of all infants against smallpox (see pp.100–03) within four months of birth. In the 20th century, vaccines against measles and mumps were developed, in 1963 and 1968 respectively, and soon became part of a regular schedule of childhood vaccinations in most countries.

The public health benefits of vaccination are incalculable in terms of lives saved and medical resources that do not have to be expended on sufferers of infectious diseases. Research scientists continue to develop vaccines for serious diseases that are difficult to treat, especially viral infections such as HIV/AIDS (see pp.242–43) and Ebola (see pp.268–69). Other targets include diseases where the method of transmission is hard to control, for instance, malaria (see pp.174–75) or the Zika virus disease, which are both spread by mosquitoes.

27,000 The number of people who contracted polio during the 1916 outbreak in the US.

73 The number of polio cases worldwide in 2015.

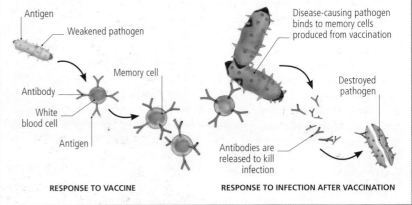

Mysteries of the Brain

Medical knowledge of the brain generally lagged behind that of other body systems, partly due to the organ's inert, featureless appearance. In the 19th century a growing awareness of the brain's role in behavior led pioneers to establish a new speciality—neurology.

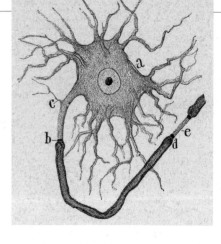

△ **Brain cell of a fish**
Advances in microscopy enabled scientists such as Santiago Ramón y Cajal (see p.97) to study nerve cells. This fish brain cell has been colored with Boveri stain (silver nitrate).

The brain's uniform structure, few obvious demarcations, and lack of moving parts give little clue to the magnitude of its functions. Access is difficult since the brain is heavily protected by the skull. Nevertheless, physicians throughout history have tried to treat its physical disorders by methods as drastic as boring holes in the skull (see pp.16–17).

The nerves, too, look like pale strings with no clear indication as to how they work. Neurological conditions such as epilepsy and migraines were often attributed to evil spirits or to divine punishment.

In the 4th century BCE, Aristotle claimed that the heart was the center of emotion and intelligence, while Roman physician Galen (see pp.40–41) associated the brain with "animal spirits" or "psychic faculties," such as reason, thought, perception, and memory.

3,000 The number of neurological patients under Charcot's care at the Salpêtrière Hospital, France, in the 1860s–70s.

With the revival of anatomy as a science in the 14th century, the gross structure of the brain—the parts visible with the unaided eye—became clearer. In 1543 Flemish physician Andreas Vesalius (see pp.74–75) depicted the brain's coverings, or meninges, its outer surface, inner chambers, nerves, and blood vessels. The chambers or "cells" were allocated functions: imagination to the anterior; reason to the middle; and memory to the posterior chamber. These cerebral ventricles, as they have since been named, actually contain cerebrospinal fluid, which has no part in mental processes.

In 1664 English physician Thomas Willis published *Cerebri Anatome* (*Anatomy of the Brain*), which offered a detailed anatomy of the brain and nerves, and introduced the term "neurology" for the study of nerves.

During the 18th century there were major advances in the understanding of the brain, but also fashions concerning mind and behavior with little scientific basis.

Phrenology (see pp.104–05), the reading of skull contours, was popular in the early 1800s. Another since discredited theory, that of "animal magnetism," was developed by German physician Franz Anton Mesmer. He believed that an unseen force or energy, subject to the laws of magnetism, flows through all living things. Those who were able to manipulate this force could use it for healing. At Mesmer's gatherings or "banquets" patients were placed in a trancelike state. "Mesmerism" is now thought to have had close links to hypnosis.

The birth of neurology

The development of anatomy and pathology through the 18th century was aided by microscopy (pp.92–93) and histology—the study of the microscopic anatomy of tissues and cells, often using methods of staining or coloring. These advances enabled 19th-century clinicians such as French professor Jean-Martin Charcot to establish neurology as a major branch of medicine.

Charcot was a talented clinician, interviewing and examining patients, diagnosing diseases, and prescribing treatment. Over a career of more than 40 years, he recorded patterns of symptoms in patients and linked his clinical findings to postmortem findings using tools of anatomy, pathology, and microscopy. He defined numerous neurological conditions: as many as 20 still bear his name.

Charcot was influential in distinguishing between neurology and psychiatry. While neurology concerns mainly the physical brain, and how problems of anatomy and physiology cause conditions such as stroke and multiple sclerosis, psychiatry developed to focus on mental health, and disorders of mood, emotions, and thoughts such as anxiety, depression, and schizophrenia that have few or no physical signs.

Charcot considered French physician Guillaume-Benjamin-Amand Duchenne, who was the first to describe and devise treatment for several nervous

◁ **A Mesmer banquet**
Mesmer held healing "banquets," at which wealthy patients held metal rods immersed in a tub of "magnetic water" and entered a trancelike state. They believed this would remedy the "imbalance" within and cure them.

△ **Master at work**
Charcot was an innovative teacher, examining, interviewing, and even hypnotizing patients during lectures. He also used visual aids such as his own paintings and medical photographs.

and muscular disorders, to be his "teacher in neurology." Charcot, in turn, inspired others, including the founder of psychoanalysis Sigmund Freud (see pp.182–83); Pierre Janet, who established psychology in France; and notable neurologists such as Gilles de la Tourette (Tourette's syndrome). Charcot himself became interested in hypnosis and its links with hysterical mental states, using it in lectures and as a possible cure.

In the late 19th century pioneering operations in the neurosurgery field included excision of a tumor of the meninges by Scottish surgeon William Macewen in 1878, and removal of a spinal cord tumor in 1887 by English surgeon and pathologist Victor Horsley.

> "To... **treat a disease...** learn how to **recognize it.**"

JEAN-MARTIN CHARCOT, FRENCH NEUROLOGIST

FRENCH NEUROLOGIST (1825–93)

JEAN-MARTIN CHARCOT

Born in Paris in 1825, Jean-Martin Charcot qualified in medicine and spent most of his working life at the Salpêtrière Hospital, Paris. Proficient in many languages, he absorbed new medical knowledge from around Europe. In 1856 he was appointed "physician to the hospitals of Paris," and became professor of Pathological Anatomy at the University of Paris in 1872. By the 1880s the Salpêtrière Hospital was Europe's leading neurology centers, with its own microscopy and photography departments. Charcot died in Paris in 1893.

Mental Illness

The idea that the mentally ill should be separated from society and treated in asylums may have seemed like progress in the 18th and 19th centuries, but the reality of their confinement and treatment continues to haunt the history of medicine.

△ **Fool's tower**
The Narrenturm, or Fool's Tower, at the Vienna General Hospital, Austria, was the first specially built mental asylum. It was constructed in 1784 with 139 cells to house inmates.

Two hundred years ago there was still very little understanding of the causes of mental illness. In early times episodes of madness had been linked to the phases of the moon (hence the word "lunatic") or seen as communications or prophesies from the gods. A supposed link between mental problems and the balance of the four humors (see pp.34–35) was first made in ancient Greece and remained popular into medieval times and beyond.

In communities with a deep sense of ancestral pride and honor, madness was a stain on the family, and sufferers were concealed from public life or even abandoned. In Europe in the medieval period, those not sheltered by their family risked appalling abuse unless they were taken in by convents, monasteries, or workhouses.

First asylums

Models for the safekeeping of the mentally ill had existed for hundreds of years, since the first facilities were provided in 8th-century Baghdad, based on the Qur'an's principle of humane treatment for those "weak of understanding." However, the approach to asylums in Europe from the early 15th century onward was based on brutality and incarceration: "treatment" included whipping, stripping, and restraining with chains. Early establishments included

▽ **Advocating humane treatment**
French physician Philippe Pinel was one of the first to insist on "moral" treatment for patients suffering from mental illness. In this painting he is shown releasing inmates from their chains at the Bicêtre Hospital asylum in Paris in 1793.

the notorious Maison de Charenton in Paris in the 1640s and the Narrenturm in Vienna in 1784.

In the 19th century the drive for the mentally ill to be placed in madhouses, or asylums, gathered momentum. In Britain the Lunacy Act and County Asylum Act 1845 insisted that local authorities take responsibility for the "mad." Asylums multiplied throughout Europe and North America in this period too.

Moral treatment

Although harsh treatment prevailed, there had been pockets of resistance since the late 18th century when, in Paris, Philippe Pinel and Jean-Baptiste Pussin stated that the mentally ill were patients and not criminals (see pp.164–65). In England, Quaker philanthropist William Tuke advocated that patients be housed in pleasant settings and treated

with minimal coercion. This emphasis on "moral treatment" traveled from Europe to the US. After visiting humane Quaker establishments in England in her campaign for reform, American teacher Dorothea Dix visited public and private mental facilities in the US, and documented appalling conditions.

Yet old habits prevailed. As asylums became overcrowded, practices such as the use of straightjackets and seclusion made a comeback. Patients became institutionalized, and asylums remained a testing ground for unscientific theories. One popular treatment saw patients swung on a harness to "calm the nerves."

New approaches

From the 1890s Austrian physician Sigmund Freud developed psychoanalysis (see pp.182–83)

(see pp.164–65)

(see pp.182–83)

◁ **Electroconvulsive therapy**
The Ectonustim 3 machine transmitted a current through the brain by means of electrodes attached to the scalp of the anesthetized patient. The current induced convulsions in the hope of alleviating mental disorders such as severe depression.

as a treatment for mental problems that lay buried deep within the unconscious mind of the patient. Freud believed that mental illness, and in particular hysteria, stemmed from repressed emotions and memories, which could be unlocked through therapy. In Freud's "talking cure," patients were encouraged to talk freely about their urges, desires, and dreams, which were analyzed by the therapist.

World War I (1914–18) saw a new approach to treating mental illness, when many thousands of soldiers who were traumatized by war were taken to specialized hospitals. Shell shock became recognized as a mental disorder affecting all ranks and classes, although numerous shell-shocked soldiers were charged with desertion.

After World War I there was a renewed enthusiasm for "physical therapies" to cure mental illness. Again, asylums were the perfect

250,000 The number of inmates in asylums in the US in 1900. In 1880 the figure was 40,000.

environment in which physicians could practice novel treatments. Lobotomy, in which surgery was used to sever physical connections between the prefrontal, frontal, and other parts of the brain, had unpredictable and sometimes disastrous results.

The first of what became known as convulsive therapies was initiated in 1934 in Budapest, Hungary, when psychiatrist Ladislas von Meduna began a regime of drug-induced seizures to treat schizophrenia. This was replaced by electroconvulsive treatment (ECT), initiated in 1938, which involved passing an electric current through the brain to trigger a seizure. By the 1960s ECT was used to treat a variety of conditions, notably severe depression. It remains in limited use, but it has been largely replaced by new drug therapies developed in the second half of the 20th century.

" **Could we in fancy place ourselves in the situation of some of these poor wretches, bereft of reason, deserted of friends, hopeless...** "

ASYLUM REFORMER DOROTHEA DIX, *MEMORIAL TO THE LEGISLATURE OF MASSACHUSETTS*, 1843

Horror of the Asylum

Hippocrates and his followers stated: "Wherever the art of medicine is loved, there is also a love of humanity." However, humanity deserted much of the medical profession in Europe during the 15th–18th centuries, when people with mental illness were locked up, abused, and even tortured in horrendous ways.

For centuries, mental and psychiatric conditions were attributed to an imbalance in the humors (see pp.34–35) or evil spirits and demonic possession. People dreaded and isolated sufferers, and many doctors believed such illnesses were incurable. From the 1400s patients were locked away from society in terrible conditions in prisons or asylums. Some were chained up, thrown occasional scraps, and left to die. Others suffered all manner of appalling "cures," such as blood-letting, to restore humoral balance. Severe traumas and shocks to exorcise the demons included being whipped or hung up by the arms or legs, and being nearly suffocated, drowned, or starved. Worse, some asylums became places of curiosity and entertainment, where people came to watch the inmates' plight, even paying for the experience.

Toward the end of the 18th century, the horrors of asylums came to the notice of reformers. In 1793 French doctor Philippe Pinel joined the staff at Bicêtre Hospital, an asylum for men in Paris. Together with the hospital governor Jean-Baptiste Pussin and his wife Marguerite, he began a series of improvements that returned humanity into the care and treatment of the mentally ill. Pinel and Pussin continued their reforms at Paris's Salpêtrière Hospital for women, introducing a rational and scientific approach. Chains were removed, living conditions improved, and prisoners became patients, encouraging a new enlightened era.

"Mental disorders are… nervous diseases… "

HENRY MAUDSLEY, FOUNDER OF LONDON'S PSYCHIATRIC MAUDSLEY HOSPITAL, FROM *BODY AND MIND*, 1870

◁ Frightful fate
"The Madhouse" is an 1835 engraving by Swiss draftsman Heinrich Merz, after a drawing by German painter Wilhelm von Kaulbach. The varied facial expressions of mentally ill patients convey their emotional states amid crowded, bleak conditions typical of the time.

At the close of the 19th century, the first vaccines were being administered and began to be incorporated into public health programs, thanks to the pioneering efforts of English physician Edward Jenner (see pp.102–03) and French chemist and microbiologist Louis Pasteur (see pp.148–49). Jenner had developed the world's first vaccine for smallpox, using the cowpox virus, in the 1790s. Pasteur, who was born the year before Jenner's death, devised vaccines for rabies and anthrax in the 1880s. However, the nature of the diseases targeted by these vaccines was not fully understood.

infected plant, filtering the liquid through paper to remove bacteria, and then rubbing the liquid onto an uninfected plant. Mayer had laid the foundations for the discovery of the first known virus.

A few years later, in 1892, Dmitry Ivanovsky repeated the principle of Mayer's experiments by applying the technique of filtration to tobacco plants with mosaic disease. Unlike Mayer, however, Ivanovsky used a more stringent method of filtration—the Chamberland filter, a porcelain tube that uses water to separate any trace of bacterial toxin from a sample. Invented by French microbiologist Charles Chamberland in 1884, and used by

that the contagion was not able to grow on its own—it needed a living host in order to replicate. Beijerinck's work established without doubt that a new type of infectious agent existed—a virus, from a Latin word meaning "poison" or "slimy liquid."

The virus particle

Whereas Beijerinck had asserted that a virus was a liquid, a study of livestock by German scientists Friedrich Loeffler and Paul Frosch, the same year, found evidence that it was actually a particle. They had discovered the world's second known virus—foot and mouth disease. By the 1920s more than

△ **Chamberland filter**
Developed for Louis Pasteur's work on vaccines in the 1880s, the porcelain Chamberland water filter was key to discovering viruses. It had pores so fine that it could filter out bacteria from any liquid sample.

Viruses and How they Work

Viral infections wreaked havoc on the populations of three continents in the 18th century, fueling efforts to deliver the world's first vaccines. Yet it took a further century to identify and understand the nature of the viruses responsible for these diseases, and how they spread.

Scientists did not comprehend the mechanics of viral activity until Russian microbiologist Dmitry Ivanovsky described the first known virus in 1892, almost a hundred years after the development of the first vaccine.

Virology begins
The history of virology, the study of viruses, began with a sick tobacco plant in the laboratory of Adolf Mayer in 1879. Mayer, a German agricultural chemist, was studying mosaic disease, which contaminated tobacco plants and destroyed entire tobacco crops. Over the following 10 years, he demonstrated that the disease could be artificially spread to other plants by taking sap from an

Louis Pasteur in his development of vaccines, the filter allowed Ivanovsky to remove all bacteria from the liquid concentrate that had been taken from a diseased tobacco plant. The filtered sample was still infectious, proving that the disease was not transmitted by bacteria.

Building on Ivanovsky's findings, Dutch microbiologist Martinus Beijerinck went one step further in 1898, concluding that not only was mosaic disease still infectious after being filtered for bacteria, but

Capsid, or shell, made of protein

Genetic material—DNA (deoxyribonucleic acid) or RNA (ribonucleic acid)

Virus shell binds to the cell membrane of host cell

1 ATTACHMENT

2 PENETRATION

Virus shell disintegrates to release its genetic material

3 REPLICATION

Nucleus of host cell

"In a flash **I understood...** a filterable virus... a **virus parasitic on bacteria.**"

FÉLIX D'HÉRELLE, CANADIAN MICROBIOLOGIST, 1917

65 different animal and human viruses had been identified, including the first human virus, yellow fever in 1901, the rabies virus in 1903, and the polio virus in 1908.

Bacteria-invading viruses

The next milestone in the history of virology came in 1915, when English bacteriologist Frederik Twort proposed that some viruses were capable of infecting bacteria and using them as hosts in which to replicate. At the Pasteur Institute in Paris, Canadian-born microbiologist Félix d'Herelle advanced the concept further by working out how to count the number of viruses that could be found in certain bacteria. He named this type of virus a "bacteriophage," or bacteria-eater.

As more viruses were discovered in the first few decades of the 20th century, attention turned to developing vaccines for some of the most devastating viral diseases—the polio vaccine (see pp.210–11), for example, which is still in use today. Modern research continues to investigate how different viruses mutate and replicate, because this is key to developing effective treatments for viral infections.

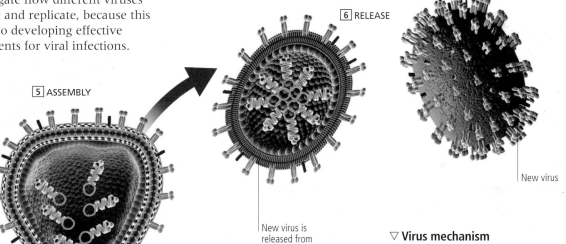

6 RELEASE

5 ASSEMBLY

New virus

New virus is released from dying cell

Cytoplasm of host cell

Capsid proteins gather around the new viral genetic material to make a new virus

Genetic material of virus enters nucleus of host cell

Virus instructs production of new viral genetic material

4 SYNTHESIS

▽ **Virus mechanism**

A bacteriophage virus acts like a parasite, injecting its genetic material into the cell of a host bacteria, which is typically found in soil, seawater, or in the stomach of an animal. Once the virus's DNA or RNA has penetrated the cell, it begins to replicate, destroying or taking over the machinery of the cell.

Fighting Rabies

Dreaded since antiquity, rabies—which spreads through the nervous system to the brain—causes terrible suffering, aggressive behavior, paralysis, and eventual death. Louis Pasteur's development of a vaccine was therefore a very welcome breakthrough.

△ **Rabies vaccine warning sign**
This German sign warns that vaccine-containing pellets have been left out for foxes, to make them immune so they would not spread rabies, so dogs must be kept away.

In the late 19th century the dreaded "mad dog" disease (as rabies was known) continued to confound physicians, who struggled to treat those who were affected. A breakthrough came in 1880, when French microbiologist Louis Pasteur (see pp.148–49) became interested in the disease. At the time France was increasingly troubled by packs of feral dogs, some of them rabid. Knowing that rabies spread through the bites of infected animals, a vet in Paris sent Pasteur saliva specimens from two dogs that had died from the disease and asked for his help.

Creating a vaccine
Working with viruses in the late 19th century was a protracted, difficult, and dangerous task. Light microscopes, which use focused light and a lens to enlarge small samples, did not provide sufficient magnification to see the rabies virus—a type of rhabdovirus called *Lyssavirus*—which is less than 0.0002mm long. Also, since viruses multiply in living cells, Pasteur and his collaborator, Émile Roux, had to carry out tests on live animals—including dogs, monkeys, and rabbits. In addition, rabies can take anything from a few days to several months to cause symptoms, depending on how quickly it spreads through the nervous system to the brain. Pasteur tested many strains of the virus, selected the fastest-acting ones, and injected them directly into the brains of test animals.

60 HOURS The time after being bitten, when French boy Joseph Meister had the first of 13 injections administered over 12 days. He became the first person to be inoculated against rabies.

Building immunity
To create a vaccine, Pasteur first needed to weaken the virus enough that it would provide immunity from rabies without causing the disease. Working with Roux and others, Pasteur tried the idea of dissecting the spinal cords of infected, freshly dead rabbits, then placing them in open flasks that contained potassium hydroxide, which acted as a drying and anti-decay agent. First, Pasteur injected healthy animals with the rabies virus present in spinal cord that had been dried for 14 days. At this late stage, the virus was weak and unlikely to do harm. He then repeated the test every couple of days using infected spinal cords that were 13 days old, then 12 days old—the idea being they would build up immunity to the virus. Finally, he injected the animals with extracts taken from a fresh infected spinal cord, which contained the most virulent virus; all the animals survived. His challenge now was to create a vaccine for humans.

Human trials
Pasteur then started trials of human vaccines, but there were two false starts: an older man who left after only one injection, and a young girl, whose disease was too advanced to be treated. But on July 6, 1885, a distraught mother brought her nine-year-old son Joseph Meister, who had been repeatedly bitten by a rabid dog two days earlier, to Pasteur for treatment. Pasteur was initially reluctant to administer the vaccine, since the boy had not yet shown symptoms of the disease and might not develop rabies—although it was likely that he would. Finally, Pasteur agreed to treat the boy. He gave Meister a series of 13 injections, starting with extracts from 15-day-old spinal cord, and building up to stronger preparations. Pasteur noted: "On the last days, I inoculated Joseph Meister with the most virulent virus of rabies." The young boy survived.

> "When **meditating over a disease,** I never **think of** finding a remedy for it, but, instead, **a means of preventing** it."
>
> LOUIS PASTEUR, FRENCH CHEMIST AND MICROBIOLOGIST, 1884

◁ **Ancient remedy**
All manner of rabies treatments had been tried throughout history, with virtually no success. This 13th-century physician applies vervain herb to the wound of a patient bitten by a rabid dog, which lies dead below.

Pasteur repeated the procedure on a shepherd, who had been attacked and severely bitten by a rabid dog, then gradually others followed. Later that year Pasteur officially reported the results from the Paris trials, and news spread worldwide.

A vaccine for the future

That December, four boys who had been bitten by a dog thought to be rabid arrived from New Jersey. A national campaign had been launched to fund their travel to Pasteur for treatment, and they returned home healthy. More rabies patients flocked to France to be treated—largely the result of the US campaign. In March 1886 Pasteur announced that he had treated 350 patients with only one loss, and by 1890 there were rabies vaccination centers in the US, Brazil, Europe, India, and China. Today the rabies vaccine, which has been improved, is on the essential medicines list of the World Health Organization (WHO) and is thought to save up to 300,000 lives annually.

▽ **Treating rabies**
As news of Pasteur's vaccine spread, lines formed to see him. Some wanted treatment for bites, others sought immunity in case of a future bite, which was the aim of Pasteur's initial research.

The Discovery of Aspirin

Throughout history, many different civilizations have used willow to alleviate aches and discomfort. In the 1800s scientists identified the active ingredient in willow and experimented with it. The white powder that resulted, aspirin, would become the most widely used drug in the modern world.

The quest to find an effective remedy for pain is as old as human civilization. Perhaps remarkably, the people of the ancient world found solutions to pain relief containing the same key ingredient as modern-day aspirin. Ancient Egyptians used willow tree extract to ease aches and pains, while Greek physician Hippocrates recommended willow-leaf tea to women to relieve the pain of childbirth. Almost two thousand years later, in the 1750s, the English clergyman Edward Stone conducted a five-year experiment that demonstrated that dried, powdered willow bark helped to cure fever. The Royal Society published his results in 1763.

Interest grew among scientists and medical practitioners in the potential of willow for pain relief.

In the 19th century science began to be seen as a true profession rather than natural philosophy, and it flourished, partly spurred on by growth in industry and commercial activity. The goal of many researchers was to find useful medicines, and efforts to pinpoint exactly how willow worked gained momentum.

Experiments with salicin

In 1828 Joseph Buchner, professor of pharmacy at Munich University, extracted a small quantity of a compound from willow bark and named it salicin. The following year, French chemist Henri Le Roux refined the process further to extract salicin in crystal form. Around the same time Swiss pharmacist Johann Pagenstecher also found salicin in the meadowsweet flower, but it was not until 1853 that French chemist Charles Frédéric Gerhardt made the breakthrough that opened the way for potential mass production of the medication.

Salicin, as it is found in willow and meadowsweet, is relatively low strength, with a mild effect on pain. Gerhardt extracted a more potent derivative of salicin, called salicylic acid, and worked out the molecular formula, which enabled him to produce it in a laboratory, at much higher concentrations than it was found in plants. However, although salicylic acid provides effective pain relief, it is hard on the stomach and can cause nausea, bleeding, and diarrhoea, so it needs to be "buffered" or neutralized in order to avoid these effects. In the course of his work on acid anhydrides, Gerhardt took the first step in addressing the side effects when he mixed acetyl chloride with salicylic acid, which created a rudimentary form of acetylsalicylic acid for the first time. Although Gerhardt showed little interest in pursuing his discovery further, other scientists did.

Two decades later, in 1876, the medical journal *The Lancet* published the results of the first clinical trial of salicin. Scottish doctor Thomas Maclagan concluded that a group

△ **Willow bark**
In 1763 it was discovered that willow bark dramatically reduced ague, a fever with symptoms similar to malaria. It was later found that the active ingredient in willow bark is salicylic acid, which forms the basis of aspirin.

of patients with rheumatism had experienced reduced fever and joint inflammation after taking the chemical compound. Maclagan had chosen to use salicin rather than the stronger salicylic acid because it was gentler on the stomach, making it more suitable for the subjects of his trial.

Final steps

The chemist who finally succeeded in creating a powerful pain medication without severe side effects was Felix Hoffmann, an employee of dye manufacturer Friedrich Bayer & Co. in Germany. Hoffmann's father suffered from rheumatism and urged his son to develop a pain remedy that was less irritating on the stomach than existing medicines based on salicylic acid. Hoffmann and his colleagues at Bayer successfully developed an easily synthesized, effective form of acetylsalicylic acid, which caused less upset to the stomach than salicylic acid. They produced the first sample of pure acetylsalicylic

40,000 TONS The quantity of aspirin consumed globally each year.

> " Aspirin is a drug that has been used for many years—it is **effective, inexpensive,** and widely **available.**"
>
> JEFFREY BERGER, AMERICAN DOCTOR, FROM *JOURNAL OF THE AMERICAN MEDICAL ASSOCIATION*, 2006

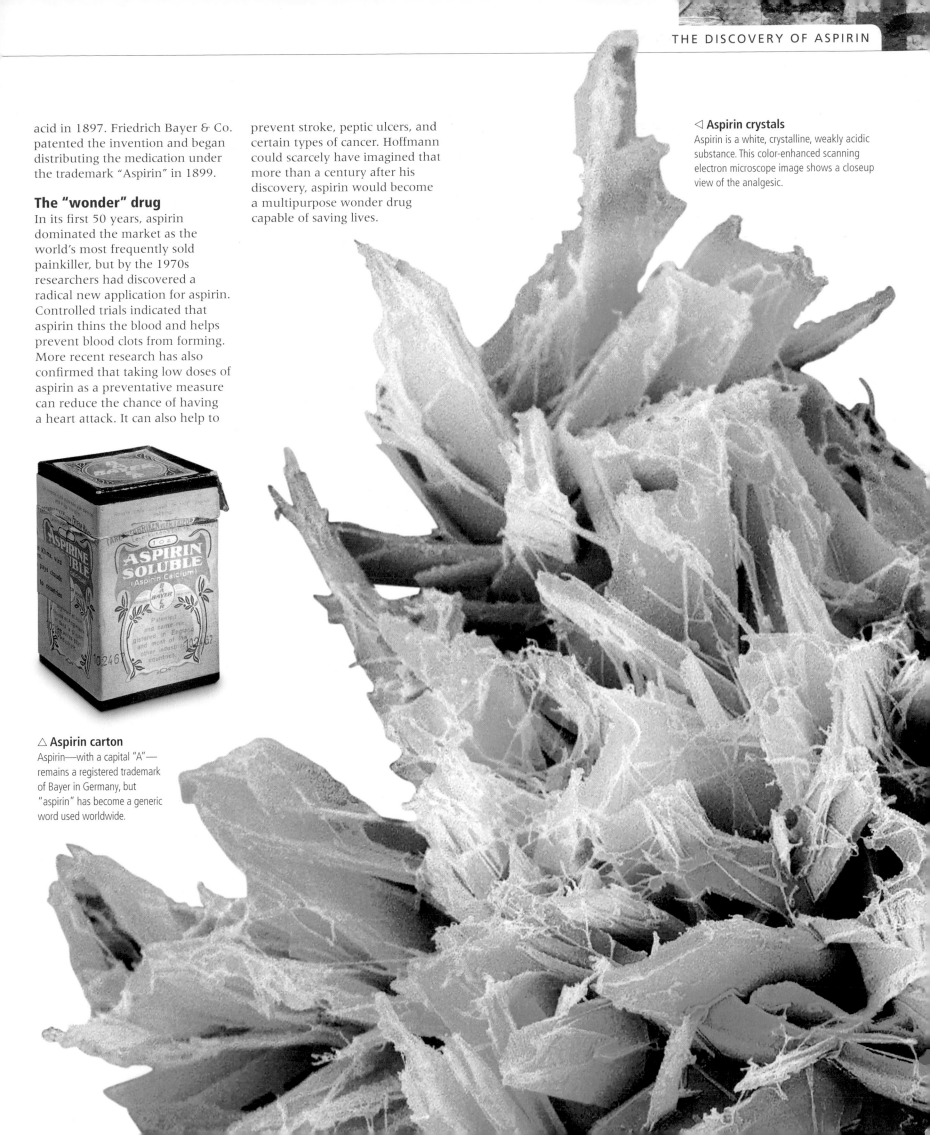

acid in 1897. Friedrich Bayer & Co. patented the invention and began distributing the medication under the trademark "Aspirin" in 1899.

The "wonder" drug

In its first 50 years, aspirin dominated the market as the world's most frequently sold painkiller, but by the 1970s researchers had discovered a radical new application for aspirin. Controlled trials indicated that aspirin thins the blood and helps prevent blood clots from forming. More recent research has also confirmed that taking low doses of aspirin as a preventative measure can reduce the chance of having a heart attack. It can also help to

prevent stroke, peptic ulcers, and certain types of cancer. Hoffmann could scarcely have imagined that more than a century after his discovery, aspirin would become a multipurpose wonder drug capable of saving lives.

◁ **Aspirin crystals**
Aspirin is a white, crystalline, weakly acidic substance. This color-enhanced scanning electron microscope image shows a closeup view of the analgesic.

△ **Aspirin carton**
Aspirin—with a capital "A"— remains a registered trademark of Bayer in Germany, but "aspirin" has become a generic word used worldwide.

X-rays

The chance discovery of X-rays by a German physicist at the end of the 19th century sparked a new age of medical imaging. Medical diagnosis was revolutionized—for the first time, physicians could look inside the body without the need for surgery.

German physicist Wilhelm Röntgen conducted an experiment on cathode rays in his laboratory on November 8, 1895. He removed all air from a glass tube, filled it with a special gas, and passed a high-voltage current through it. As he did this, the tube emitted a fluorescent glow. Next, Röntgen darkened the room and shielded his tube in a casing of thick black cardboard to exclude all light. To his surprise, he noted that even though his tube was completely encased a nearby screen that was coated with a fluorescent chemical glowed. Röntgen studied this for several weeks and concluded that the glow must result from an undiscovered kind of ray—a type of radiation that differed from visible light. He named it "X-ray"—"x" being the mathematical term for an unknown quantity.

Röntgen subsequently tried to block the path of the X-rays to his screen with a selection of denser materials, including wood, copper, and aluminum, but the rays penetrated all of them.

Looking inside the body

However, when he held a lead disk in front of the tube he was amazed to see the bones of his own hand glowing on the screen—it was the first ever radiographic image. He then placed his wife's hand in the path of the rays over a photographic plate, capturing the world's first X-ray image. The bones were clearly visible, while the soft tissue was barely noticeable. Six weeks later, Röntgen published a paper titled *Über eine Neue Art von Strahlen* (*On a New Kind of Rays*).

Röntgen's discovery caused a public sensation. The implications of being able to look inside the human body were immense, and

> "I did not **think,** I **investigated.**"
>
> WILHELM RÖNTGEN, IN AN INTERVIEW FOR *MCCLURE'S MAGAZINE*, 1896

X-rays were soon being used to diagnose a wide range of conditions. Within a year, the world's first radiology department was set up at a hospital in Glasgow, Scotland, and produced the first X-rays of a kidney stone and a coin stuck in a child's throat.

The first machines were basic and emitted weak radiation, so patients had to remain still for more than 30 minutes for images to be captured. It also became apparent that X-rays caused burns and hair loss. But by the early 1900s scientists had also discovered that controlled doses of X-ray radiation could be used positively to fight cancers and skin diseases. X-rays proved useful in wartime—during World War I military doctors used X-ray machines to locate bullets and shell fragments in soldiers' bodies.

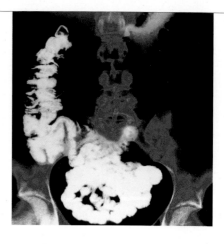

△ **Barium X-ray imaging**
The insoluble salt barium sulfate shows up on X-rays in the same manner as metal or bone. It can be administered orally into the gastrointestinal tract, which is not visible on standard X-rays, so that its lining, size, and shape can be examined.

However, in 1904 American inventor Thomas Edison's assistant, Clarence Dally, who worked extensively with X-rays, died of cancer. His death caused scientists to begin to take the risks of X-ray radiation more seriously.

Further developments

More work was needed to fully comprehend the nature of X-rays. In 1912 German physicist Max von Laue decided to transmit X-rays through crystals, and in the process demonstrated that X-rays, like light, were subject to diffraction (interaction after being split). The diffraction pattern showed how a crystal's atoms were arranged—a technique crucial to the analysis of molecular structure. X-ray crystallographic techniques were later used to study the structure of proteins and were employed by researchers worldwide. This work has lead to incalculable advances in chemistry and molecular biology.

Although X-rays continue to be used for medical diagnosis, they are employed in a wide range of other fields, from biotechnology, genetics, and astronomy, to scanning luggage for security.

Röntgen was a modest man who loathed all the attention that was heaped on him after his discovery. In 1901 he was awarded the first Nobel Prize for physics, but he bequeathed the prize money to scientific research and deliberately never patented the X-ray, ensuring that the public could benefit.

GERMAN PHYSICIST (1845–1923)

WILHELM CONRAD RÖNTGEN

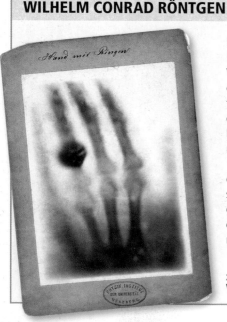

Born into a family of cloth merchants in Lennep, Prussia (now Germany), Wilhelm Röntgen spent part of his childhood in the Netherlands. Far from being a brilliant pupil, he was expelled from school, and found his vocation only after he was taken under the wing of an inspiring tutor.

Although he is best known for his discovery of X-rays, Röntgen studied several areas of physics, including gases, heat transfer, and light. He died of intestinal cancer (not thought to be related to his work with X-radiation).

X-RAY IMAGE OF RÖNTGEN'S WIFE'S LEFT HAND, WITH HER WEDDING RING VISIBLE

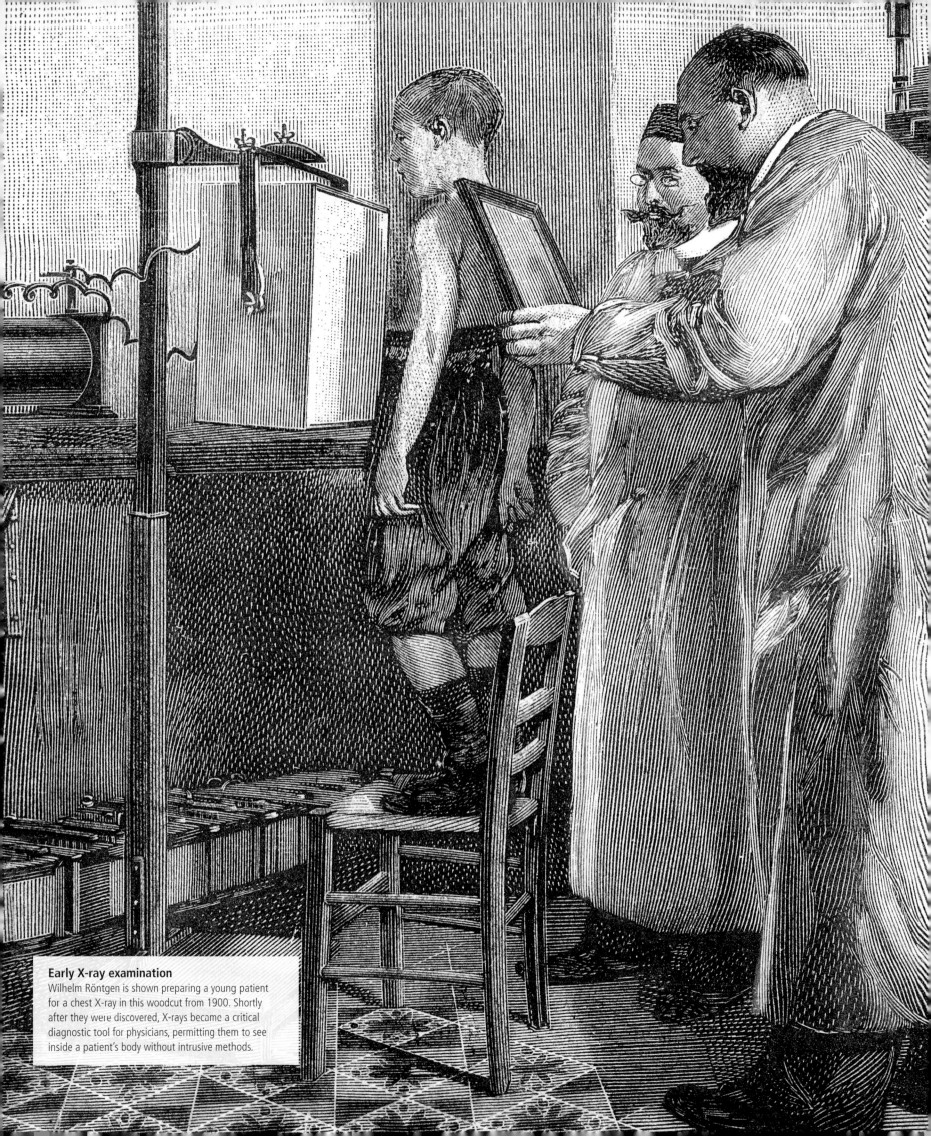

Early X-ray examination
Wilhelm Röntgen is shown preparing a young patient for a chest X-ray in this woodcut from 1900. Shortly after they were discovered, X-rays became a critical diagnostic tool for physicians, permitting them to see inside a patient's body without intrusive methods.

The Struggle Against Malaria

One of the deadliest diseases, malaria has shaped much of world history. It has dictated patterns of migration and settlement, decided wars, and shattered peace. The ongoing search for a malaria vaccine is one of the most intensive in medicine.

Malaria is caused by several kinds of single-celled parasites belonging to the *Plasmodium* genus. It is transmitted when a female *Anopheles* mosquito, having fed on the blood of an infected human, bites a healthy individual. The main symptoms of malaria are flulike and include a high temperature (fever), shaking, chills, headaches, muscle aches, and fatigue. Vomiting, nausea, and diarrhea may also occur. In severe cases, malaria may lead to kidney failure, confusion, seizures, coma, and sometimes death. Symptoms usually start 7 to 30 days after infection, although they can take up to one year to develop. In some forms, the illness recurs for many years, because the parasite can remain dormant in the liver cells.

Herbal treatments for the disease are mentioned in two texts dating back more than 2,000 years: the Chinese *Huang-di Neijing* (*Yellow Emperor's Classic of Internal Medicine*) and the Indian *Susruta Samhita*. The latter asserted that the illness was associated with

220 **MILLION The estimated number of cases of malaria each year.**

insect bites. The ancient Greek physician Hippocrates noted the effects of malaria, and the ancient Romans called it "swamp fever" because they believed that the disease was caused by the noxious fumes of foul-smelling, swampy areas—an idea that became known as the miasma theory (see pp.120–

> ▷ **Malaria parasite**
> The red blobs seen here are egg clusters of the malaria parasite in the mosquito gut. Each cluster produces thousands of infectious, actively moving parasites, which travel to the mosquito's salivary glands and are injected into people when it bites.

21). In the medieval period, this notion remained popular and the name "malaria" came into use, from the Italian for "bad air."

Treatments and causes

One of the first effective malarial treatments was cinchona bark (see pp.88–89), brought back from South America to Europe in the 1630s. The bark's active ingredient was identified as quinine—still a mainstay of malaria medication. However, the cause of the infection was not known until a series of discoveries starting in 1880, when the French army surgeon Charles Laveran found microscopic parasites in the blood of a malaria sufferer.

Around 1886 Italian physician Camillo Golgi showed that there are different kinds of malaria and that fever and chills coincide with the release of the parasite in the blood. In 1890 Italian researchers Giovanni Grassi and Raimondo Filetti identified several kinds of malarial parasites. The same year, Ronald Ross showed that

> **"The belief is growing on me that the disease is communicated by the bite of the mosquito."**
>
> RONALD ROSS, IN A LETTER TO SCOTTISH PHYSICIAN PATRICK MANSON, 1896

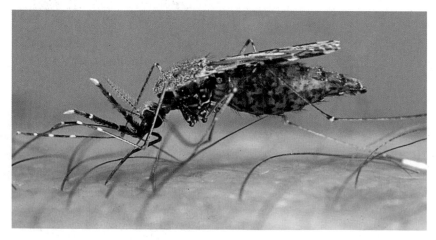

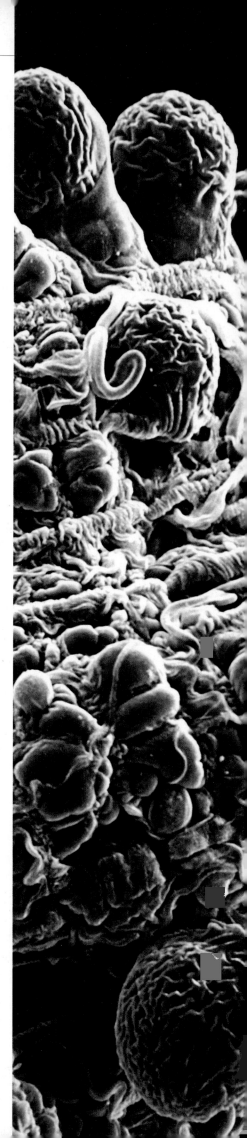

◁ **Malaria vector**
The mosquito shown here is the South American malaria vector mosquito (*Anopheles albimanus*). Although there are approximately 430 *Anopheles* species, only 30 to 40 of these transmit malaria. *Anopheles* mosquitoes are found throughout the world, except Antarctica.

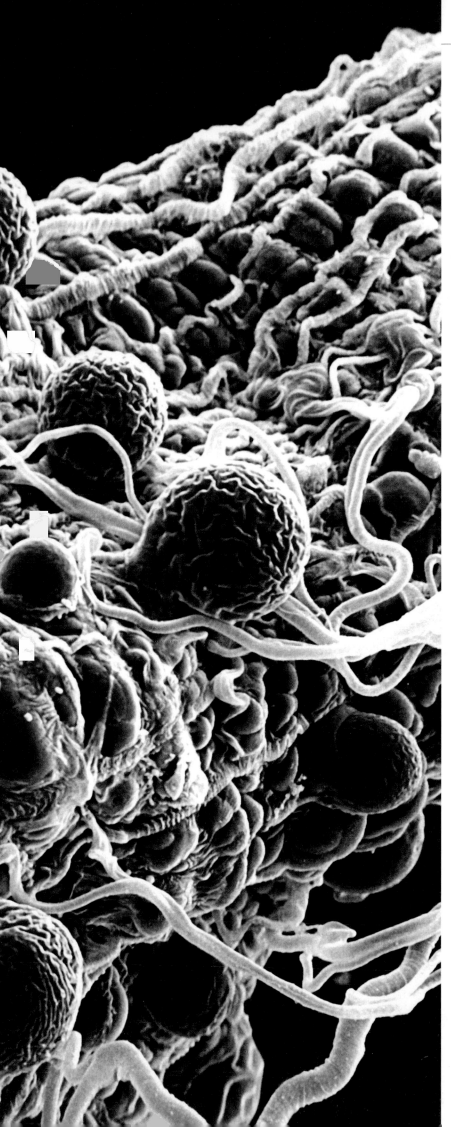

Wooden bellows

Tin-plate nozzle

◁ **Antimalarial spray**
This device was used to spray the insecticide powder Paris Green—a poisonous mix of copper and arsenic. The insecticide was common in the 1940s, until it was discovered to be fairly toxic to plants and damaging to human health.

mosquitoes that bite humans take up the parasite and transmit it between individuals. In 1898–99, Grassi claimed, correctly, that only female mosquitoes of the *Anopheles* genus are the vectors (transmitters) of human malaria.

Tackling the problem

In 1904 the US took over the construction of the Panama Canal after the French had to stop, largely due to massive illness caused by malaria and yellow fever. The US Army initiated a program to drain swamps where mosquitoes bred, to use insecticides, and to protect their workers with mosquito nets, screens, and medicine. As a result, the hospitalization rate for canal workers fell drastically.

In 1939 Swiss chemist Paul Müller discovered that dichloro-diphenyl-trichloroethane (DDT) was a powerful insect-killer, and it quickly

450,000 The approximate number of deaths each year from malaria.

became a global weapon against insect pests and vectors, but when its harmful environmental effects came to light in the 1960s and 1970s it was phased out. Meanwhile, in 1955 the World Health Organization (WHO) set up a campaign to wipe out malaria, using prevention (such as mosquito nets), insecticides, and drug treatments. In the 1980s simple tests were developed to diagnose malaria, allowing outbreaks to be addressed rapidly. In 1981 Chinese pharmacologist Tu Youyou showed that artemisinin was an effective antimalarial treatment.

However, malaria is a complex and persistent disease—further strains and species of mosquito vectors were discovered in the 20th century, and some strains have become resistant to drug treatments. Although many nations are now malaria-free, the infection remains endemic in about 100 countries.

BRITISH PHYSICIAN (1857–1932)

RONALD ROSS

Born in India, Ronald Ross studied medicine at St. Bartholomew's Hospital, London, UK. He joined the Indian Medical Service in 1881 and became interested in malaria in 1892. In 1899 Ross returned to the UK to teach at the Liverpool School of Tropical Medicine and worked as a medical troubleshooter for the government during World War I. He was also the first director of the Ross Institute for Tropical Diseases, London, founded in recognition of his work in 1926.

Transfusion Breakthrough

Today, blood transfusions are everyday procedures, responsible for saving millions of lives. However, it took many failed attempts and false starts—and an important discovery at the turn of the 20th century—before blood tranfusions became a practical reality.

After British physician William Harvey's account of blood's continual circulation (see pp.84–85) was published in 1628, medical minds began to consider the possibility of transferring blood between living beings—both from animals to humans and between humans. However, a major problem observed in the early experiments was that blood tends to clot the moment it is exposed to air. In 1654 Italian physician Francesco Folli wrote that he had managed to transfer blood directly between two patients, using slim tubes inserted into the donor's and the recipient's vessels. However, he did not record his results.

Early developments

In 1665 British physician Richard Lower showed how blood could be transferred between two dogs by joining their blood vessels. In 1667 French physician Jean-Baptiste Denys described using lamb's blood to treat a feverish patient. The same year, Lower and his colleague Edmund King transfused blood from a lamb into an ailing patient; the man survived and said his condition was much improved. Further experiments followed, mainly in France, Italy, and England, but the results were so unpredictable that governments and religious authorities banned the practice.

In 1828 London-based obstetrician James Blundell revived the idea, to treat new mothers who suffered

▷ **Animal–human blood transfusion**
The apparent similarity between the blood of humans and other mammals led to experimental transfusions in the 17th century. Lamb's blood was often used, with the additional aim of conferring youth and vitality to the human recipient.

> "A **single pint** can **save** three **lives...** create a million smiles."

AMERICAN POSTER TO RAISE AWARENESS FOR WORLD BLOOD DONOR DAY, 2012

from excessive bleeding after childbirth. The donor was often a close family member, and the blood flowed directly between donor and recipient. Others developed the procedure by using apparatus such as funnels, syringes, and valves. Again, the results were inconsistent. Attempts to delay clotting using chemicals, in order to do away with the necessity of having a donor next to the patient at the time of transfusion, were also unsuccessful.

The A-B-C-O of blood

In 1875 German physiologist Leonard Landois described the process of mixing blood plasma—the liquid without cells—from one animal with the red cells from another. He observed that this often caused the red cells to clump together (a mechanism known as agglutination) and even burst.

AUSTRIAN-BORN PHYSICIAN (1868–1943)

KARL LANDSTEINER

Born in Baden bei Wien near Vienna, Austria, Landsteiner qualified in medicine at Vienna University in 1891. Five years later, he joined the Vienna Hygiene Institute, where he carried out much of his research into blood. After World War I, he moved to the Rockefeller Institute for Medical Research, New York. He received the Nobel Prize in Physiology or Medicine in 1930 for his "discovery of human blood groups." Landsteiner died of heart failure in New York City in 1943.

In 1895 Karl Landsteiner became interested in immunity and how the body defends itself using antibodies to "fight" alien matter such as invading germs. He focused his studies on blood serum—that is, blood plasma from which cells and clotting substances have been removed. In 1900 Landsteiner began a long and complex series of experiments to see if agglutination took place every time human blood samples were mixed, and noticed

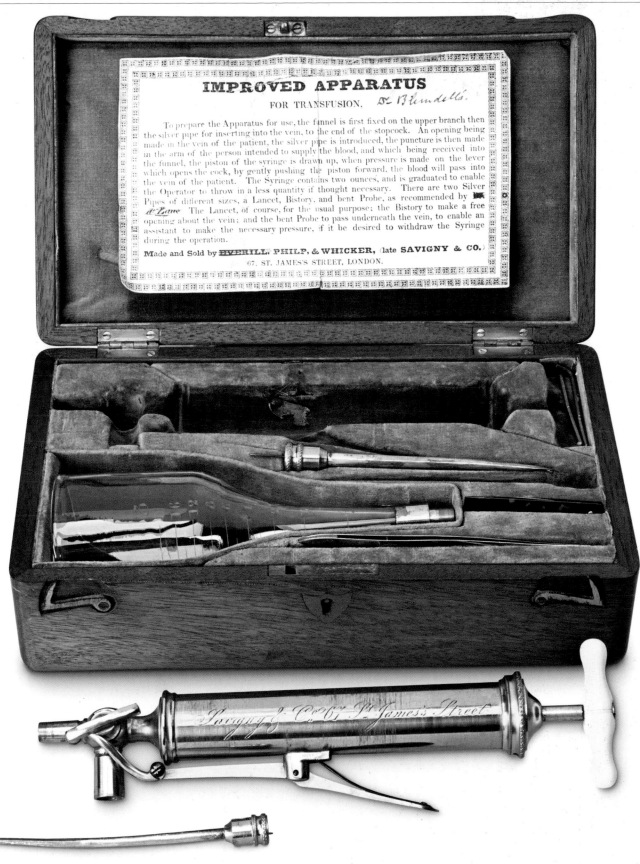

IMPROVED APPARATUS
FOR TRANSFUSION. Dr Blundell's.

To prepare the Apparatus for use, the funnel is first fixed on the upper branch then the silver pipe for inserting into the vein, to the end of the stopcock. An opening being made in the vein of the patient, the silver pipe is introduced, the puncture is then made in the arm of the person intended to supply the blood, and which being received into the funnel, the piston of the syringe is drawn up, when pressure is made on the lever which opens the cock, by gently pushing the piston forward, the blood will pass into the vein of the patient. The Syringe contains two ounces, and is graduated to enable the Operator to throw in a less quantity if thought necessary. There are two Silver Pipes of different sizes, a Lancet, Bistory, and bent Probe, as recommended by Dr d'Tanne. The Lancet, of course, for the usual purpose; the Bistory to make a free opening about the vein; and the bent Probe to pass underneath the vein, to enable an assistant to make the necessary pressure, if it be desired to withdraw the Syringe during the operation.

Made and Sold by EVERILL, PHILP, & WHICKER, (late SAVIGNY & CO.)
67, ST. JAMES'S STREET, LONDON.

◁ **Blundell's apparatus**
James Blundell began transfusions for mothers who bled profusely after giving birth. Blood was drawn from a vein in a donor's arm and then injected directly into an artery in the recipient's arm before it could clot.

the MNSs blood group system, based on antigens found on the surface of red blood cells. By 1937 Landsteiner and US forensic expert Alexander Wiener discovered the rhesus (Rh) factor antigen in blood. Further research has identified more than 30 blood group systems.

Storing blood for transfusion

Although transfusions had become much safer by World War I, the problem of blood clotting while in

108 MILLION The number of blood donations collected globally in 2012—approximately 50 percent from rich countries, home to only 18 percent of the world's population according to the WHO.

storage persisted. In 1914 Belgian physician Albert Hustin found that sodium citrate and glucose worked as anticoagulants. By 1915 German-born scientist Richard Lewisohn, spurred on by wartime casualties, calculated the amounts needed to prevent clotting without posing a risk to the recipient, so by 1916 blood could be stored, taken to battlefield medical units, and transfused ito patients, thus saving thousands of lives. This practice is now routine in every hospital.

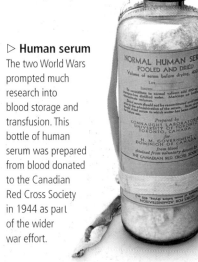

▷ **Human serum**
The two World Wars prompted much research into blood storage and transfusion. This bottle of human serum was prepared from blood donated to the Canadian Red Cross Society in 1944 as part of the wider war effort.

that it did not occur in all cases. In 1901 he described his findings and stated that every individual has blood from one of three different groups, which he called A, B, and C. Landsteiner observed that agglutination occurs when antibodies in blood serum react with the substances called antigens on the surface of the red blood cells. So mixing group A blood from one person with A from another did not result in agglutination and likewise for B and B. Also, he found that group C red cells did not clump when added to either A or B serum as they have no antigens. Group C blood, now know as O, has both anti-A and anti-B antibodies in the plasma and as a result blood from people in this group can be given to any recipient. In 1902 a fourth group called AB was identified.

Rhesus factor

Landsteiner moved to New York and began a collaboration with Russian-born blood specialist Philip Levine and in 1927 they identified

4
ERA OF SPECIALIZATION
1900–1960

ERA OF SPECIALIZATION
1900–1960

1900

1901
Alois Alzheimer writes the first account of a form of dementia that will come to be known as Alzheimer's disease.

1905
Fritz Schaudinn and Erich Hoffmann identify the causative bacterium of syphilis, *Treponema pallidum*.

« *Treponema pallidum*

1901
Karl Landsteiner announces that blood exists in different forms, or groups, initially termed A, B, and C.

1905
Eduard Zirm carries out the first successful cornea transplant.

1903
Willem Einthoven constructs the first practical electrocardiograph machine, or ECG.

❯ Einthoven's string galvanometer—the first ECG

1906
Claudius Regaud discovers a side effect of using X-rays is sterility, which leads him to investigate and begin their use in radiotherapy against cancers.

1910

1910
Hans Christian Jacobaeus carries out early laparoscopic (minimally invasive or "keyhole") surgery on a human patient.

1916
Progress in anticlotting and storing blood allows front-line transfusions for soldiers in World War I.

❮ School gymnasium in the US converted into a hospital during the Spanish flu pandemic

1910
Paul Ehrlich discoveres the first effective treatment for syphilis, Salvarsan.

1918
The influenza (Spanish flu) pandemic spreads. One of the deadliest disease outbreaks in history, it kills around 100 million people.

1920

1921
Bacillus Calmette–Guérin (BCG) vaccine, developed over many years by Albert Calmette and Camille Guérin, comes into use against tuberculosis.

1924
Hans Berger records the first human electroencephalogram, EEG, showing the electrical activity of the brain.

1921
Margaret Sanger and her colleagues found the American Birth Control League, in the campaign to put women in charge of contraceptive use.

1926
Alexander Glenny greatly increases the efficacy of diphtheria toxoid vaccine, although problems remain.

1921
Edward Mellanby shows that lack of newly discovered vitamin D causes rickets.

1921–22
Frederick Banting and Charles Best use pancreas extracts (containing insulin) to treat diabetes in dogs; the method is then applied successfully to treat humans.

❯❯ Banting and Best with the first dog that survived on insulin

1923
George N. Papanicolaou devises the Pap test, or Pap smear, for cervical screening.

1927
Karl Landsteiner and Philip Levine identify the M, N, and P blood groups.

1924
An effective toxoid vaccine is introduced for tetanus.

1928
Lax laboratory practices allow Alexander Fleming to discover the antibiotic penicillin.

Progress in the field of medicine continued in the 20th century with more new vaccines; better, implantable prosthetics; and rapid advances in blood transfusions and other aspects of emergency medicine spurred on by two world wars. World War II also stimulated mass production of penicillin, and further research yielded more antibiotics. The roles of hormones in health and illness became clearer with the introduction of insulin-containing pancreatic extracts to control diabetes in 1922. The lengthening list of medical specialities included geriatrics, primary care, and oncology, with developments in cancer screening, chemotherapy, and radiotherapy.

1930

1940

1950

»

1935
Two early polio vaccines are trialed in the US but fail terribly causing illness, paralysis, and even death.

» Patients at polio treatment and rehabilitation center

1940
The first artificial hip is implanted; the design and materials will be much improved in the 1960s.

⊻ One of the early asthma inhalers

1955
George Maison invents the pressurized metered-dose "aerosol" inhaler (to deliver the same measured amount each time); it is suitable for conditions such as asthma.

1937
Daniel Bovet discovers the antiallergy properties of antihistamines.

1941
The US Blood Donor Service and the American Red Cross Blood Bank are established during World War II.

1950
Richard Lawler carries out the first successful kidney transplant.

1955
R. Adams Cowley begins to promote the term "golden hour" as a working concept in emergency medicine.

1935
The first sulfonamide antibacterial is marketed under the trade name Prontosil.

⊻ Skin-testing kit for allergies

1944
Pyrilamine, discovered by Daniel Bovet, is introduced as an antihistamine medication.

1952
Charles Hufnagel implants the first mechanical heart valve—a ball-in-cage device that he designed himself.

TALIDOMIDA: ¡Justicia!

⋀ Protest against thalidomide

1952
Britain's Royal College of General Practitioners (RCGP) is founded, acknowledging the specialist status of primary care and family doctors.

1957
"Wonder drug" thalidomide is marketed for numerous conditions. But soon its use by women during early pregnancy is linked to babies being born with malformations and disabilities.

1937
Max Theiler and Hugh Smith produce "17D," the first effective vaccine against yellow fever.

1937
The rhesus (Rh) factor blood group is discovered by a team including Karl Landsteiner and Alexander Wiener.

1942
The first antihistamine drugs are developed.

1943
Willem Kolff constructs and tests the first kidney dialysis machines, with little success—this comes two years later.

1945
First widespread use of vaccines against influenza.

1946
After many years of specialist applications, the Pap test enters more general use in hospitals.

1953
Francis Crick and James Watson announce they have worked out the structure of the "molecule of life" DNA—it is a double helix.

1958
Åke Senning inserts the first implantable heart pacemaker—invented by Rune Elmqvist.

AUSTRIAN NEUROLOGIST Born 1856 Died 1939

Sigmund Freud

"**Dreams** are the royal road to the **unconscious.**"

SIGMUND FREUD, *THE INTERPRETATION OF DREAMS*, 1900

F ew medical specialities owe as much to one person as psychoanalysis does to Sigmund Freud. From a form of treatment Freud first used to treat a young woman in the 1880s, psychoanalysis has become a philosophy and a theory of psychotherapy; and it has had a lasting influence on 20th-century thinking. It has spread throughout Western culture into literature, cinema, and theater, and has transformed notions of identity, memory, childhood, and sexuality.

Early years
Freud was born in Pribor (now in the Czech Republic) and his family moved to Vienna when he was 3 years old. He studied at the University

▷ **Father of psychoanalysis**
Freud founded psychoanalysis, a new approach to mind and behavior that greatly affected Western civilization. He is regarded as one of the most influential, and controversial, figures of the 20th century.

▽ **Mind map**
Freud suggested that the human psyche is similar to an iceberg, with the id (the primitive drives) hidden in the unconscious. The ego, which deals with conscious thoughts, regulates both the id and the superego (the critical and judging voice).

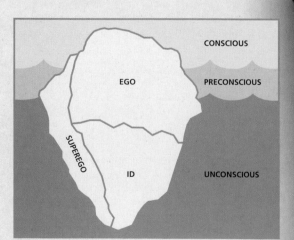

CONSCIOUS

EGO

PRECONSCIOUS

SUPEREGO

ID

UNCONSCIOUS

TIMELINE

- **1885–86** Trains under Jean-Martin Charcot at the Salpêtrière Hospital in Paris, France, where he studies hysteria and the use of hypnosis.

- **1887–1902** Returns to Vienna, Austria. Corresponds with German physician Wilhelm Fliess in Berlin. These letters, published posthumously, reveal his views as he was developing his theories.

- **1888** Drops the use of hypnosis for treating hysteria and turns to free association.

- **1895** Publishes *Studies on Hysteria*, with his friend Josef Breuer, introducing the concept that symptoms of hysteria were symbolic representations of traumatic memories, possibly of a sexual nature.

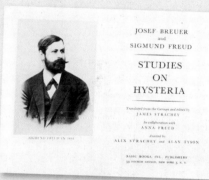

JOSEPH BREUER AND SIGMUND FREUD'S *STUDIES ON HYSTERIA*, 1895

- **1896** Introduces the term psychoanalysis.

- **1900** Publishes *The Interpretation of Dreams*, containing the heart of his theory.

- **1905** *Three Essays on the Theory of Sexuality* charts for the first time the stages of development of the sexual drive in humans from infancy to adulthood.

- **1908** The first meeting of psychoanalysts is held in Salzburg, Austria. Carl Jung and Freud are invited to lecture in the US.

- **1909** Writes his case studies, including his first analysis of a child, Little Hans, 5.

- **1915–17** Sets out the full range of his theories and observations, in a set of 28 lectures delivered at the University of Vienna, outlining his core concepts, including the libido, free association, and his theories of the unconscious.

- **1923** Publishes *The Ego and the Id*; is diagnosed with cancer.

- **1933** Adolf Hitler becomes dictator of Germany; Freud's books are among 25,000 volumes burned in Berlin for being "un-German."

- **1938** Leaves for London, and dies a year later.

of Vienna under German physiologist Ernst Brücke. Later, he became interested in hypnotism and traveled to France in 1885 to study under French neurologist Jean-Martin Charcot (see p.161).

He returned to Vienna and began collaborating with Austrian physician Josef Breuer, who was studying hysteria. Breuer was treating a case of hysteria by placing the patient, Anna O. (see pp.250–51), into a trance

Sometimes these ideas emerged in a disguised form, such as a slip of the tongue, now known as a Freudian slip, or in dreams. To help the release of such repressed thoughts, Freud later developed a technique called "free association," whereby patients could talk of whatever came into their mind and in doing so, provide insights into their unconscious and any repressed emotions or memories. Addressing these would set the patient on the

△ **Analyst's couch**
This room in London where Freud spent his later days houses his couch, where patients would lie down, and talk freely to him, while Freud sat unobserved behind them.

at around 3–5 years of age, when a child is sexually attracted to the parent of the opposite sex, and feels a rivalry with the parent of the same sex. The fear and guilt aroused by such feelings leads the child to repress them, affecting subsequent stages of personality development. However, later psychoanalysts, including Swiss psychotherapist Carl Jung, underplayed this role of sexual drive.

> "There is a **psychological technique** which **makes it possible** to **interpret dreams.**"
>
> SIGMUND FREUD, FROM *THE INTERPRETATION OF DREAMS*, 1900

and encouraging her to talk. He noted that during these sessions, she recalled traumatic events, felt the emotions associated with them, and temporarily lost the symptoms of her condition. This led Freud to theorize that the mind was divided into three levels of consciousness. He concluded that people's behavior was influenced more by their unconscious—buried motives, fears, and wishes—than their conscious rational thoughts.

Freud saw repression as a means by which feelings that are too unbearable are transferred from the conscious mind to the unconscious.

road to recovery. Several breakaway theories have been developed since, but the essence of the "talking cure"—where patients talk out their issues in order to attain well-being—persists.

In 1897 Freud began exploring his own dreams, believing that they had symbolic meaning. He proposed that unconscious wishes have their origins in early childhood, and are linked to the sexual development of a child. He identified a series of psychosexual stages in the development of a child's personality. These included the Oedipus complex, which occurs

Mapping the psyche

In *The Ego and the Id* (1923), Freud divided the personality into three separate but interacting parts—the id, the ego, and the superego. He saw the id as representing the primary source of psychic energy, the ego as using this energy to cope with external reality, and the superego as the controlling and parental influence over the id, making demands on the ego to follow moral goals. Neurosis, Freud believed, was a symptom of the conflict between the three.

Freud's legacy endures. His methods, although modified, are still widely used, amid an ongoing debate about efficacy and whether the "talking cure" is scientific.

The **Development** of the **ECG**

An electrocardiogram (ECG) is a recording of the heart's electrical activity. Today this is usually done by placing sensors on the skin, but early ECGs required room-sized equipment and patients had to place their hands and feet in saltwater to ensure a good electrical connection.

Around 1786 Italian scientist-physician Luigi Galvani noticed that a dead frog's leg would twitch when pieces of metal were applied to it. He believed that this phenomenon was intrinsic to life, and called it "animal electricity." A few years later, Galvani's rival Alessandro Volta showed that an electric current was made by the combination of various metals with the leg, and it was this that stimulated muscle contraction. These findings encouraged further research into the presence and effect of electricity in all manner of living things, from worms to humans.

Early apparatus

Galvani's name lives on as the galvanometer (a device to measure electric current developed in 1820) is named after him. Improvements in the galvanometer's design and sensitivity meant that in 1827 Italian physicist Leopoldo Nobili, from Florence, managed to detect tiny electrical currents, again in dissected frogs. In 1838 one of Nobili's students, Carlo Matteucci, a physics professor in Pisa, Italy, connected a frog's heart to its leg muscle and noticed that it twitched with each heartbeat, indicating that the heart itself actually gave out some kind of electrical activity.

In the mid-19th century, the true nature of electricity and its relationship with magnetism was still a mystery. In spite of this, all manner of electrical batteries, generators, and other machines were being invented, and some of these were marketed as electrical or electromagnetic treatments for therapeutic use on patients. These devices sent small amounts of "tingling" electrical current or

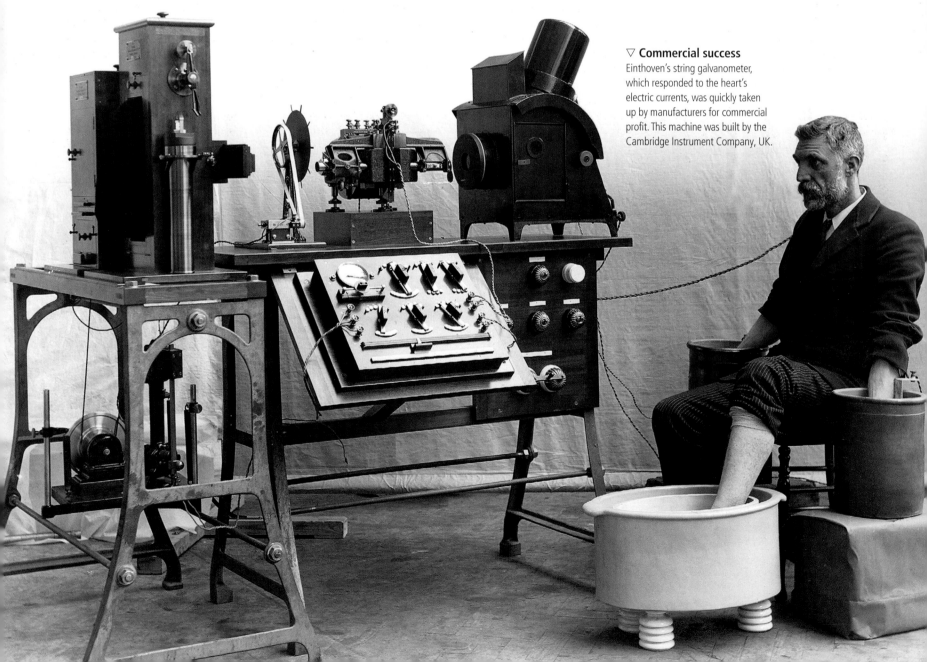

▽ **Commercial success**
Einthoven's string galvanometer, which responded to the heart's electric currents, was quickly taken up by manufacturers for commercial profit. This machine was built by the Cambridge Instrument Company, UK.

"An **instrument** takes its true value... from the **work** it really **does.**"

WILHELM EINTHOVEN, IN A LETTER TO ENGLISH
CARDIOLOGIST THOMAS LEWIS, 1922

even larger, more painful shocks through the body. None of the machines came into general use at the time, although they did eventually lead to the invention of the heart defibrillator (see p.206).

Reading the heart

In 1843 German physiologist Emil du Bois-Reymond detected a small electrical potential, or voltage difference, in resting animal muscles and saw how it changed when the muscle contracted. He called this phenomenon "action potential.". By 1856 direct readings from the exposed hearts of animals indicated varying electric currents with each beat. Additionally, reports began to appear describing patients whose breathing and/or heartbeat had stopped but were resuscitated when shocks of 300 volts or more were

0.25
MV The voltage of a P wave of an ECG—equivalent to 1/400 of an electrical volt.

applied to their chests. In 1887 at St Mary's Medical School, London, British physiologist Augustus Waller published an article titled *A Demonstration on Man of Electromotive Changes Accompanying the Heart's Beat*, which described what is regarded as the first human ECG. It used leads connected to sensors on a patient's hands and feet, rather than directly to an exposed heart. However, the procedure was complex and not suited to practical use.

In 1890 British physician George J Burch, based in Oxford, published several articles showing how variations in electricity too rapid to be recorded by a galvanometer could be worked out using

calculations and graphs, which gave an insight into the true wave pattern of the heart's electrical activity. The following year British physiologists William Bayliss and Edward Starling, at University College London, enhanced the technique and linked these electrical changes to phases of the heart's contraction and relaxation.

Dutch physiologist Willem Einthoven had seen Waller demonstrate his early ECG at the First International Congress of Physiologists in Basel, Switzerland in 1889. In 1893 Einthoven coined the term "electrocardiogram," and announced further progress in a report called *New Methods for Clinical Investigation*. Over the following years Einthoven developed the machinery, recording methods, and analysis of ECGs for both healthy and diseased hearts.

Putting ECGs to use

From about 1910 heart specialists began using ECGs to diagnose heart conditions such as atrial fibrillation (fast, erratic "trembling" of the upper chambers), angina (disorder

WILLIAM EINTHOVEN

Born in Java in the Dutch East Indies (now Indonesia), William Einthoven studied medicine at the University of Utrecht in the Netherlands. He was appointed professor at Leiden University in 1886.

Einthoven made the ECG a practical reality by combining several different innovations. In 1895, using an improved galvanometer and new correction formulas, he identified five peaks and troughs, or "waves," in the heart's electrical activity, which he called P, Q, R, S and T—the letters following O, which is the origin, or bottom left corner, of a graph.

In 1901 Einthoven invented a new apparatus—the string galvanometer, which included a very thin wire of silver-coated quartz positioned between powerful electromagnets. Changing currents in the wire caused movements that could be seen when a projection microscope directed them

onto a strip of passing photographic paper for continuous recording. In 1906 Einthoven published the first series of normal and abnormal ECGs for 10 heart conditions. He was awarded the Nobel Prize in Physiology or Medicine in 1924 for his discovery of the mechanism of the ECG. He died in 1927 in Leiden.

caused by restricted blood flow to the heart), and acute myocardial infarction (commonly known as MI or a heart attack). As this knowledge was obtained without invasive procedures, it allowed for much better treatment, and even prevention, of existing disorders. The drawback was that the early machines were cumbersome and required a dedicated room. The first portable ECG machines,

powered by a vehicle battery, arrived in 1928. However, they weighed more than 44 lb (20 kg) and were therefore unwieldy. The invention of increasingly small transistor electronics led to desktop versions by the 1960s and, more recently, compact electronic systems and microchips have allowed for the development of ECG recorders that can easily fit in the hand.

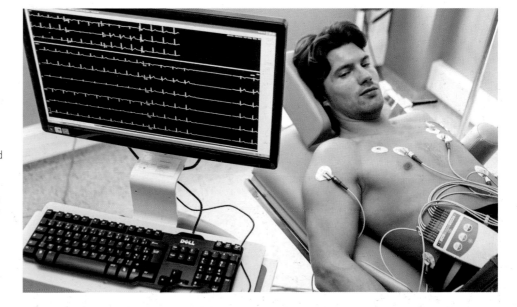

▷ **Modern ECG**
Sensors can be attached to patients to give an instant reading or a person can be fitted with an ambulatory device—an ECG machine small enough to wear. The "portable" device can store recordings of the heart for up to two weeks as the wearer goes about their normal daily activities. The readings are then downloaded to a computer for analysis—as shown here.

A **Cure** for **Syphilis**

In 1495 an unfamiliar disease swept through Europe, which was transmitted sexually or from mother to child in the womb. Causing painful sores, madness, and death, syphilis exacted a terrible toll for over four centuries. Then, in 1909 an effective treatment was found, relegating the "Great Pox" to a minor, yet serious, disease.

An extremely virulent disease, syphilis claimed the lives of thousands. Victims of the disease suffered disfiguring and painful pus-filled sores and skin ulcers, and in extreme cases soft tumors that ate into the flesh and bones. Many died of the disease, and those who did not were left with scars and were often disfigured. Sufferers were stigmatized, doubly so when it became apparent that the disease was sexually transmitted.

Questionable origin

A number of theories exist regarding the origin of syphilis. At first it was believed to have come via mercenaries who had fought in

▽ **Pleasure and pain**
This 17th-century engraving is a wry comment on the sexual nature of contracting syphilis. It points out the contrast between the "one pleasure" the victim may have, with the "thousand pains" that will follow.

the French invasion of Naples, Italy. In France it was called the "Italian disease," and in Italy, it was thought to be of French origin—no nation wished to be known as its birthplace. Another popular theory maintains that syphilis was brought back to Europe by sailors returning from Columbus's first voyage to the Americas in 1493. This is supported by the widespread evidence that shows the presence of diseases of the treponemal family—which includes syphilis—in ancient American skeletons.

By 1502 syphilis had spread to Central and Northern Europe. However, by this time it had apparently mutated into a slightly milder form. It was common in European life, affecting about 10 percent of the adult population at any one time. Physicians devised many treatments but none was reliable. Of all the available remedies, mercury was the preferred drug. It was believed that the sweating it induced would

expel the toxins of the disease from the body. Unfortunately the severe side effects of mercury, such as mouth ulceration, tooth loss, and bone deterioration were often confused with the original symptoms of syphilis. So, many patients suffered from needless additional pain as a result.

Identifying syphilis

Attempts to identify the cause of the disease were hampered by its early confusion with leprosy, which causes similar disfiguration. From the 16th to the 19th centuries, syphilis was also confused with gonorrhea, another sexually transmitted disease. It not until 1837 that French venereologist Philippe Ricard confirmed that syphilis and gonorrhea were two distinct infections.

Ricard also identified three stages of syphilis: the primary stage, in which canker sores (small, painless, nodular growths or ulcers) appear at the site of contact; the secondary stage in which a more general inflammation, flulike symptoms, and a rash occur; and a tertiary stage, in which soft tumors may appear, and the nervous system

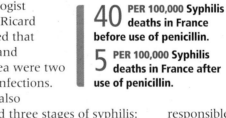

40 PER 100,000 **Syphilis deaths in France before use of penicillin.**

5 PER 100,000 **Syphilis deaths in France after use of penicillin.**

△ ***Treponema pallidum***
Syphilis is caused by a thin, spiral-shaped organism that spreads through sexual contact. The name "pallidum" comes from the very pale colour of this bacillus. Its paleness made it very difficult to observe and hence to discover.

can be affected, causing blindness and insanity. Ricard also observed that the third stage of syphilis may even occur after years of remission. By 1876 doctors had also identified cardiovascular syphilis, which spreads through the blood, but they were still no nearer to finding the microorganism responsible for the disease, let alone a cure for it. In the absence of a cure, a public health approach was adopted to try to control the spread of syphilis. In 1864 the Contagious Diseases Act was passed in Britain calling for the regular examination of prostitutes, and their necessary detention and treatment if they were found to be infected.

"A disease so cruel, so distressing... nothing more terrible or disgusting, has ever been known on this earth.**"**

JOSEPH GRÜNPECK, FROM *THE PUSTULAR EPIDEMIC "SCORRE," OR THE FRENCH SICKNESS*, 1496

The search for the cure

In 1905 German zoologist Fritz Schaudinn finally identified the microorganism that caused syphilis (*Treponema pallidum*), and the race to find a cure began.

The decisive step toward a cure was made by German immunologist Paul Ehrlich. He began looking at different variants of arsenic, seeking what he called a "magic bullet"—a term he coined to describe a drug that would target a disease without affecting any other parts of the body. In doing this he was initiating the concept of chemical therapies, now known as chemotherapy.

In May 1909 Sahachiro Hata, a Japanese researcher working in Ehrlich's laboratory, was testing the 606th arsenical compound and found that it worked on *Treponema pallidum*. Human testing of the drug—named Salvarsan—rapidly followed, and within a year, it was publicly available. This was the first drug to target a particular pathogen.

Salvarsan, and its variant Neosalvarsan—introduced a few years later—remained the principal anti-syphilitic drugs for two decades, until 1943, when it was found that the new antibiotic, penicillin (see pp.198–99), was even more effective.

The invention of the Wasserman test in 1906, named after its inventor, German bacteriologist August Paul von Wassermann, also aided treatment. It tested blood to see if the syphilis bacterium antibody was present, and with this test it was possible to identify victims of the disease even if there were no symptoms.

Controlled but not cured

Syphilis infection rates began to fall rapidly once penicillin became widely available and reached a low point in the 1950s. Since then, the number of syphilis cases around the world has fluctuated, with a general increase during the 2000s. The disease remains a serious public health problem, with more than 110,000 deaths worldwide in 2010, while patients with tertiary syphilis continue to suffer debilitating long-term damage.

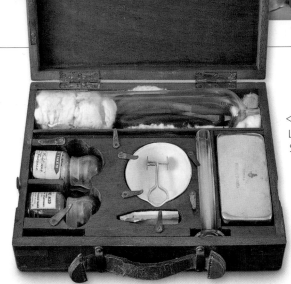

◁ **The Salvarsan kit**
Laboratory tests found the drug Salvarsan to be so effective against the syphilis bacteria, and there was such a high demand for the medicine, that Paul Ehrlich was forced to make it commercially available before any further tests. This Salvarsan kit is from the year 1910.

▷ **Encouraging treatment**
The drug Salvarsan had a number of toxic side effects and patients had to continue treatment for two years. Only around 25 percent of patients actually completed the course, which greatly reduced its effectiveness.

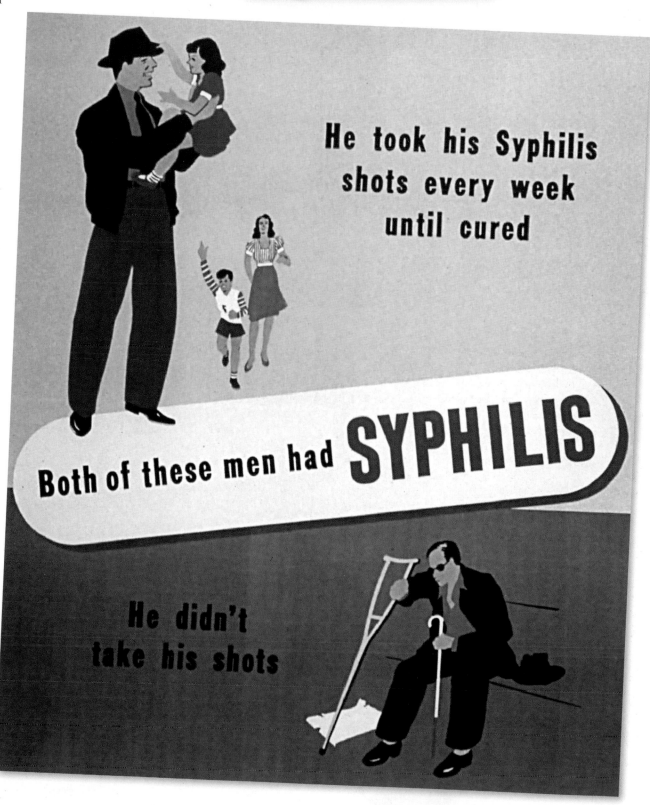

He took his Syphilis shots every week until cured

Both of these men had SYPHILIS

He didn't take his shots

Minimally Invasive Surgery

Surgeons in ancient times had no alternative but to make large openings in the body to access the operating area. Improvements in equipment and techniques gradually reduced the size of these cuts. In the 20th century new technology enabled surgery through tiny "keyhole" incisions.

Bodies from the Neolithic period show clear evidence of a form of surgery known as trepanning (see pp.16–17), which involved cutting a small hole in the skull. The ancient Greeks and Romans are known to have practiced basic surgical procedures too. Although advances were made over the ensuing centuries, most notably in the areas of antisepsis and pain relief, the premise remained the same—cutting the body open to perform surgery, then stitching it up again.

A surgical revolution

Carrying out surgery without significantly cutting up the body—a procedure known as minimally invasive surgery (MIS)—first became a reality in the 20th century. Unlike traditional open surgery, which requires at least one large cut to access the operation site, minimally invasive surgery involves making a small incision just large enough to insert a miniature light source and imaging device (usually fiberoptic), as well as

△ **The "Lichtleiter" (light conductor)**
The first endoscope with internal illumination was invented by German surgeon Philipp Bozzini in 1806. It consisted of an aluminum tube containing a candle and a mirror, which reflected light from the candle into the body, helping physicians see the organs better.

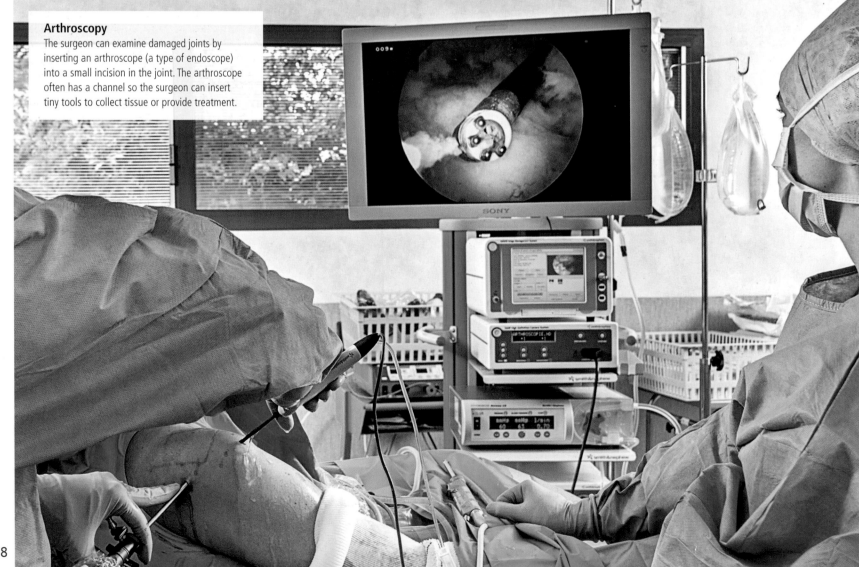

Arthroscopy
The surgeon can examine damaged joints by inserting an arthroscope (a type of endoscope) into a small incision in the joint. The arthroscope often has a channel so the surgeon can insert tiny tools to collect tissue or provide treatment.

small surgical tools. This is also known as keyhole surgery, since the incisions are similar in size to a keyhole (¼–½ in/0.5–1.5 cm wide). The imager is connected to a high-definition monitor, through which the surgeon and supporting medical team can view the area that is to be examined and treated. The advantages of such surgery include less pain for the patient, quicker recovery time, minimal scarring, and a reduced risk of infections and other complications.

Minimally invasive surgery is now practiced regularly in hospitals throughout the world, having replaced traditional surgery for a variety of procedures such as gallbladder and kidney removal, tumor removal in the head, neck, lungs, bladder, and uterus, and repair of hernias and heart defects.

Looking inside the body

Minimally invasive surgery has only been made possible by the invention and development of the modern endoscope. An endoscope is a long, thin, flexible tube, with a powerful light and a miniature imager attached, which is inserted through a natural opening, such as the mouth or anus, or via a small

GERMAN SURGEON (1866–1945)

GEORG KELLING

Born and educated in Dresden, eastern Germany, Georg Kelling started studying medicine in 1885 at the University of Leipzig and later—due to his military service—at the University of Berlin.

Guided by leading scientists, Kelling got his medical doctrate in 1890 then became a surgeon at Dresden hospital, where he specialized in gastrointestinal conditions. In his attempts to better understand the problems he was treating, he performed the world's first laparoscopy—a procedure he called "celioscopy." He performed surgery on the abdomen of a living dog, using a technique called "insufflation" to inflate the abdomen before inserting a cystoscope—a tubular device with a magnifying lens and a light source—into the abdominal wall. Kelling's innovation laid the foundation for the modern era of minimally invasive surgery. He is also credited with the invention of the esophagoscope—an endoscope to inspect the food pipe (esophagus). He and his wife were both killed in the World War II Allied bombing of Dresden in 1945.

incision made in the skin. The camera sends live images to a screen in the examination room or operating room so that medical professionals can see exactly what is happening inside the body.

The idea of using endoscopic techniques is not new—the writings of Hippocrates in the 4th century BCE (see pp.36–37) indicate that the ancient Greeks used tools to carry out internal inspections. However, the first significant advances did not take place until the 19th century.

One of the principal challenges in endoscopy was to provide a light source that could illuminate the dark recesses inside the body. In 1806 German army surgeon Phillip Bozzini invented the "Lichtleiter" (see opposite). However, the device was difficult to work and became very hot, so it was not used on patients in Bozzini's lifetime. By 1853 French surgeon Antoine Desormeaux had made variations to the Lichtleiter (renaming it the "endoscope") and it was used for the first time on a patient. However, like Bozzini's version, the risk of burning the patient meant it was not considered practical for long-term use.

A considerable breakthrough in modern endoscopy came in 1878, when German urologist Maximilian Carl-Friedrich Nitze presented the first working cystoscope—a long, tubular device with built-in electric light and magnification, used to view inside the bladder. The invention of the incandescent light bulb in the 1870s and 1880s allowed further improvements to the cystoscope.

Developments in surgery

In 1901 German surgeon Georg Kelling used the Nitze cystoscope to perform the first abdominal keyhole operation—a procedure that would later become known as laparoscopy. The pioneering operation was carried out on the abdomen of a dog, and Kelling later used the same technique on two human patients.

Kelling's groundbreaking surgery led to a number of new devices and techniques for minimally invasive surgery. In 1938 Hungarian medical intern Janos Veress created a spring-loaded needle for draining fluid from surgical sites and sucking out air and fluid from the chest. In 1970 American doctor Harrith Hasson developed a technique that enabled an even smaller incision site for laparoscopic surgery. A decade later, the 1980s ushered in the age of video laparoscopy, with high-quality miniature cameras or other imagers used for the first time.

One of the greatest advances in minimally invasive surgery in recent years has been the use of robots (see pp.254–55). This involves a surgeon directing surgery on a computer console, while looking at a high-definition monitor, with the robot carrying out instructions. Surgery has never been safer for the patient, and remarkable technological developments continue.

> ## "Laparoscopy requires a different **skill set** than open surgery."
>
> M. CENK ÇAVUŞ OĞLU, DIRECTOR, MEDICAL ROBOTICS AND COMPUTER INTEGRATED SURGERY LABORATORY, CASE WESTERN RESERVE UNIVERSITY, OHIO, 2006

1910 The year that Swedish surgeon Hans Christian Jacobaeus performed the first laparoscopic surgery on a human.

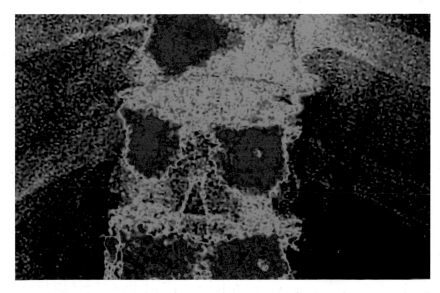

◁ **Vertebroplasty**

During vertebroplasty, a procedure that is used to treat fractured bones in the spine and spinal compression, the surgeon makes a small hole in the skin and injects a bone cement mix (shown here in red) into the affected bone.

Diabetes and **Insulin**

A disease of the endocrine glands in the pancreas, diabetes was known to physicians for thousands of years. However, it was not until the discovery of the hormone insulin in 1921 that it was possible to treat people with this condition.

The first reference to diabetes appears in an Egyptian papyrus dating back to c.1500 BCE, where it is described as a disease of "too great emptying of the urine." The condition of excessive urination, or polyuria, was also observed by ancient Greek physician Aretaeus of Cappadocia, who noted the excessive thirst that the disease caused. In the 6th century Indian physician Susruta identified the characteristic sweet or honey taste of diabetic urine. However, the awareness of these symptoms was not accompanied by any remedy, and diabetics usually died young.

The sweetness of diabetic urine was rediscovered in the 17th century by British physician Thomas Willis—who gave the disease the name diabetes mellitus after the Latin word *mel* meaning

"honey." However, it was not until 1776, when Liverpool-based British physician Matthew Dobson discovered that the evaporated urine of a diabetic left a sugarlike residue, that physicians began to understand that the condition was associated with an excess of blood sugar. This residue was confirmed to be glucose in 1815.

Link with pancreas

The cause of diabetes, however, remained unknown. In 1673 Swiss scholar Johann Brunner found that a dog suffered from polyuria after removal of its spleen and its pancreas—an organ in the upper left abdomen that helps digestion by secreting enzymes that assist in the breakdown of nutrients in food. The procedure was repeated in 1889 by German physiologist Joseph von Mering and Lithuanian pathologist Oskar Minkowski, who found that removal of the pancreas led to diabetes in dogs. However, they also found that when a portion of the organ was replaced, the disease subsided.

In 1884 Friedrich von Frerichs, a German biochemist, found that one-fifth of diabetic patients had lesions or other damage in the pancreas, again identifying it as

◁ **Canine insulin**
American-born scientist Charles Best (left) and Canadian physician Frederick Banting removed the pancreas from dogs, making them diabetic; injecting the dogs with insulin allowed them to survive. In 1923 only Banting received a Nobel prize for this work.

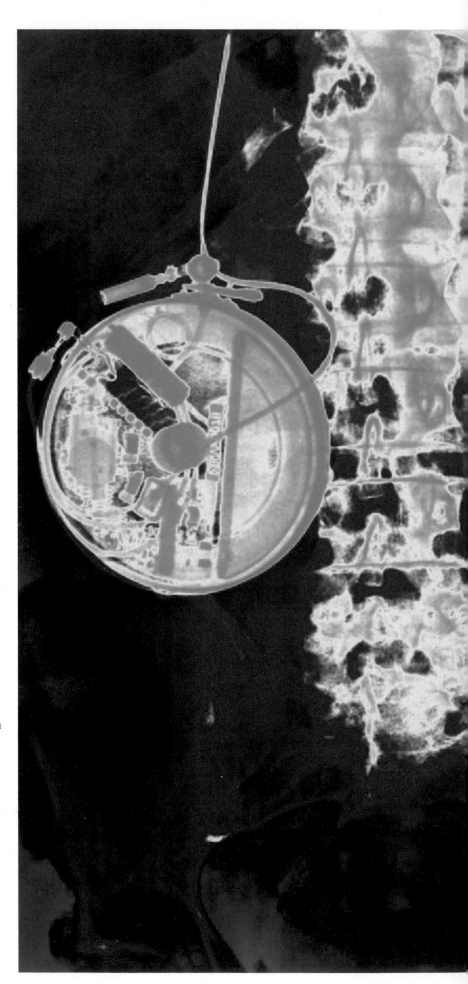

crucial to the development of the disease. Then, in 1893, French pathologist Gustave-Édouard Laguesse made the critical link between a hormone secreted by Islets of Langerhans, a glandular system within the pancreas, and the regulation of blood sugar levels. But the hormone itself remained elusive and therapies for diabetes concentrated on controlling a patient's blood sugar level—a task made easier in 1841 by German chemist Hermann von Fehling's invention of a blood glucose test.

Physicians tried diet to moderate the effects of diabetes, partly in response to the discovery in 1861 by British physician Frederick Pavy, that low-carbohydrate diets could reduce glucose levels. But results were mixed and patients also suffered nerve and eye damage, circulatory disorders—which sometimes led to amputation—and, in extreme cases, coma.

Extracting insulin

In 1921 Canadian orthopedic surgeon Frederick Banting and his assistant Charles Best derived an extract—the hormone insulin—from the excised pancreases of dogs. When the insulin was given to rabbits, their blood sugar level

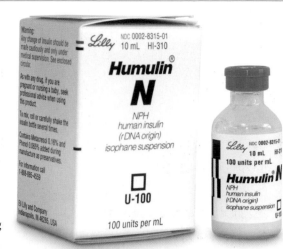

◁ **Human insulin**
The first human insulin was created in 1981 using genetically modified *E.coli* bacteria. By the 1990s both fast- and slow-acting versions were available, enabling more effective management.

decreased. Insulin therapy was first used on a human—a 14-year-old diabetic boy—in Toronto General Hospital in January 1922. After purifying the insulin further, it lessened his symptoms. It was now clear that the insulin prevented

24 hours became available in 1936. Initially insulin was synthesized from animal proteins, but in 1981 human insulin was synthesized.

The greater need now is for patient education and long-term care to help diabetics avoid the

> "Being a **melting down** of the limbs and flesh into urine… the patient **never stops making water…**"
>
> ARETAEUS OF CAPPACDOCIA, ANCIENT GREEK PHYSICIAN, 2ND CENTURY CE

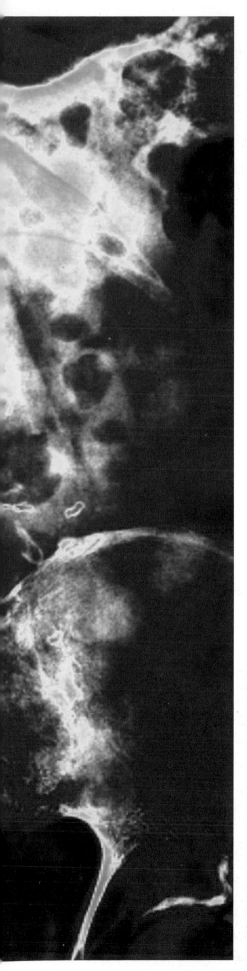

◁ **Artificial pancreas**
Pancreatic transplants in diabetic patients began in 1966, but had limited success. This led to the development of a subcutaneous insulin infusion pump, or "artificial pancreas", in 1978. By 2000, these had over 200,000 users worldwide.

excessive buildup of sugar and within days its use spread.

In 1936 British scientist Harold Himsworth made further strides in understanding the mechanism behind diabetes when he described the difference between Type 1 and Type 2 diabetes (see panel, below).

New challenges

Improvements were made over the decades in insulin production. Longer-acting insulin that lasted

complications associated with the disease. By 2014 around 400 million people worldwide were estimated to have diabetes, causing up to 2 million deaths annually; 90 percent of these people had Type 2 diabetes. These numbers have quadrupled since 1980. With affluent lifestyles expected to increase the number of people with Type 2, and as yet no cure, diabetes is one of the world's greatest medical challenges.

CONCEPT

DIABETES (TYPE 1 AND 2)

In Type 1 diabetes, the body cannot produce sufficient insulin to break down blood glucose. It is caused by damage to the insulin-secreting endocrine glands in the pancreas— possibly as a result of a viral infection. The excess sugar can cause nerve damage. In Type 2 diabetes, lifestyle factors such as obesity result in an excess of blood sugar, which over time, makes the body resistant to insulin, and therefore, permanently unable to process glucose effectively.

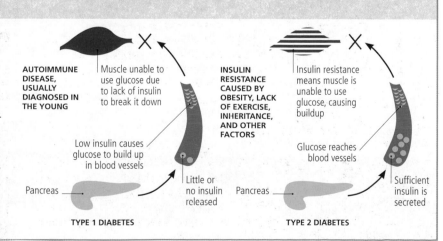

AUTOIMMUNE DISEASE, USUALLY DIAGNOSED IN THE YOUNG

Muscle unable to use glucose due to lack of insulin to break it down

Low insulin causes glucose to build up in blood vessels

Pancreas

Little or no insulin released

TYPE 1 DIABETES

INSULIN RESISTANCE CAUSED BY OBESITY, LACK OF EXERCISE, INHERITANCE, AND OTHER FACTORS

Insulin resistance means muscle is unable to use glucose, causing buildup

Glucose reaches blood vessels

Pancreas

Sufficient insulin is secreted

TYPE 2 DIABETES

War and Medicine

During World War I (1914–18), medicine made progress on several fronts. But it struggled to keep pace with the harm caused by new armaments and weaponry, in particular the dreadful and indiscriminate toll of gas, chemical, and germ warfare.

With germ theory (see pp.146–47) and the causes of infections well established, World War I became the first major conflict where more soldiers, and other military, died in battle than died from nonviolent causes, such as contagious diseases and starvation. This war incurred high casualties, with close to 20 million dead, and a similar number wounded, two-thirds in action.

During the war, efforts were made to limit infections such as tetanus and typhoid. New technology also brought momentous changes. Motorized transportation allowed the wounded to reach newly mobile medical facilities swiftly, ushering in a new era in emergency care (see pp.256–57). The first extensive use of X-rays (see pp.172–73) enabled medical personnel to locate bullets and shrapnel in the body for fast removal. Blood storage and transfusions, developed in the last decade, also advanced speedily.

With so many men involved in fighting, women's roles in the general workforce proliferated. They contributed to the war effort, serving as ambulance drivers and messengers near the frontline, orderlies and nurses in field hospitals, and caregivers and rehabilitation specialists back home.

The end of war in 1918 coincided with the global spread of the influenza pandemic (see pp.196–97). Indeed, the war may have helped it spread, through crowding, malnourishment, and opportunities for the virus to mutate. By 1920 this pandemic had claimed perhaps twice as many lives as World War I.

"**Wars...** have furthered the progress of the **healing art.**"

EMIL GEIST, AMERICAN PHYSICIAN, FROM *JOURNAL OF THE AMERICAN MEDICAL ASSOCIATION*, 1919

▷ Gas attack
The use of poison gases during World War I posed a health threat to soldiers and civilians alike, prompting research into respiratory medicine and the treatment of chemical burns. Here German nurses treat victims of an allied gas attack at the Western Front.

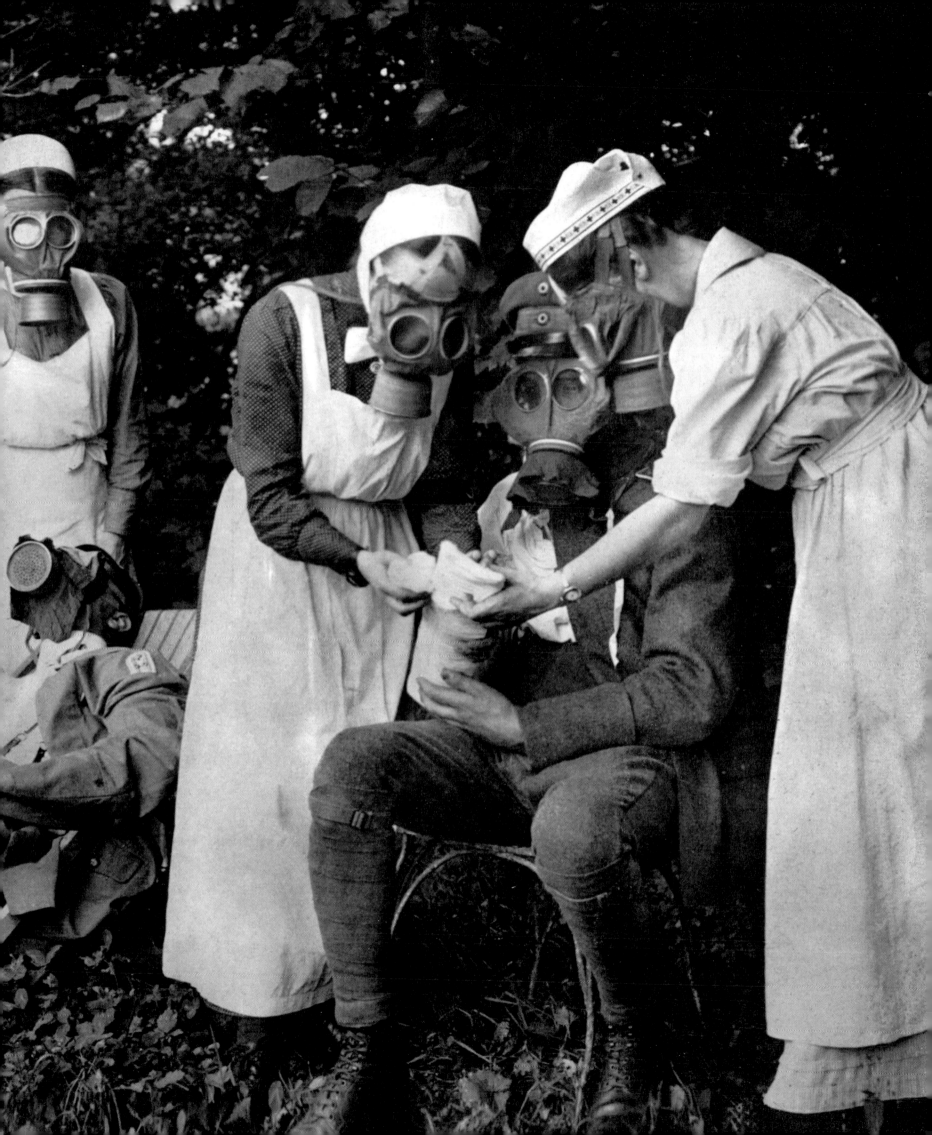

Battlefield Medicine in World War II

Medical innovation raced ahead during World War II. Progress was made in the fields of infection prevention, drug therapy, treatment for medical shock, rapid trauma surgery, and the speedy transportation of casualties by employing a highly organized system.

It is often said that wartime sees more deaths from disease than from military action. During World War II (1939–45), millions lost their lives to infections such as malaria (see pp.174–75). Supply shortages and enemy blockades hampered the availability of the antimalarial drug quinine, increasing the spread of the infection among Allied troops (the US, UK, France, and others). A synthetic version had been invented in Germany in the 1930s, so those tablets were made available to the troops instead. The tablets had a very bitter taste, and occasionally caused headaches and vomiting; however, they were also effective and Allied soldiers had orders to take their doses.

Typhus, which is spread by lice, was another infection risk, mainly in Europe and North Africa. The US Typhus Commission was formed in 1942 to research the prevention and treatment of the disease. The commission supplied three million doses of the vaccine, used insecticides such as DDT (dichlorodiphenyltrichloroethane) to treat military personnel, their kits, and camps, and circulated information about conditions that encouraged typhus such as poor sanitation and decaying refuse.

Penicillin was successfully being used as an antibiotic toward the end of the war (see pp.198–99). An earlier generation of antibiotics called sulfonamides (see pp.200–01) was also used. Many soldiers were given a first-aid pouch containing antibacterial sulfa powder

△ **Popular painkiller**
Processed from the poppy plant, morphine was widely used as an analgesic during World War II. To prevent accidental overdose or addiction, a syrette—a small tube with a measured dose that could be injected at the battlefront—was used to administer the drug. It was then attached to the casualty's collar to show the dose taken.

for treating open wounds and a bandage for protection. Frontline first aiders and combat medics carried both tablets and the powder.

Life-saving blood
Rapid advances were made in stored-blood transfusion (see pp.176–77), which had first been used during World War I. Techniques were developed that meant blood could be separated into its constituents, such as the liquid portion, or plasma, and red cells. The plasma lasted longer in storage, was easier to transport, and for many purposes was as effective as whole blood. The next major development was dried plasma, which could be

reconstituted with distilled water. As the war progressed, US researcher Edwin Cohn devised a process to purify serum albumin—the main protein in blood, and one that is essential for healthy blood volume, blood pressure, and tissue supply. Transfusions of these blood products (known as fractions) saved thousands from medical shock due to severe blood loss, which if left untreated was fatal.

New blood donation and transport routines were also huge life-savers. The 1940 "Blood for Britain" campaign encouraged civilians to donate blood, and the American Red Cross collected blood from donors in New York City to export

△ **Vital jobs**
After returning home, many disabled war veterans continued to support the war effort, for example, by producing artificial limbs that they themselves tested and improved. War stimulated huge improvements in reconstructive and plastic surgery (see pp.238–39).

> **"If I could reach all America… I would… thank them for blood plasma and whole blood."**
>
> DWIGHT D. EISENHOWER, SUPREME COMMANDER OF THE ALLIED FORCES IN EUROPE DURING WORLD WAR II, AND PRESIDENT OF THE US (1953–61)

its plasma to the UK. The US Blood Donor Service and the American Red Cross Blood Bank were established in 1941.

Battlefield care systems

By the end of the war, the Allies had established a massive military machine for medical care, which extended all the way from the battlefield to hospitals back home. Combat medics provided first aid just behind the front line. Mobile aid stations, located further into Allied territory, received casualties on stretchers. At the clearing stations, still further back—staff checked field dressings, provided pain relief and blood transfusions, and other emergency care. The use of triage assessment (see pp.256–57) ensured that those needing further attention were moved on to mobile field hospitals.

In August 1945 the US established their Mobile Army Surgical Hospital (MASH) units. These played major roles in the Korean War (1950–53) and Vietnam War (1955–75), when helicopter air ambulances (see pp.256–57) revolutionized casualty transport. MASH-type units were used by many other nations and led to the Combat Support Hospitals (CSH) that are still in use today.

▷ **MASH in Korea**
A sergeant selects blood for transfusion to a patient at a MASH unit during the Korean War. A seriously wounded soldier had a 97 percent chance of survival at such hospitals during the conflict.

▽ **On the battlefield**
Front-line blood, plasma, and serum albumin transfusions helped stabilize casualties for transport to proper facilities. Here, a wounded soldier is shown receiving a transfusion before being evacuated to a hospital ship during the Allies' Normandy landing in 1944.

Influenza and the Pandemic

The influenza pandemic of 1918–19 crossed international barriers to become one of the most widespread disasters of the 20th century. It infected some 1 billion people, killing an estimated 50 million in a single year, and wiping out 6 percent of the world's population.

Private Albert Gitchell reported to military medics at the US Army base at Fort Riley, Kansas, feeling unwell on March 11, 1918. He said he had a bad cold, was achy and feverish, with a burning throat and bad cough. He was quarantined in a tent for soldiers suffering from infectious diseases. However, by lunchtime, 107 soldiers were unwell, all complaining of the same symptoms. The disease spread rapidly—by the end of the week 522 men were affected. Far from being a common cold, these men were suffering from a virus called influenza, or "flu."

Influenza was not a new phenomenon, but this strain of the virus—H1N1—was particularly virulent and lethal. It caused an array of violent symptoms rarely seen with the disease, such as bleeding from mucus membranes, especially the nose, ears, stomach, and intestines. The disease also made its victims vulnerable to bacterial infections and many of them died of pneumonia.

From epidemic to pandemic

As soldiers made their way to the battlefields of France during World War I, and then back home, the virus traveled swiftly to all corners of the globe. It became a pandemic, an infectious disease that spreads over a wide geographical area. It later became known as "Spanish flu" because the Spanish press was the first to report it widely.

Doctors tried all known methods of treatment to deal with the virus, but there was no cure. Total isolation—which involved cutting the infected people off from the outside world—was successful if done soon enough. People were urged to avoid gathering in large crowds, but this proved difficult to enforce. On September 28, 1918, 2,000,000 people gathered in Philadelphia, Pennsylvania, for the fourth Liberty Loan Drive—a parade to raise money for the war effort. The city lost more than 12,000 people to Spanish flu in the month that followed. In Britain thousands gathered at Trafalgar Square in London to celebrate the Armistice announcement on November 11, 1918, causing many more flu-related casualties.

▷ **Makeshift hospital**
This US school gymnasium was converted into a hospital during the Spanish flu pandemic of 1918. Wearing gauze masks was mandatory. Drapes and sheets can be seen here separating patient beds.

The Spanish flu pandemic of 1918 was over by 1919, vanishing as quickly as it had arrived. A virus that usually makes people ill for a few days had claimed an estimated 50 million lives. More people had

CONCEPT

HOW INFLUENZA VIRUSES MUTATE

The flu virus uses two main proteins—hemagglutinin (HA), and neuraminidase (NA)—to get into, replicate, and infect cells inside the body. The virus can change in two ways.

An antigenic drift is a slight mutation that can occur in the HA and NA, and humans are partially immune to it. This mutation occurs as the virus spreads from person to person.

Antigenic shift, a complete change of HA, NA, or both, is more dangerous because it infects animals and humans. The virus now has a new HA or NA glycoprotein that has never been exposed to a human immune system, which leads to pandemics.

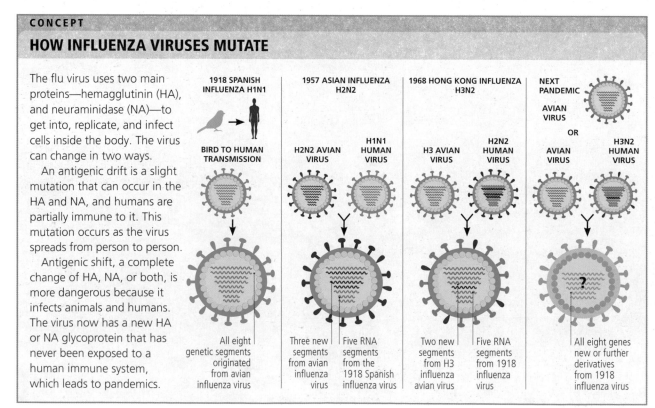

1918 SPANISH INFLUENZA H1N1	1957 ASIAN INFLUENZA H2N2	1968 HONG KONG INFLUENZA H3N2	NEXT PANDEMIC AVIAN VIRUS OR

BIRD TO HUMAN TRANSMISSION — H2N2 AVIAN VIRUS / H1N1 HUMAN VIRUS — H3 AVIAN VIRUS / H2N2 HUMAN VIRUS — AVIAN VIRUS / H3N2 HUMAN VIRUS

All eight genetic segments originated from avian influenza virus — Three new segments from avian influenza virus / Five RNA segments from the 1918 Spanish influenza virus — Two new segments from H3 influenza avian virus / Five RNA segments from 1918 influenza virus — All eight genes new or further derivatives from 1918 influenza virus

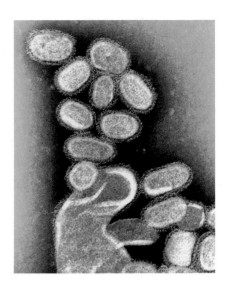

" No one… picked my brains about influenza so expertly as he did. "

SIR FRANK MACFARLANE BURNET, AUSTRALIAN PHYSICIAN, ON HAVING MET A YOUNG JONAS SALK, 1943

▷ **Influenza to H1N1**
These particles have been recreated from the virus strain that caused the 1918 Spanish flu pandemic. Scientists hope to identify the traits that made it so lethal and develop new vaccines.

died in one year than perished in a century of the bubonic plague in the medieval period.

Searching for a cure
Through the 1920s and 1930s researchers looked for the origin of the pandemic, without success. Then in 1997 US scientists obtained the virus's genetic material in lung tissue from 1918 preserved by US army doctors. They concluded that the virus passed from birds to pigs, then jumped another species barrier to humans. They also believed that the strain of flu was so lethal because it rapidly filled victims' lungs with fluid, so they drowned.

In 1938 US physicians Jonas Salk and Thomas Francis developed the first vaccine against flu. This was used to protect US military forces from the disease in World War II.

While flu vaccines have become increasingly effective, it is not possible to create a lifelong immunization because flu viruses mutate constantly (see panel, opposite). Scientists predict that another flu pandemic is inevitable.

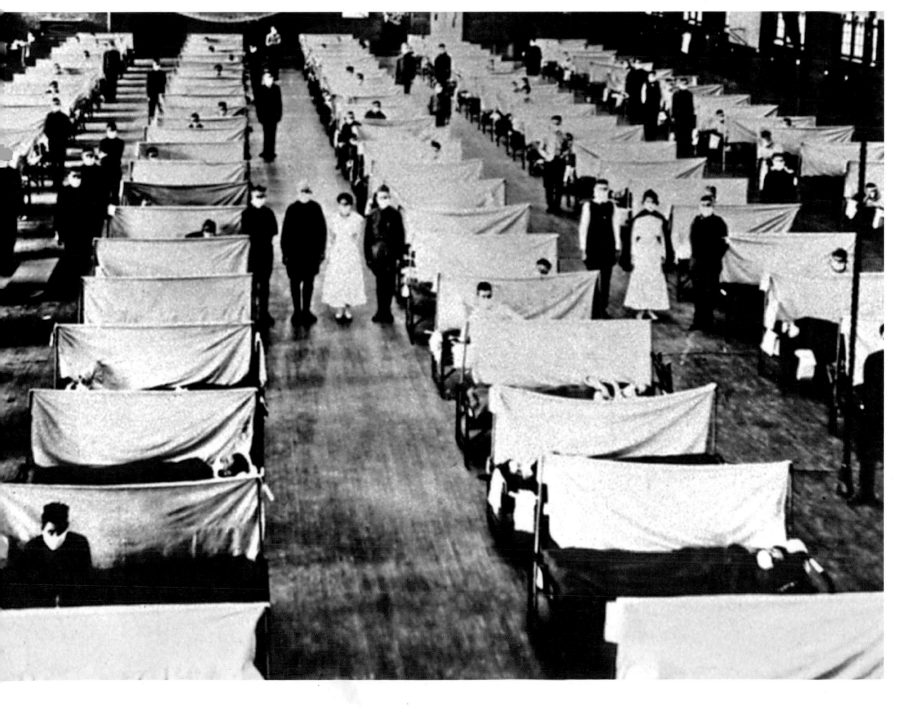

The Discovery of Penicillin

The discovery by Alexander Fleming in 1928 that a mold called *Penicillium* had the power to inhibit the growth of disease-causing organisms opened the way to a new era in which antibiotics—medicines derived from penicillin and similar substances—could finally cure infectious diseases.

Mold—in the form of bread gone bad—had traditionally been used to treat infected wounds. British apothecary John Parkington had recommended its use as far back as 1640, yet physicians did not know how mold worked, could not control its operation, and had no way of making it effective against infections.

It was only after work on germ theory (see pp.146–47) in the mid-1800s by French microbiologist Louis Pasteur (see pp.148–49) that scientists discovered the mechanisms by which bacteria spread diseases, and began to devise ways to counter them. In 1871 British physiologist John Sanderson observed that spores of the microscopic fungus *Penicillium* seemed to inhibit the growth of bacteria. In 1877 Pasteur and German microbiologist Robert Koch observed that airborne spores impeded the growth of the anthrax bacillus. French bacteriologist Jean Paul Vuillemin named the phenomenon "antibiosis" in 1889. Scientists began to make attempts to exploit this property for its therapeutic purposes.

British surgeon Joseph Lister had tried a crude form of antibiotic using mold to treat surgical infections in the 1870s. In 1895 Italian researcher Vicenzio Tiberio

> **10** The number of people who could be treated with the penicillin available in 1942.
>
> **600** **BILLION** The number of penicillin doses produced by US drug companies in 1945.

injected *Penicillium* extract into typhoid-infected mice and found it had some effect in inhibiting the disease. In 1897 French researcher Ernest Duchesne published a paper entitled *On the Antagonism between Mold Fungi and Bacteria* in which he described his belief that *Penicillium* could impede bacterial growth. But this early research was taken no further, and the active chemical agent responsible for antibiosis remained unidentified.

Turning point

The breakthrough came in 1928 at St. Mary's Hospital, London, UK, where pharmacologist Alexander Fleming was studying the effect of lysozymes—enzymes that attack bacterial cell walls—and needed to culture *Staphylococcus* bacteria for his research. He went away for a month's vacation and found on his return that his culture dishes had been stacked in a sink. While most of the petri dishes were covered with *Staphylococcus*, on one there was a moldlike substance that seemed to inhibit the growth of bacteria around it. Intrigued, Fleming cultured this mold in a broth and found that he could replicate its antibacterial property.

Extracting the drug

Fleming published his findings the following year. The mold was identified as *Penicillium notatum*, and the substance secreted by it was named penicillin. Efforts to isolate penicillin proved frustrating. From 1930 to 1932 Harold Raistrick,

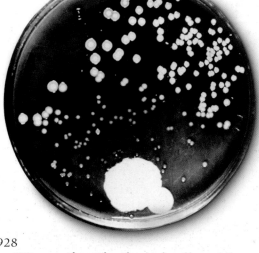

△ **Alexander Fleming's culture plate**
The upper half of Fleming's original culture plate from 1928 is covered with *Staphyloccus* bacteria. The growth of the *Penicillium* mold has retarded the development of bacteria toward the bottom, and few bacterial colonies exist there.

professor of biochemistry at London's School of Hygiene and Tropical Medicine, made progress toward extracting penicillin, but it was chemically unstable and the process of extraction destroyed much of it.

In the late 1930s a team at Oxford University, led by Australian-born pathologist Howard Florey and his German-born colleague Ernest Chain, overcame the problem by freeze-drying the broth, which kept it stable for long enough to extract the penicillin. In 1940 Chain and Florey improved the process by using a carbon-based solvent to extract and purify penicillin. By doing this, they were able to manufacture enough of the

SCOTTISH PHARMACOLOGIST (1881–1955)

ALEXANDER FLEMING

The son of a Scottish farmer, Alexander Fleming studied medicine in London, UK, under British bacteriologist Almroth Wright, a pioneer of vaccine therapy. After serving in World War I, he returned to medical research, producing important work on lysozymes—enzymes for which little therapeutic use could be found at the time. He became professor of bacteriology at London University in 1928. Fleming's accidental discovery of the antibacterial properties of penicillin led to him being awarded the 1945 Nobel Prize in Physiology or Medicine, alongside researchers Howard Florey and Ernest Chain, who developed it.

> **"[The discovery of] penicillin started as a chance observation. My only merit is that I did not neglect the observation."**
>
> ALEXANDER FLEMING, IN HIS NOBEL PRIZE LECTURE, 1945

Penicillin's source
The penicillin-making mold that Fleming identified was a variety of fungus known as *Penicillium notatum* (now called *Penicillium chrysogenum*). It is common in humid environments in temperate and subtropical regions, and its spores travel easily through air.

substance to test it on a human—a policeman who was suffering from blood-poisoning. At first, the results were startling and the infection began to disappear. However, the effect of the penicillin was short lived and it had to be reinjected every three hours and when the supplies ran out, the patient died.

Interest from drug companies

Nonetheless, the results were promising enough to raise interest in the pharmaceutical industry. War-ravaged UK lacked the

funds for the research, but drug companies in the US developed it and were soon able to manufacture penicillin in large quantities. It was trialed on war wounded in North Africa in 1943 and by the time of the Normandy landings in 1944 it was being used routinely on battle casualties. After the war, it was employed on a massive scale and saved thousands of lives. Doctors, who had previously had few options to fight bacterial disease and infections, now had a powerful new weapon.

▷ **Penicillin in use during World War II**
Penicillin injections were first given to wounded soldiers in May 1943. By June 1945 the US was producing enough of the antibiotic to treat over 250,000 battle casualties. The death rate of soldiers with chest wounds fell by around two-thirds after the drug was introduced.

Before penicillin's widespread use in the 1940s, antibiotics were available in the form of the sulfonamides, or sulfa drugs. These work by disrupting bacterial production of folic acid, which affects bacteria's ability to make genetic material so they cannot grow and breed. Sulfonamides were developed in the 1930s in Germany, after it was observed that synthetic dyes from distilled coal tars may have antibacterial effects. In 1932 German pathologist Gerhard Domagk discovered that one chemical, sulfanilamide, countered bacterial infection in mice. He tested the chemical on humans, including his daughter when she fell ill with an infection. Marketed in 1935 under the name Prontosil, this was the first commercially available general antibiotic. For his work, Domagk received the Nobel Prize in Physiology or Medicine in 1939.

Ways of attack

The detail of how atoms are arranged in different drugs is vital for understanding how they work, as well as for discovering new types. Penicillin belongs to a class of antibiotics known as beta-lactams. These antibiotics have a beta-lactam ring (a square molecular structure that has one nitrogen atom and three carbon atoms, one of which is attached to an oxygen atom). They work by interfering with the way a

Antibiotics in Action

The discovery of penicillin in 1928 (see pp.198–99) and its eventual introduction sparked a global effort to discover more kinds of antibiotic drugs. There are now more than 20 groups, or classes, categorized by features such as their structure, how they attack bacteria, and which types of bacteria they affect.

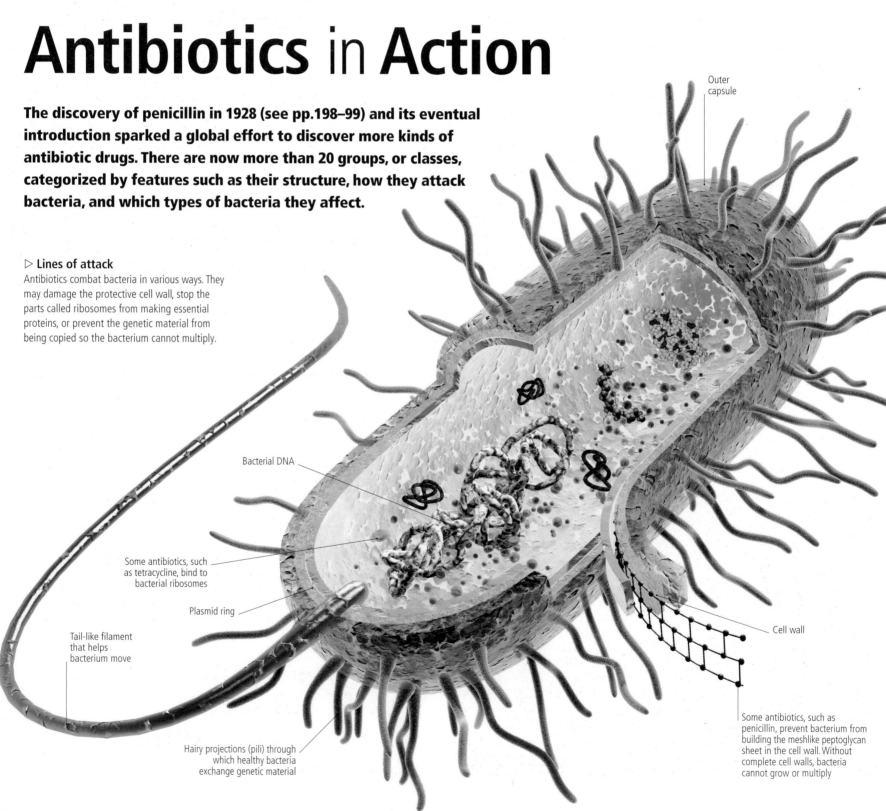

▷ **Lines of attack**
Antibiotics combat bacteria in various ways. They may damage the protective cell wall, stop the parts called ribosomes from making essential proteins, or prevent the genetic material from being copied so the bacterium cannot multiply.

Outer capsule

Bacterial DNA

Some antibiotics, such as tetracycline, bind to bacterial ribosomes

Plasmid ring

Tail-like filament that helps bacterium move

Hairy projections (pili) through which healthy bacteria exchange genetic material

Cell wall

Some antibiotics, such as penicillin, prevent bacterium from building the meshlike peptoglycan sheet in the cell wall. Without complete cell walls, bacteria cannot grow or multiply bacterium makes its outer coat—the cell wall. Pencillin targets conditions such as blood poisoning, and local wound and skin infections. Ampicillin became available for general use in 1961, widening the use of beta-lactams to target new kinds of bacteria, such as those causing pneumonia and bacterial meningitis. One of the most common beta-lactams is amoxicillin, developed in England in the 1960s, then released into the market in the1970s.

A subgroup of cell-wall disrupting beta-lactams is the cephalosporins. They were extracted from a fungus—*Cephalosporium acremonium* (now known as *Acremonium*)—growing in sea water near a sewage outlet on the Mediterranean island of Sardinia. Italian Giuseppe Brotzu discovered the fungus in 1945 while researching why there were fewer cases of typhoid fever in the city than elsewhere. He noticed that the fungus worked against *Salmonella typhi*, the bacteria that caused typhoid. The cephalosporins had long, complicated trials and many

> **150** **The number of different antibiotics that have come onto the market since Prontosil in 1935.**

different versions were launched, but finally came into medical use in the 1960s. There are now five or more successive generations of cephalosporins with dozens of individual kinds.

Further classes

The aminoglycosides are another class of antibiotics that work by interrupting the action of ribosomes—the molecular "factories" in bacteria that make the proteins used for cell structures and the enzymes that control cell reactions. Streptomycin, the first to be discovered (and still widely used), was made by another bacterium, *Streptomyces griseus*. Streptomycin was found in soil samples in 1943 by US student and microbiologist Albert Schatz, who was searching for an antibiotic that could combat tuberculosis. Schatz was working with eminent US-based scientist Selman Waksman, who had coined the term "antibiotic" the previous year, and whose team identified several other antibiotics. In 1952 Waksman was awarded the Nobel Prize in Physiology or Medicine. Other aminoglycosides include neomycin, discovered in 1949, and gentamicin, found in 1963.

Protein manufacture by ribosomes is affected by the tetracyclines, the first of which was also discovered in a soil sample, and is made by

▽ **Testing antibiotics**
A specific bacterium is grown on a gel, and disks of antibiotics are added. The antibiotics on the left plate have killed the bacteria around the disk, leaving clear areas—zones of inhibition. The antibiotics on the right plate were not effective.

▷ **Antibiotics and viruses**
Most antibiotics have little or no effect on viral diseases. A virus has no cellular machinery and processes of its own. It uses those of its host cell, which antibiotics do not disrupt.

another type of bacterium, *Streptomyces aureofaciens*. It was identified in 1948 by 76-year-old US plant biologist Benjamin Duggar, who had already made many prominent contributions to botany and plant physiology. During its testing phase, the drug is credited with saving the life of a five-year-old boy, Tobey Hockett, who had a ruptured and infected appendix. Known as tetracycline for its four-ring structure, it was later called chlortetracycline to distinguish it from the many other antibiotics that followed.

There are numerous other classes of antibiotics that attack different kinds of bacteria in various ways, including amphenicols, such as

chloramphenicol, found in 1947; macrolides, an example being erythromycin from 1949; glycopeptides such as vancomycin, isolated from a soil sample from Borneo in 1952; rifampicin, isolated in 1957; quinolones; streptogramins; nitroimidazoles; lipopeptides, discovered in 1987, and oxazolidinones, in use since 2000. Each member of each class of antibiotics has its own special uses and side effects. The many kinds of antibiotics are neccessary because bacteria can develop resistance to existing drugs. Patterns of treatment are constantly changing to stay ahead of the development of antibiotic-resistant bacteria (see pp.258–59).

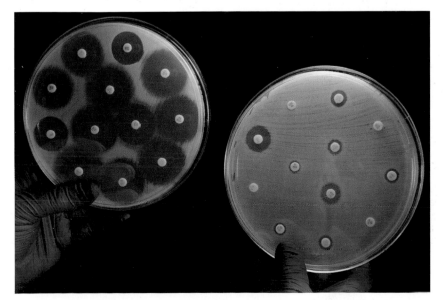

" From the moment he is born…
**man is subject to…
numerous microbes.** "

SELMAN WAKSMAN, ON RECEIVING HIS NOBEL PRIZE, 1952

The **Evolution** of **Syringes**

A syringe is a tube with a plunger, which lets liquid enter when it is pulled up or leave when pushed down. The invention of the hypodermic syringe in 1855 added a needle, which allowed medicine to be injected subcutaneously (beneath the skin).

1 **Silver tube with plunger** Arab physician al-Zahrawi used this type of device to extract bladder stones. 2 **Pewter syringe and nozzle** Syringes such as this one were used to deliver enemas to patients. 3 **Mechanical syringe** Mechanically operated syringes such as this could be used to extract fluids from patients. 4 **Disposable syringe** Invented by New Zealand pharmacist Colin Murdoch, this syringe overcame the problems of cross-infection caused by reuse. 5 **Tuberculin syringe** This syringe, with its multiple-puncture "Sterneedle gun" design, was used to test for tuberculosis. 6 **Glass syringe in case** The invention of these sterilizable, precision-made glass syringe bodies in the 1940s allowed contents to be seen and the plunger to be fitted accurately. 7 **Auto-injector** Designed for self-administration, auto-injectors or "EpiPens" come with single-use dosages. They are particularly useful for allergy sufferers experiencing anaphylactic shock. 8 **Insulin pen and cartridge** These pens allow diabetics to carry insulin conveniently and to measure doses accurately. 9 **Packet of hypodermic needles** Made in Germany, these needles were probably used by the German army during World War I. 10 **Butterfly cannula** This device allows a precision grip close to the needle, making intravenous injection and blood collection easier. 11 **Coated needles in tin box** Used by British doctors during World War II, these handy tins contained needles that were coated in paraffin wax to keep them in good condition. 12 **Trocar** Used to drain fluid-filled swellings, trocars consist of a puncturing device inside a tube.

1 SILVER TUBE WITH PLUNGER (10TH CENTURY)

Pewter
plunger barrel

2 PEWTER SYRINGE AND NOZZLE
(17TH –18TH CENTURIES)

Ivory
handle

LONDON

3 MECHANICAL
SYRINGE (17TH–
18TH CENTURIES)

Trocar was inserted into
the skin after an incision
was made with a lancet

Needle case

2
ml
2.5

4 DISPOSABLE
SYRINGE (1956)

Barrel held
detachable
six-point
needle block

5 TUBERCULIN SYRINGE (1960)

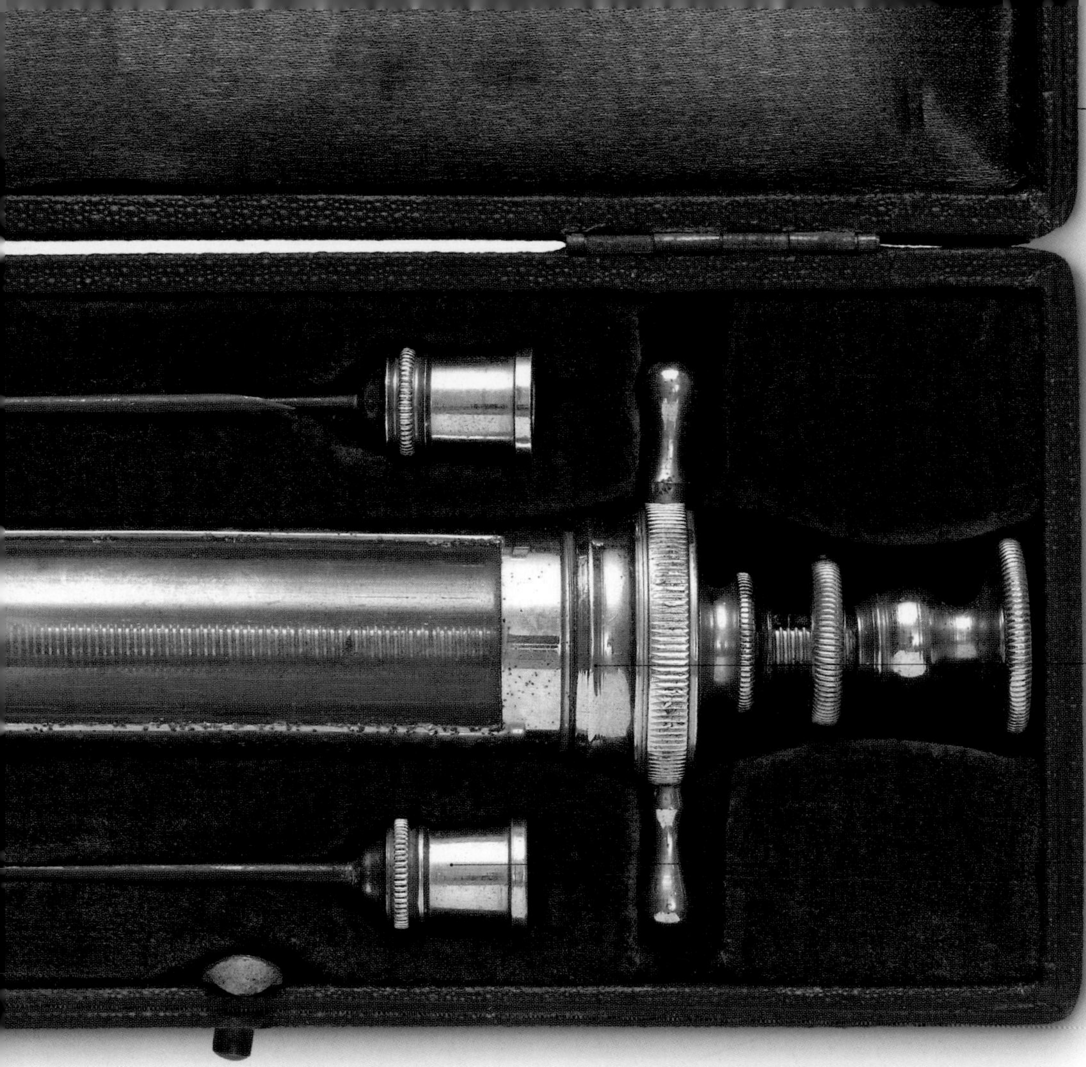
Metal and glass plunger
Metal barrel to attach needle to plunger

[6] GLASS SYRINGE IN CASE (1940s)

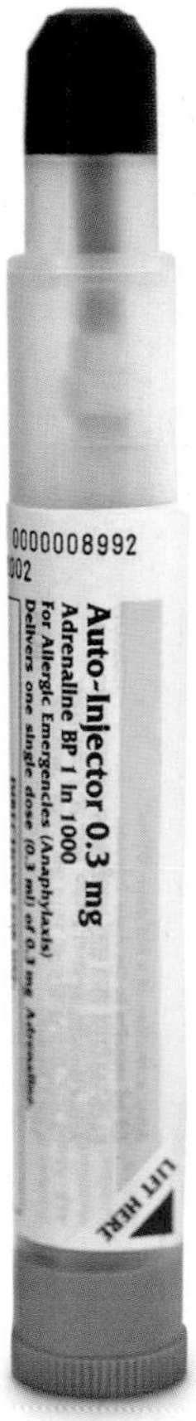
0000008992
Auto-Injector 0.3 mg
Adrenaline BP 1 in 1000
For Allergic Emergencies (Anaphylaxis)
LIFT HERE

[7] AUTO-INJECTOR (LATE 20TH CENTURY)

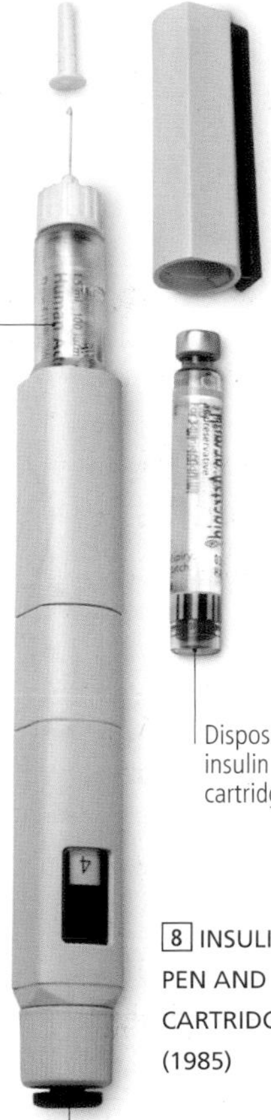
Dose calibration window
Disposable insulin cartridge
Press plunger

[8] INSULIN PEN AND CARTRIDGE (1985)

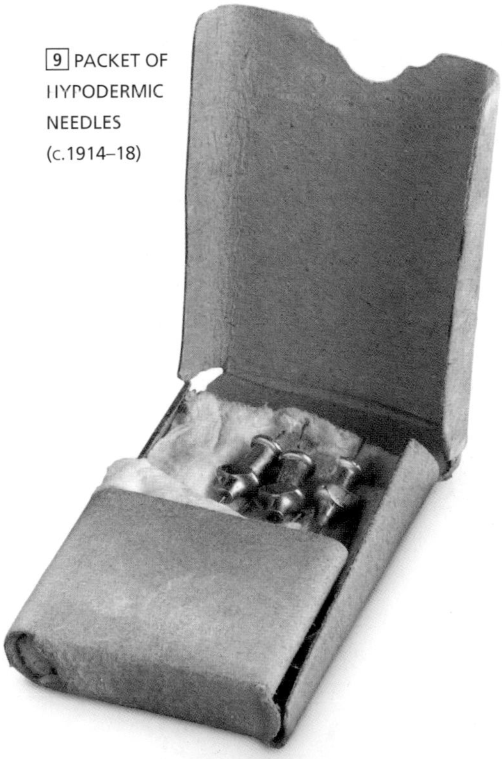

[9] PACKET OF HYPODERMIC NEEDLES (c.1914–18)

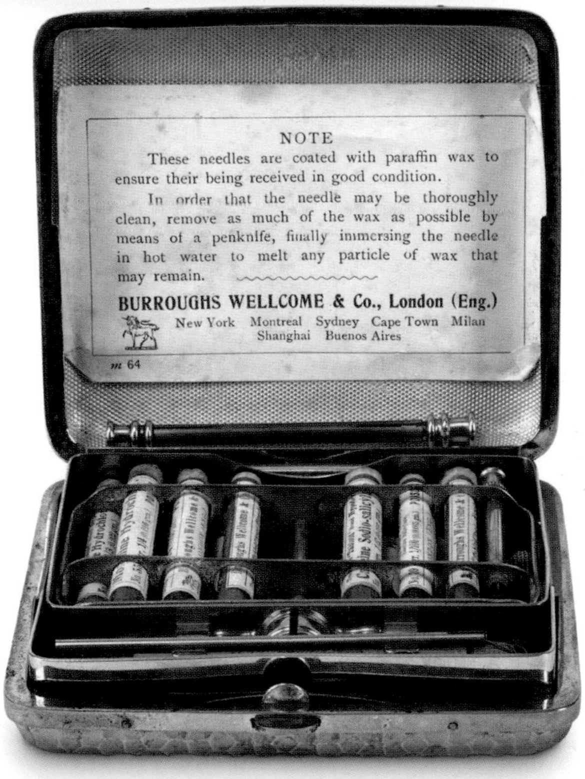
NOTE
These needles are coated with paraffin wax to ensure their being received in good condition.
In order that the needle may be thoroughly clean, remove as much of the wax as possible by means of a penknife, finally immersing the needle in hot water to melt any particle of wax that may remain.
BURROUGHS WELLCOME & Co., London (Eng.)
New York Montreal Sydney Cape Town Milan Shanghai Buenos Aires
m 64

[11] COATED NEEDLES IN TIN BOX (c.1939–45)

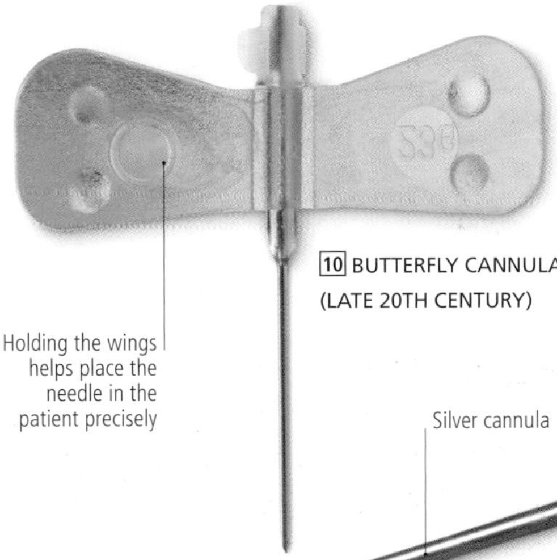
Holding the wings helps place the needle in the patient precisely

[10] BUTTERFLY CANNULA (LATE 20TH CENTURY)

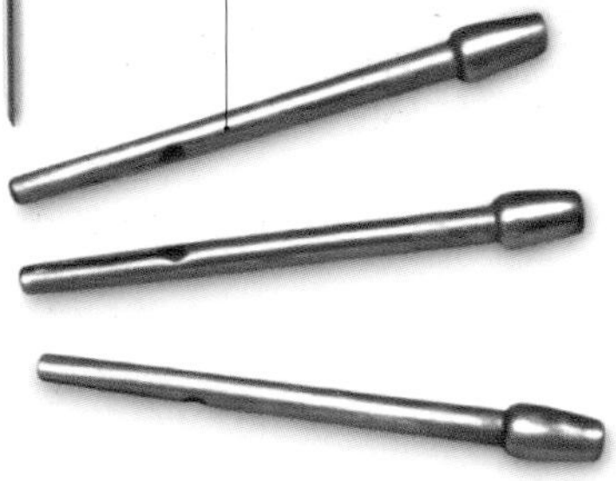
Silver cannula

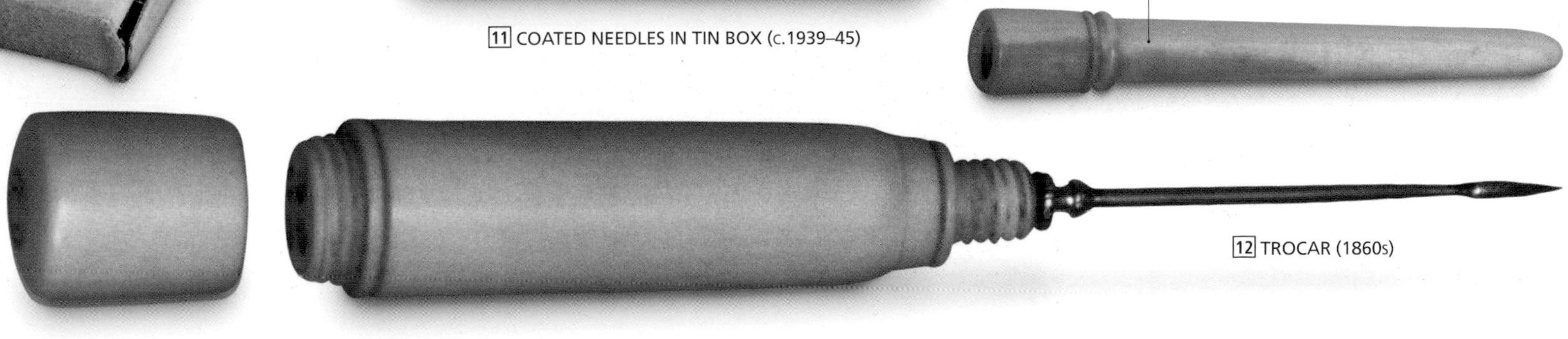
Ivory casing

[12] TROCAR (1860s)

Women's Health

The 20th century saw great progress in the field of women's health. The discovery of estrogen and progesterone, and the means to detect cervical cancer—previously a common killer—dramatically improved women's quality of life and life expectancy.

At the beginning of the 20th century women were still a mystery to the largely male-dominated medical profession. Even Sigmund Freud (see pp.182–83), founder of psychoanalysis, admitted that he found women difficult to understand. Female hysteria, which had been a notion since ancient times, gained ground as a medical diagnosis in the 19th and early 20th centuries. At a time when female nervous disorders were perceived as directly connected to a woman's reproductive organs, operations such as ovariotomies (removal of the ovaries) and hysterectomies (removal of the uterus) began to be carried out as extreme cures.

Early examinations

The speculum has been used in gynecological inspections since Roman times, although it was modified in the 19th century to resemble the instrument still used today. At the time, the use of the device set off a vigorous debate in the medical community. Doctors thought that such an intrusive procedure was indecent and might corrupt women. Yet the speculum enabled doctors to examine the cervix and perform cervical biopsies.

In the 1930s gynecological pathologist Walter Schiller carried out studies on cervical cancer at the University of Vienna. He followed the development of lesions of the cervix (the opening of the womb), which he believed indicated the presence of cancerous cells. Schiller concluded that cervical cancer usually develops slowly, and if identified early it can be treated before spreading to other tissues. Schiller devised a simple screening test that involved painting the cervix with diluted iodine, and promoted the idea of routine testing to reduce mortality

rates. To treat the disease, he advocated a radical hysterectomy followed by radiation therapy.

Despite Schiller's milestone in cervical cancer treatment, cure rates were only about 30 percent, and his screening technique using iodine was discovered to be too nonspecific. At around the same time, a more effective screening technique was developed in the form of a test based on examination of cervical smears (samples taken from the cervix): exfoliative cytology.

Birth of the Pap test

Exfoliative cytology at the opening of the cervix became known as the Pap test, or Pap smear, after George

Papanicolaou, who developed it. In 1925 Papanicolaou began a joint study with the Woman's Hospital of the City of New York and Cornell University Medical College's Department of Anatomy. Looking at changes in various tissues during the reproductive (menstrual) cycle he was able to detect cancerous cells.

GREEK–AMERICAN PATHOLOGIST (1883–1962)

GEORGE PAPANICOLAOU

A Greek-born American physician, Papanicolaou worked at the Cornell University Medical College, New York, from 1913 to 1962. It was during an examination of vaginal smears of guinea pigs to study their sex cycle that he found cancer cells originating in the cervix. In 1943 he published *Diagnosis of Uterine Cancer by the Vaginal Smear*. His method later came to be known as the Pap test. His work led to the acceptance of cytology as a basis for diagnosing cervical cancer.

▷ **Cells in normal cervical smears**
The Pap test has saved the lives of millions of women. As hospitals began using the test, training on how to recognize certain cells and changes became imperative for accurate diagnosis. These drawings by George Papanicolau depict different types of normal cervical cells.

> **"The first observation** of **cancer cells** in the … **cervix:** one of the **greatest thrills** … [of] my scientific career."
>
> GEORGE PAPANICOLAOU, ON EXAMINATION OF A SMEAR OF THE UTERINE CERVIX

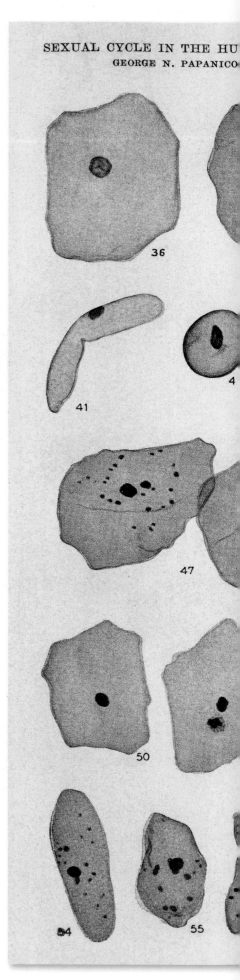

SEXUAL CYCLE IN THE HU
GEORGE N. PAPANICO

36
4
41
47
50
54
55

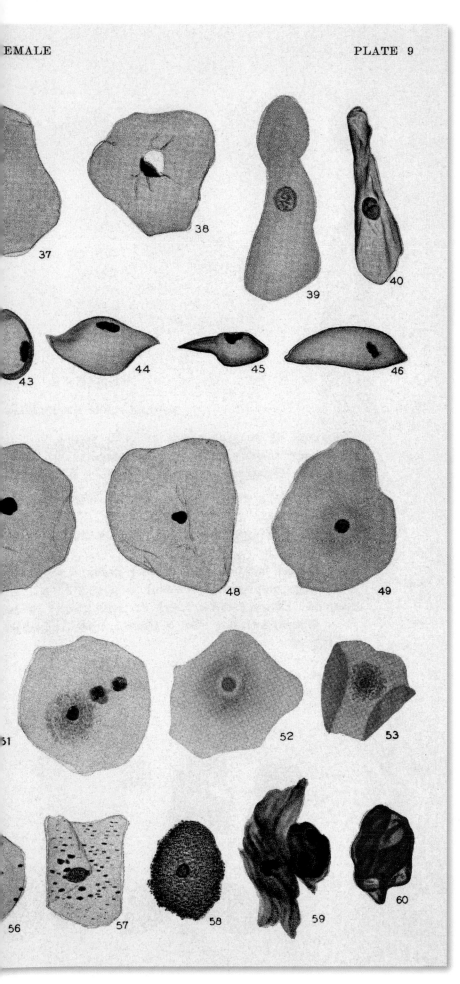

EMALE   PLATE 9

In the 1940s, when the Pap test was first introduced, cervical cancer was a major cause of death among women. The early detection of the disease that the Pap test allowed greatly reduced the global death rate of patients. The same technique was also used for studies on the menstrual cycle, amenorrhea (abnormal absence of menstruation), sterility, and hormone therapy.

60 **PERCENT Decline in the incidence of cervical cancer in the US from the introduction of the Pap test into everyday clinical practice, in the 1940s, to 1992.**

Discovery of hormones

The discovery and isolation of the female sex hormones estrogen and progesterone revolutionized women's health too. In 1905 British physiologist Ernest Starling referred to certain glandular secretions as "hormones," and shortly after this endocrinology—the study of hormones—advanced rapidly. American scientists Edgar Allen and Edward Doisy first isolated estrogen in 1929. By the mid-1930s pharmaceutical companies were manufacturing estrogen products to help with symptoms of menopause. The discovery of progesterone

△ **Screening for breast cancer**
This colored mammogram of a woman's breast shows a cancerous tumor, highlighted in blue. Regular self-examination and screening through mammography can help ensure the early detection of breast cancer, allowing for effective treatment before metastasis occurs.

in 1934 led to its use to prevent miscarriage and treat infertility, and the idea of hormonal contraception became a reality (see pp.224–25). Hormone replacement therapy (HRT) first became available in 1942, with the drug Premarin, as the replacement of lost estrogen, or both estrogen and progesterone, by HRT had been found to bring relief from symptoms of menopause such as hot flashes.

Further developments

In the 20th century scientific advances in women's health expanded rapidly. Developments included the early detection of breast cancer through mammograms (introduced in the 1960s), improved family planning, in vitro fertilization (IVF, see pp.240–41), safer delivery of babies, and better methods of pain relief in labor and childbirth. Partly due to greater understanding of women's health, female life expectancy in, for example, the US, increased from 48 years in 1900 to 78 years in 1980.

IN PRACTICE

PROGESTERONE

The female sex hormone progesterone did not become viable as a therapeutic agent until almost two decades after it was first discovered. Isolated by American anatomy professor George Corner, American gynecologist Willard Allan, and German biochemist Adolf Butenandt in 1934, it was extremely difficult, and expensive, to extract from natural sources. Its synthetic forms were developed by Bulgarian-American chemist Carl Djerassi for Syntex in 1951, and American chemist Frank Colton for Searle in 1952.

Heart Disease

The most common cause of death worldwide, heart disease includes a range of malfunctions and degenerations of the valves and muscles of the heart. Modern understanding of the heart's mechanisms has enabled limited treatment of heart disease, but is yet to reveal a cure.

Ancient physicians recognized the heart as a critical organ for well-being—Greek philosopher Aristotle even regarded it as more important than the brain. However, it was not until English physician William Harvey (see pp. 84–85) discovered the circulation of blood in 1628 that physicians began to understand how modifications in that circulation, caused by damage to the heart, might be fatal to a patient.

Advances in knowledge

Early understanding of heart disease was gained through anatomical dissections, such as those carried out by Italian physician Giovanni Mara Lancisi. His 1707 work *De Subitaneis Mortibus* (*On Sudden Death*) examined the effects of cardiac dilatation—the enlarging of the heart cavity and stretching of the heart muscle—on the heart's valves. In the 1720s German professor of physiology Friedrich Hoffmann theorized that the narrowing of arteries observed in

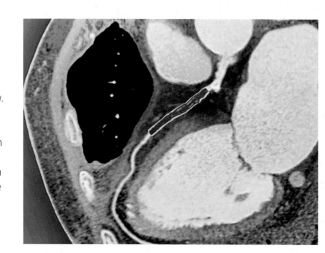

some patients might be the cause of disease and death. This would be confirmed nearly two centuries later, in 1912, by American cardiologist James B Herrick.

Late 18th-century cardiology focused on the nature and cause of angina pectoralis—severe chest pains caused by the blocking of the arteries—which was first identified as a disease by English physician William Heberden in 1768. When Scottish surgeon John Hunter was dissected in 1793 following his death after a severe angina attack, his coronaries were found to be ossified (or hardened)—further confirming suspicions of the disease's cause.

Physicians produced increasingly detailed accounts of the state of hearts suffering from coronary artery disease—caused by atherosclerosis, or blocking of the arteries—such as English physician James Hope's 1831 work *A Treatise*

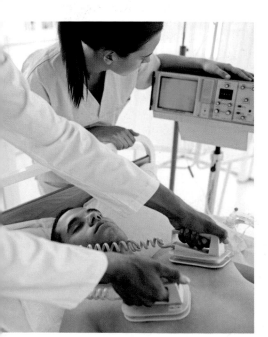

◁ **Restarting the heart**
A heart that has stopped or become irregular during a cardiac arrest can be revived with a defibrillator—a device that delivers electric impulses to restart heart contractions. It was invented by American electrical engineer William Kouwenhoven in 1930.

on the Diseases of the Heart and Great Vessels. However, doctors could do little to treat the disease.

Advances were made in detecting the signs of heart disease, such as the invention of the stethoscope by French physician René Laënnec in 1816 (see pp.114–15). In 1855 German physician Karl Vierordt devised the sphygmograph, which allowed a graph of pulse activity to be traced. Others gradually refined the device until, in 1890, Scottish cardiologist James Mackenzie invented a means to distinguish between the pulse in the veins and in the arteries. This permitted more sophisticated monitoring of heart irregularities.

Lifestyle causes

After Herrick's 1912 discovery, doctors understood the role of atherosclerosis in causing heart disease. However, it was only gradually that the role of patients' lifestyles became apparent. In 1948 the United States Heart Institute began the Framingham Heart Study, intended to identify behaviours that made heart disease more likely. Smoking, high alcohol intake, lack of exercise, obesity, and diabetes emerged as high risk factors.

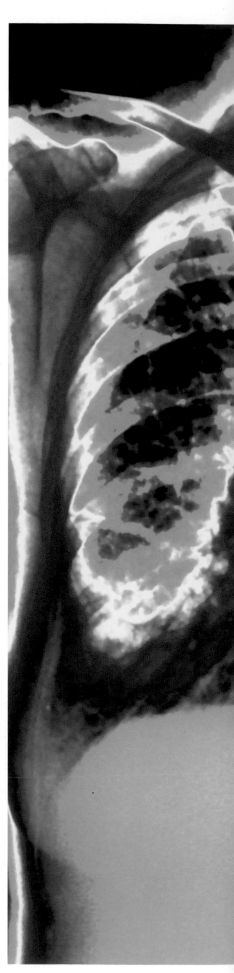

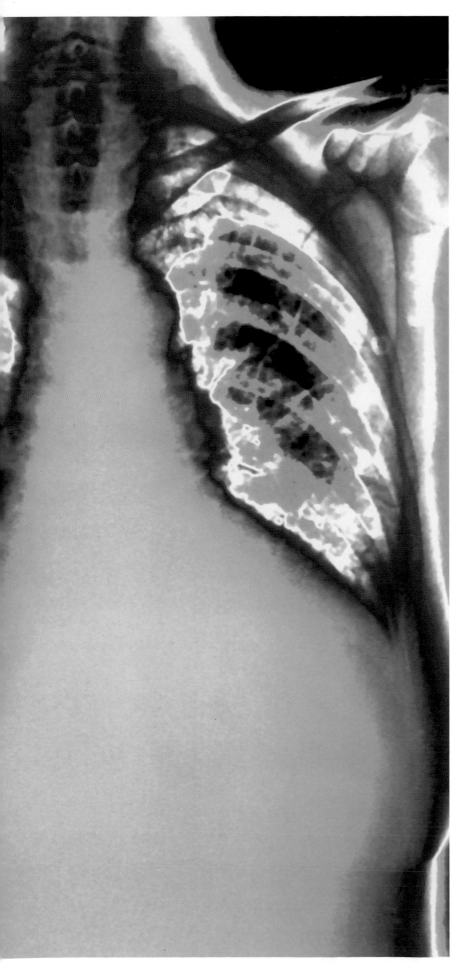

IN PRACTICE

CARDIAC BYPASS SURGERY

In bypass surgery, an artery blocked by atherosclerosis is bypassed using a vein or artery, normally from the patient's leg, arm, or chest. The vessel is grafted between the aorta (the main vessel in the arterial network) and a position on the coronary artery (the artery supplying blood to the heart) beyond the point where narrowing has occurred. This allows coronary blood to flow to the heart muscle while avoiding (bypassing) the blocked area. The first clinical coronary artery bypass was carried out in 1960 by German-born surgeon Robert Goetz. Today, bypassing several arteries to carry out triple (seen here), quadruple, or even quintuple bypasses is routine.

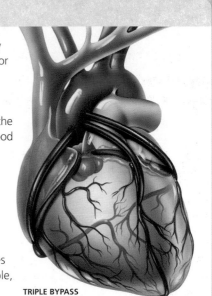

TRIPLE BYPASS

As societies became prosperous, these behaviors became more prevalent—to the extent that 90 percent of coronary artery disease is now considered to be preventable through early lifestyle modification.

Surgical solutions

Heart damage caused by coronary artery disease, congestive heart failure (due to weak or damaged heart valves), or myocardial infarction (heart attack) is irreversible. However, the advent of heart surgery has provided some hope for patients. The first operation on a heart valve— conducted to correct the abnormal narrowing of the mitral valve, which joins the two left chambers of the heart—was carried out by English surgeon Henry Souttar in London in 1925. Surgery on a baby to correct congenital heart abnormalities was first performed at Johns Hopkins Hospital in Baltimore, Maryland, in 1944. From 1952 open heart surgery— in which the heart is exposed for surgical repair, while an artificial pump temporarily keeps blood flowing—allowed a wide range of surgical procedures to be attempted. Radically, from 1967, it became possible to perform a heart transplant by replacing a damaged heart with a donor organ (see pp.234–35).

Coronary artery disease, seen in Egyptian mummies dating back to c.1000 BCE, continues to be a severe challenge for physicians 3,000 years later. In 2013 more than 17 million people died of the disease, making it the leading cause of death in most industrialized countries today. Prevention measures include altering diet and lifestyle, and medications such as statins.

> "It is sometimes possible, I think, to **prevent the disease,** but never to **cure it.**"

JEAN-NICOLAS CORVISART, FRENCH CARDIOLOGIST, FROM *ESSAY ON THE DISEASES AND ORGANIC LESIONS OF THE HEART AND GREAT VESSELS*, 1806

Allergies and Antihistamines

Greater understanding of allergies in the 20th century led physicians to explore their prevention, treatments, and cures. However, the fast-rising prevalence of asthma (see pp.214–15) and other allergic conditions means they have become a "21st-century epidemic."

Allergies are nothing new: records of allergic reactions date back thousands of years. Chinese Emperor Shennong (c.2700 BCE) was the first to treat respiratory distress, now known as allergic asthma, with ephedra, a shrub related to pines and firs. However, little was understood about allergies until the early 1800s, when English physician John Bostock described hay fever or seasonal allergic rhinitis as a disease that affected the upper respiratory tract. It is now known that allergic reactions arise as a result of unwanted immune responses to harmless substances called allergens, and can range from the seasonal and common (such as hay fever—the world's most widespread allergy) to the severe or life-threatening (such as reactions to medications, wasp or bee stings, or foods such as peanuts, shellfish, and dairy products).

Allergies and anaphylaxis

As vaccines against some of the world's most virulent diseases were developed, it was noted that some caused inexplicable reactions. Unharmed by the initial injection, some patients would react violently to the second. In 1902 French physiologist-physicians Paul Portier and Charles Richet coined the term "anaphylaxis" on discovering this life-threatening response to certain medications. Anaphylactic shock occurs within minutes of allergen exposure, causing swelling (of, for example, the eyes, throat, or hands), difficulty breathing or swallowing, a sudden drop in blood pressure, and even a loss of consciousness.

Research progressed further in 1906 when Viennese pediatrician Clemens von Pirquet noted that exposure of the body to a particular substance resulted in the production of antibodies. He called this response "allergy," from the Greek *allos* ("other"), and *ergia* ("capacity to react"). The symptoms were caused by the body's attempts to fight off a disease agent. Applying this idea, hay fever can be explained as fluid production to protect the nasal passages against attack by an "invader"—actually harmless pollen.

Histamine was first recognized and suggested as the cause of allergic reactions by British physiologist Henry Dale and British virologist Patrick Playfair Laidlaw in 1910. Histamine is released from cells when the body is irritated by outside substances. The body then tries to expel the perceived invader, causing the symptoms typical of an allergic reaction (see panel, left). By 1932 histamines were confirmed as causative agents in allergic response.

Discovery of antihistamines

In the 1930s Swiss-born Italian pharmacologist Daniel Bovet began a search for compounds that would relieve the symptoms of allergies. In 1937 Bovet discovered the first antihistamine substance, which, in countering the effect of histamine, became effective in treating allergic reactions. This discovery led to the development of the first antihistamine drugs in 1942. Further milestones were reached with the detection of allergens such as pollen, dust mites, peanuts, and latex. In 1967 Japanese

△ **Bovet's breakthrough**
For his ground-breaking work in developing the first antihistamine, Daniel Bovet received the Nobel Prize in Physiology or Medicine. In 1947 he set up the Therapeutic Chemist Laboratory at the Superior Institute of Health, Rome, where he worked with his wife, the scientist Filomena Nitti.

▷ **Skin-testing kit**
This early 20th-century skin-testing kit contains a variety of potential allergens. The substances responsible for allergies in some people are put onto or injected into the skin, to see if they cause a reaction, typically redness or inflammation.

CONCEPT

HOW ANTIHISTAMINES WORK

An allergic reaction occurs when the body assumes that something harmless, such as pollen or dog hair, is a dangerous invader. On detecting an allergen, the inflammatory response in tissues involves mast cells releasing histamines. These bind to histamine receptors, causing reactions such as localized swelling, itchiness, or watery eyes. There are four types of histamine receptors, H1–H4. It is mainly through the H1 receptors that histamines cause the symptoms of hay fever. Antihistamines block histamines, and prevent them from binding to histamine receptors. This prevents receptor activation and the events leading to an allergic reaction.

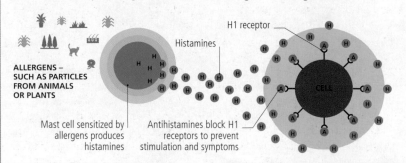

H1 receptor

Histamines

ALLERGENS – SUCH AS PARTICLES FROM ANIMALS OR PLANTS

CELL

Mast cell sensitized by allergens produces histamines

Antihistamines block H1 receptors to prevent stimulation and symptoms

---

> " What is **food for some** may be fierce **poison for others**."
>
> LUCRETIUS, ROMAN PHILOSOPHER, IN *DE RERUM NATURA (ON THE NATURE OF THINGS)*, 56 BCE

ALLERGEN TEST SOLUTIONS—DUNCAN.
KEY TO MULTIPLE TESTS

GRASS POLLENS
A
Meadow
Foxtail
Soft Brome
Vernal
Cock's Foot
Rye
Meadow

B
False Oat
Yorkshire Fog
Sheeps Fescue
Dog's Tail
Wild Oat
Timothy

TREE POLLENS
A
Hazel
Poplar
Birch
Elm
Ash
Oak

B
Hornbeam
Beech

FLOWER POLLENS
A
Hyacinth
Hawthorn
Gt. Plantain
Pinks
Carnation
Rose

B
Stocks
Marguerite
Privet
Golden Rod
Michaelmas
Daisy
Orris Root
House Dust

EXTRANEOUS
INHALANTS
Horse
Dandruff
Feather
Poultry
Dog Hair
Cat Hair
Cow Hair
Rabbit Hair
Sheeps wool

CONTROL
SOLUTION

MIXED
CEREALS
Wheat
Oat
Rye
Barley
Corn
Rice

MIXED
FISH &
SHELL
FISH
Cod
Salmon
Crab
Oyster
Lobster
Sardine

MIXED
FRUITS
Strawberry
Orange
Tomato
Peanut
Apple
Banana

MIXED
MEATS
Chicken
Beef
Pork
Lamb
Veal
Mutton

MIXED MILK
& EGG GROUP
Cowsmilk
Cheeses
Egg White
Egg Yolk
Chocolate
Cocoanut

MIXED
VEGETABLES
Beans
Celery
Peas
Rhubarb
Mustard
Spinach
Potato

Mast cell

△ **Mast cell**

Mast cells contain histamine, an important substance for fighting infection. When an allergic reaction occurs, the mast cells release a high quantity of histamine into the body.

immunologists Kimishige and Teruko Ishizaka discovered the role of IgE (immunoglobulin E) class antibodies in allergic reactions. In response to repeated exposure to an allergen, the allergic individual produces IgE antibodies, which cause mast cells (a type of white blood cell) to release chemicals, such as histamines, into the bloodstream. Drugs that interact with this process can offer effective allergy treatments.

The cause of some allergic reactions can be difficult to diagnose. Allergy testing kits were first developed in 1894 with the "functional skin test." Skin-prick testing is now widely used to determine certain food, insect venom, and drug allergies. A drop of liquid allergen is placed on the skin, then a pinprick is made through the drop; an itchy, red bump then appears within 15 minutes if the patient has an allergy to that substance. Home-testing kits are also available, but some question their accuracy.

The number of people suffering with allergies has increased in recent years. The rate of those with food allergies doubles roughly every 10 years, and the incidence of peanut allergy alone tripled between 1997 and 2008. The rise in eczema and asthma has gone hand-in-hand with this increase, and it appears that Western, developed countries have higher incidences than developing countries. Research continues into the causes of the increase, although it has been linked to urbanization, environmental factors, air pollutants, and diet.

Polio: A Global Battle

For millennia, the highly contagious viral disease poliomyelitis (polio) has caused paralysis, deformity, and even death. The worst epidemics occurred in the first half of the 20th century, but the disease began to fade with vaccination from the mid-1950s, and polio may be the next disease to be eradicated globally.

The polio strains of *Enterovirus*, a group of viruses that affect the digestive tract, spread from person to person through mucus and other nose and mouth secretions, and via contaminated water and food. Up to 98 percent of cases have either no symptoms, or mild ones, such as fever, vomiting, and diarrhea. However, depending on the virus strain, and the age and health of the sufferer (children tend to be more affected by the disease than adults), in 1–2 percent of cases polio proved more serious. The virus can move from the digestive system into the nerves, especially the spinal cord, which can lead to deformity, wasting of muscle, and paralysis, especially of the lower limbs. Polio can also affect the muscles of the throat and chest, making swallowing and breathing difficult.

△ **Victims from the past**
Perhaps the earliest image of polio is an ancient Egyptian scene, the "polio stele," from about 1,400 years ago. It shows a man, probably a priest, with a deformed leg and foot.

History of polio
While the history of polio may be traced back to antiquity, the disease was not much written about until the 19th century—perhaps because its vague and often mild symptoms were difficult to identify. It was first described as a distinct disease in 1840 by German physician Jakob Heine, and in 1874 it was given its current name by another leading German physician, Adolph Kussmaul.

By the early 1900s outbreaks of polio began to occur in Europe and North America. The epidemics became increasingly serious and widespread. Severe cases resulted in permanent paralysis, major breathing problems—treated by placing the patient in a metal chamber called an "iron lung" that pulled air in and out of the lung—

▷ **Rehabilitation**
In 1927 US President Franklin D. Roosevelt set up the Georgia Warm Springs Foundation (now Roosevelt Warm Springs) for the treatment and rehabilitation of polio patients. Here patients in the 1950s read their mail from home.

◁ **Salk administering vaccine**
In 1953 Jonas Salk gained much publicity for the battle against polio when he famously injected himself and his family with his latest vaccine, saying, "It is safe, and you can't get safer than safe."

and even death. In 1916 the US recorded almost 28,000 cases and more than 6,000 deaths. While the reasons for this sudden upsurge in the disease remain unclear, the hygiene hypothesis suggests that as living conditions improved, people's natural immunity decreased.

The US led the way in polio research. In 1935 two early vaccines underwent extensive trials—one at New York University

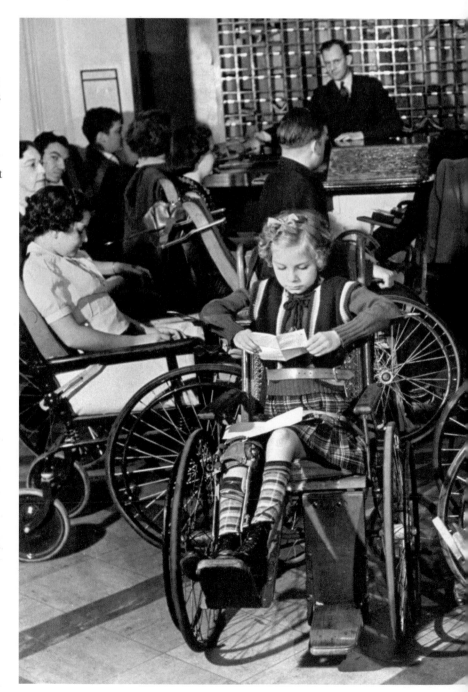

"The **people [own the polio vaccine]**... There is no patent. Could you patent the sun?"

JONAS SALK, AMERICAN PHYSICIAN AND MEDICAL RESEARCHER, 1955

and the other at Temple University, Philadelphia. Both of the trials used attenuated (weakened) viruses and failed badly, with many volunteers becoming very ill, and some died. Three years later US President Franklin D. Roosevelt, who had contracted polio in 1921, became the focus of the major fundraising "March of Dimes" campaign, when people were asked to donate 10 cents each to help fight the disease.

In 1941 the US-based researcher Albert Sabin and his colleagues discovered the polio virus in the digestive tract, suggesting that it entered the body via the mouth. In 1949 new and less expensive methods of growing the virus for vaccine development were discovered, and by 1950 further trials resulted in some success.

▷ **Poster for polio vaccine**
More than 50 years ago, the US Communicable Disease Center, now Centers for Disease Control and Prevention (CDC), introduced the national symbol of public health, Wellbee. Personifying well-being, this mascot's first assignment was to increase awareness regarding Sabin's oral polio vaccine in Atlanta and across the US.

Polio vaccine

In 1952 the US suffered its worst ever polio epidemic—there were more than 57,600 recorded cases, 21,000 of which involved paralysis. American virus expert Jonas Salk's team began new polio vaccine trials by administering the killed virus, which gave immunity but was not long-lasting. In 1954 large-scale trials began. More than one million children were injected with the Salk vaccine. It proved up to 90 percent effective against paralytic polio.

At the same time, Albert Sabin was developing an oral vaccine against polio using live, attenuated viruses. After another major trial, this vaccine was deemed successful too. Easier to administer than the Salk vaccine, and mimicking the natural virus's entry into the body, mass vaccination began around the world from the early 1960s.

Further improvements to polio vaccines, and global campaigns to administer them, made outbreaks of polio rare by the 1990s. However, it lingered on in some regions, such as India (which was finally declared polio-free in 2011), Afghanistan, Pakistan, and Syria. The World Health Organization (WHO), in conjunction with governments worldwide, launched the Global Polio Eradication Initiative in 1988. Renewed in 2013, with injectable vaccines, the initiative aims to banish the infection by 2018.

▽ **Vaccine being administered**
Vast numbers of volunteer health workers have been involved in the polio eradication campaign. Here infants in Ghazni, Afghanistan, receive the vaccine in droplet form.

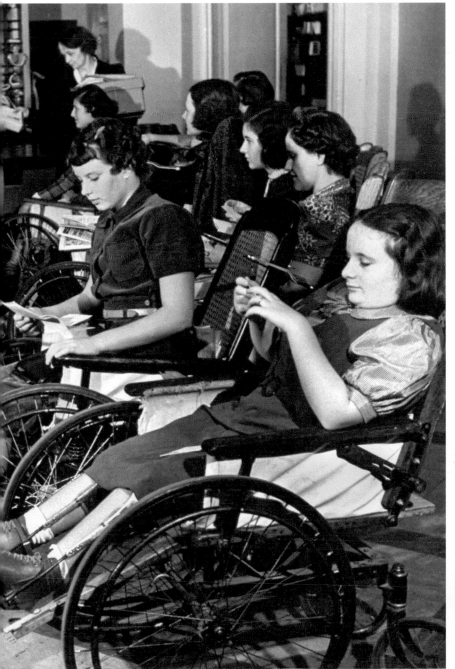

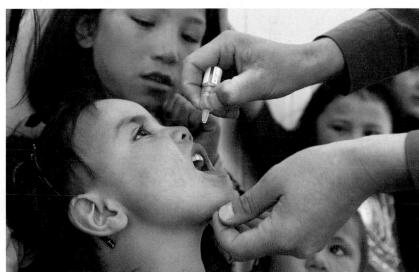

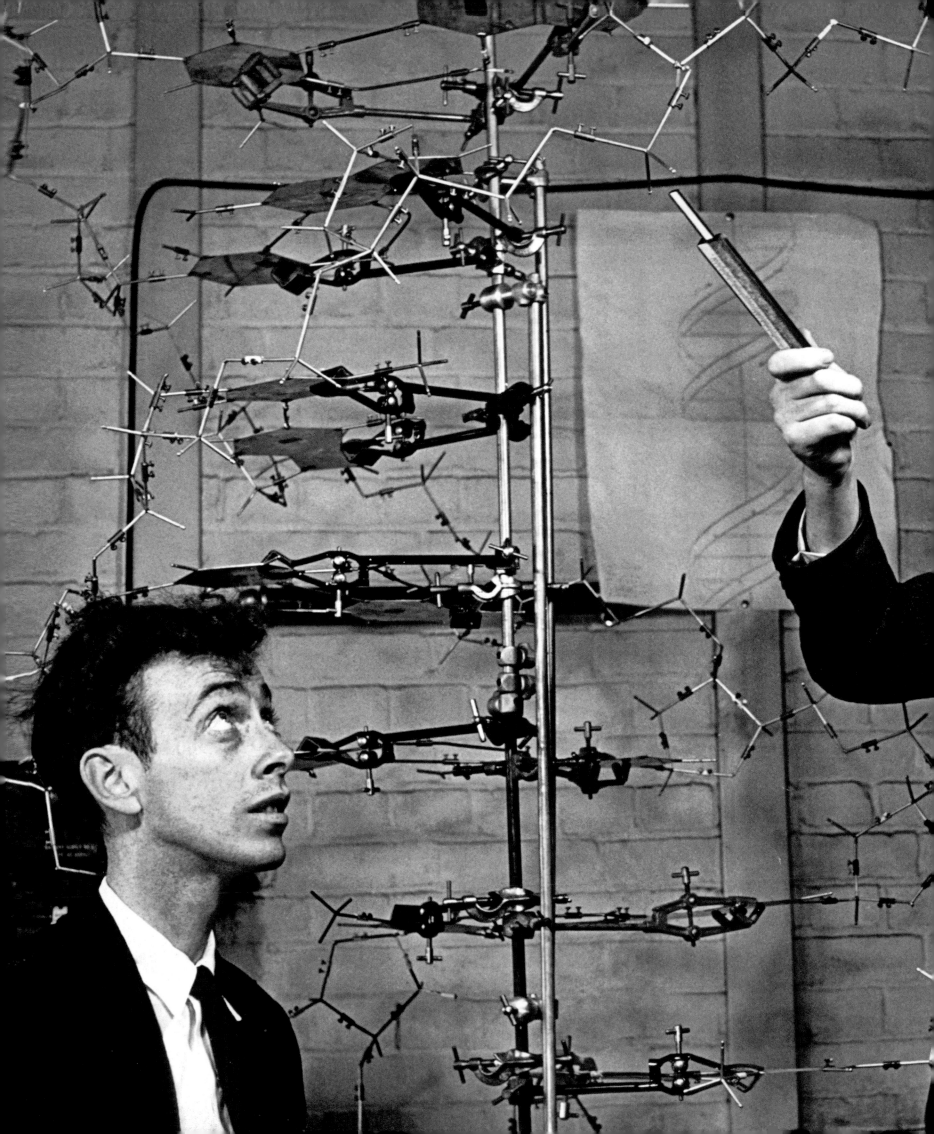

The **Structure** of **DNA**

One of science's greatest discoveries occurred in Cambridge, England, in 1953, when Francis Crick and James Watson proposed a "double-helix" structure for the genetic substance DNA. Their findings opened up vast new areas of biological research, and the potential to find causes and cures for innumerable medical conditions.

The momentous breakthrough by British scientist Francis Crick and American biologist James Watson was part of widespread, active research on DNA (deoxyribonucleic acid). It had already been established in 1944 that DNA carried genetic information. Crick and Watson knew that it somehow had to copy itself, to pass from one cell to the next, and from one individual to its offspring. Their key contribution lay in identifying DNA's twisted-ladder structure in which the "rungs" were made up of substances called bases, which were linked in specific pairs. The order of the bases formed the code for genetic information. The pairs could be unlinked so that each side of the ladder could build a new matching half, thereby producing two DNAs from one. Subsequent research unlocked the genetic code, provided insight into how genes work (see p.246) and patterns of heredity, and offered potential solutions to innumerable medical conditions, from inherited diseases to infections and cancers.

Watson and Crick were greatly helped by X-ray images taken by British biophysicist Rosalind Franklin. The breathtaking scale and possible applications of their discovery led to Crick, Watson, and their colleague Maurice Wilkins receiving the Nobel Prize in Physiology or Medicine in 1962. Franklin could not be included in the list because she had succumbed to ovarian cancer in 1958, and Nobel rules prevent posthumous nominations.

"We could 'play God' with the **molecular underpinning** of all of life."

JAMES WATSON, AMERICAN BIOLOGIST, FROM *DNA: THE SECRET OF LIFE*, 2003

◁ Watson and Crick's DNA model
Watson (left) and Crick often used paper and metal cutouts, balls, sticks, string, and laboratory glassware to model the angles and arrangements of atoms in molecules such as DNA shown here.

Inhalers and Nebulizers

Asthma and other respiratory conditions have been treated for thousands of years by inhaling medicine into the lungs. By the end of the 18th century, specific inhalation devices were invented, which have now evolved into sophisticated inhalers that can deliver precise doses of medication.

The benefits of breathing in the smoke of certain herbs, or the vapor from an infusion of them, were discovered at least 4,000 years ago. At first this was simply a matter of breathing in the fumes of a fire or steam from a cooking vessel, but more specific inhalation therapies evolved in many ancient civilizations. In Egypt, for example, herbs were put onto hot stones, and the resultant vapor was inhaled, while Indian ayurvedic medicine recommended smoking pipes that contained an herbal preparation of the roots of datura plants, which are now known to act as bronchodilators that open up constricted airways. Similar medicinal pipes were also used in Central and South America to smoke a number of different herbs.

Early inhalers

The first specific inhalation device was possibly designed in ancient Greece, and has been attributed to the physician Hippocrates (see pp.36–37). This inhaler was simply a cooking pot, with a hole in its lid through which a straw could be inserted. The pot would be filled with an infusion of herbs, spices, or other

▷ **First pressurized inhaler**
The "pulverisateur," invented by Jean Sales-Girons, won the 1858 silver prize of the Paris Academy of Science. By pressing on the handle, liquid is pumped from the reservoir and forced through a nozzle to form a spray.

"The appliance **reduces the solution to a mist** so fine it actually floats in air."

FROM INSTRUCTIONS FOR ASTHMANEFRIN HAND-BULB NEBULIZER, 1940

medicines and heated, and the vapor that arose would be inhaled through the straw. Despite its simplicity, this Greek invention became the model for the first modern inhalers, which appeared in the late 18th century.

At that time, the Industrial Revolution was well under way in England, and innovations of all kinds were in the air—as was a great deal of atmospheric pollution from burning coal by industry. So it is perhaps not surprising that it was an English physician, John Mudge, who, in 1778, revived the idea of an inhalation vessel, modifying a pewter beer tankard along the same lines as the Greek device. Holes in the handle of the tankard allowed air to be drawn through the liquid in the tankard and inhaled through a mouthpiece fitted to the lid. Mudge's invention soon caught on, and in the 19th century ceramic versions of this inhaler became common. While Mudge had advocated an infusion

of opium for his device, bronchodilating datura plants began to be imported from India in the early 1800s, and by the middle of the century, this became the standard drug for inhalation therapy. However, many users of these ceramic inhalers simply breathed in the steam from plain warm water.

Invention of atomizers

A breakthrough in the development of inhalers was made in the middle of the 19th century, with the invention of the atomizer by French doctor Auphon Euget-Les Bain. As much a product of the French perfume industry as an advance in medical technology, this device used air pressure to form a mist of droplets from a liquid, which could then be inhaled. This technology was used to develop nebulizers—machines that "atomize" liquid medicine into a mist or spray to be inhaled via nose, mouth, or both. Jean Sales-Giron's portable inhaler was

334
MILLION
The number of people in the world suffering from asthma, according to the 2012 Global Burden of Diseases study.

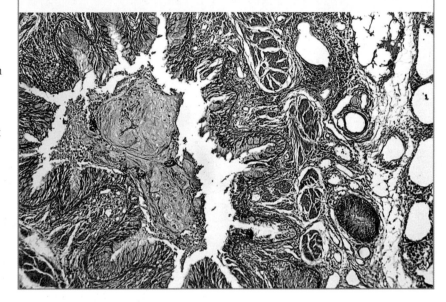

BRONCHIAL ASTHMA

The characteristic difficulty in breathing experienced during an asthma attack is due to the constriction and obstruction of the airways caused by inflammation. As well as swelling of the bronchi, muscles around the airways contract, and excess mucus is produced, restricting breathing still further.

The immediate effects can be alleviated by the inhalation of drugs to dilate the airways. The causes of asthma are complex, including genetic and environmental factors. Attacks can be triggered by pollution, pollen, animal hair (see pp.208–09), or changes in air temperature. Exercise, stress, or anxiety can also bring on an attack.

operated by a hand pump, while a steam-powered inhaler was developed by German doctor Emil Siegle. These atomizers were originally designed to deliver a mist of mineral waters from spas, and later included medicinal ingredients. However, at the turn of the 20th century the discovery that adrenal extract worked as a bronchodilator prompted the use of epinephrine in aerosols to treat asthma. The success of the drug led to widespread use of nebulizers, which became more portable and convenient to use with compressed air produced by squeezing a bulb. Another important innovation of the 1930s was the use of an

electric compressor that could be regulated more easily.

Further developments

Ensuring an accurate dosage of medication was still difficult, however. In 1948 researchers at Riker Laboratories in the US developed the Aerohaler, which gave a measured dose of isoprenaline or isoetharine powder when air was inhaled through it. The turning point came in the 1950s with the development of the pressurized metered-dose inhaler (pMDI), which used a valve on a pressurized container that delivered a single measured dose of atomized liquid.

Unfortunately, the pMDI devices relied on chlorofluorocarbon (CFC) propellants in the pressurized cartridge, which were banned in 1987 in an effort to protect the ozone layer. CFCs have been superseded by hydrofluoralkenes (HFAs). Also available are dry powder inhalers (DPIs), which were first developed in the 1970s.

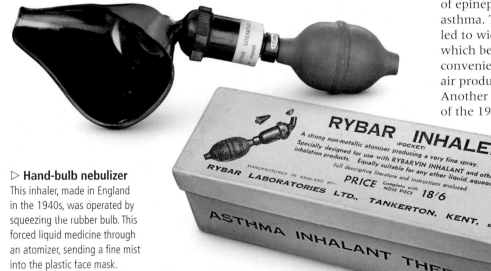

▷ **Hand-bulb nebulizer**
This inhaler, made in England in the 1940s, was operated by squeezing the rubber bulb. This forced liquid medicine through an atomizer, sending a fine mist into the plastic face mask.

Early MRI machine
One of the pioneers of MRI, Dr. Raymond Damadian (standing) demonstrates his body scanning machine, which used a "super magnet" to detect cancerous cells. The first MRI body scan in 1977 was of the thorax of his colleague Laurence Minkoff (seated), and took nearly five hours.

Scanning Machines

The discovery of X-rays in 1895 made it possible to see inside the body without surgery. Further advances in medical imaging came in the second half of the 20th century, with the development of various methods of scanning that provide more detailed, three-dimensional images.

Attempts in the early 20th century to improve the level of information provided by X-ray images (see pp.172–73) led to the invention of the first scanning technology—tomography, from the Greek *tomos*, meaning a slice or section. Initially, tomography involved moving an X-ray source and detector simultaneously over a patient, which created a single, blurred image with one plane in focus. By the mid-20th century, several X-ray images of a "slice" of an object or body from various angles could be assembled to create a composite image.

Computed tomography

The first computed tomography (CT) or computerized axial tomography (CAT) scanning machines appeared in the early 1970s, and were slow and cumbersome. Today, however, modern CT scanners can take thousands of X-ray readings in seconds, and interpret them to create a consolidated computer image almost immediately. As the patient passes through the hoop-shaped machine, the scanner rotates around the frame, sending and detecting beams of X-rays through the body. The data is then processed digitally to produce a three-dimensional (3D) image.

CT scanning became a mainstay of medical imaging soon after it was introduced. The principle of tomography, however, was not restricted to X-ray imaging. At the time the first CT scanners appeared, research was already underway into an alternative to X-rays—using radio waves and magnetic fields.

Magnetic resonance imaging

When a body is subjected to a very strong magnetic field, particles known as protons are forced into alignment. When they return to their original positions, the protons emit detectable radio signals. The protons of different tissues send out different signals, enabling the clear imaging of bones and soft tissues, including tumors.

Magnetic resonance imaging (MRI, see pp.232–33) was pioneered in the 1970s, and is now widespread. Unlike X-rays and CT scanning, MRI does not involve exposing the patient to radiation. However, the patient must remain still during an MRI scan, lying on a table that passes through the machine, which is effectively a hollow cylindrical magnet.

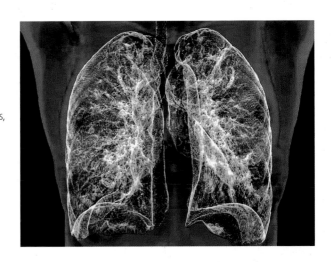

▷ **CT scan of healthy lungs**
The three-dimensional capabilities of CT scanning are especially useful in examining the internal structure of lungs, which cannot be as clearly seen using simple radiography.

Other imaging techniques

Developed in the 1950s, ultrasound scanning transmits high-frequency sound waves into the body and detects their echoes, using the same principle as the sonar used by submariners. Ultrasound machines can now produce real-time moving images, and have the advantage of being more mobile than other scanning technologies, with hand-held devices available.

A new generation of imaging machines is now being developed, based on discoveries in nuclear medicine. Molecular imaging techniques such as SPECT (single-photon emission computed tomography) and PET (positron emission tomography) use gamma rays, and are sometimes used in a hybrid machine along with CT or MRI scanning to take highly detailed images of the body.

◁ **Pulsed doppler**
Modern ultrasound scanners can assess the blood flow in veins and arteries by sending and receiving pulses of high-frequency sound waves, and taking into account the Doppler effect—changes in wave frequency that occur with direction of movement.

BRITISH ELECTRICAL ENGINEER (1919–2004)

SIR GODFREY HOUNSFIELD

Godfrey Hounsfield was a pioneer in the field of computed tomography in the 1960s and 1970s. Fascinated by electronics from childhood, Hounsfield worked on electronics and radar with the Royal Air Force during World War II before studying electrical engineering at Faraday House in London. In 1949 he joined Electrical and Musical Industries (EMI), where he developed the idea of building a cross-sectional image of an object from X-ray readings taken from various angles. He was a joint winner with physicist Allan McLeod Cormack of the 1979 Nobel Prize for Physiology or Medicine.

The **Pharmaceutical Industry**

The roots of the pharmaceutical industry can be traced as far back as the medieval period, to small apothecaries offering traditional remedies. Today, the industry has become a global phenomenon, a multi-billion dollar enterprise that produces thousands of new drugs and has revolutionized human health.

The last 200 years have seen enormous growth in the development of medicines. Advances in the field of chemistry during the 19th century led to the isolation of active ingredients, such as morphine extracted from opium, and quinine from the bark of the cinchona tree. In the 19th century, small firms such as Merck in Germany, Hoffmann-La Roche in Switzerland, Burroughs Wellcome in England, and Smith Kline in the US started making and selling these drugs wholesale. Other firms that are now well-known pharmaceutical companies began as makers of organic chemicals, producing dyes and other substances used in the textile industry. These include Bayer in Germany, and Pfizer in the US. Bayer's first move into manufacturing medicine came with the development of aspirin (see pp.170–71)—one of the most successful pharmaceutical products ever—which it began distributing in 1899.

The end of the 19th century also saw the development of vaccines, such as those for diphtheria and tetanus (see pp.158–59), which created further business for fledgling pharmaceutical firms. In 1909 German scientist Paul Ehrlich and his assistant Sahachiro Hato found that the compound Salvarsan was deadly to syphilis bacteria. Demand for the drug was so high that Ehrlich made it available commercially with no further tests (see pp.186–87).

The Germans and the Swiss became the dominant forces in the expanding pharmaceutical industry prior to World War I. However, in 1917, the aspirin trademark and its US assets were seized from German company Bayer and the US subsidiary of Merck was split off from its German parent company. Germany's position as leader of the pharmaceutical industry was compromised and other companies, particularly in the US, took advantage of this. The pharmaceutical industry was about to experience phenomenal global growth, boosted by breakthroughs such as the isolation of insulin to treat diabetes (see pp.190–91) and the discovery of penicillin in 1928 (see pp.198–99).

1 TRILLION US DOLLARS The total level of pharmaceutical revenue worldwide in 2014.

The thalidomide scandal

However, not all new drugs were entirely beneficial. There was little rigorous testing, and the toxicity of some drugs was discovered too late. One example of this came with the use of thalidomide in the 1950s and 1960s. Developed by West German company Chemi Grünenthal, thalidomide was introduced as a sleeping pill, and later given to pregnant mothers to ease morning sickness. After a short time on the market in Europe, obstetricians observed that babies were being born with deformities to their limbs. Research rapidly identified thalidomide as the culprit and it was withdrawn in 1961. Thalidomide led to a public outcry over the insufficient testing of drugs, and as a result major regulatory reforms were enforced.

Blockbuster drugs

A number of world-changing drugs were developed in the post-war era. The contraceptive pill, introduced in 1960 (see pp.224–25), transformed society and the lives of women. Valium (diazepam)—used mainly to treat anxiety disorders, seizures, and alcohol withdrawal symptoms—was brought to the market by

◁ **Thalidomide victim**

A victim of thalidomide leaves court in a wheelchair, while a protester stands with a banner reading "Thalidomide, Justice!" during the trial of the drug's manufacturer in Madrid, Spain. This was one of a spate of cases against the makers, Grünenthal.

"The thalidomide tragedy took place 50 years ago in a **world completely different** from today."

HARALD STOCK, GRÜNENTHAL'S CHIEF EXECUTIVE OFFICER, IN AN OFFICIAL APOLOGY, 2012

Pill production

After World War II, drugs were being mass produced and manufacturers began compromising on quality and safety. This technician did not observe hygiene measures, such as gloves and a mask, when working with these pill-coating machines, each with a capacity of 400,000 tablets.

Roche in 1963, followed by the introduction of the monoamine oxidase inhibitor (MAOI) class of antidepressants. The widely used drugs acetaminophen and ibuprofen were developed in 1956 and 1969, respectively. The 1970s brought a period of great development in the field of cancer treatment and medication. Angiotensin Converting Enzyme (ACE) inhibitors arrived in 1975, improving cardiac health, while in 1977, Tagamet (cimetidine), a medication for stomach ulcers, became the first ever "blockbuster" drug—earning its creators the Nobel Prize and its makers more than $1 billion a year. This sparked a new trend, as pharmaceutical companies competed to become the developers of the next "big thing" to generate huge profits.

Pharmaceuticals has grown to become one of the world's largest industries, but its reputation has taken a battering along the way. With some drugs costing upward of $100,000 for a full course, and with the cost of manufacturing comprising just a fraction of this, drug companies have been accused of profiteering, although the industry disputes this, citing huge research and development (R&D) costs. Companies have also been accused of pulling out of research into drugs that help the poorest of communities, even though the industry is worth over $300 billion per year. Despite the industry's massive size and influence today, many challenges remain, including finding a cure for diseases such as malaria and cancers.

▽ **Drug design**

A researcher uses a computer to model the binding of an anticancer (chemotherapy) drug to an enzyme. Computational biology enables the development of drugs, and understanding how they function.

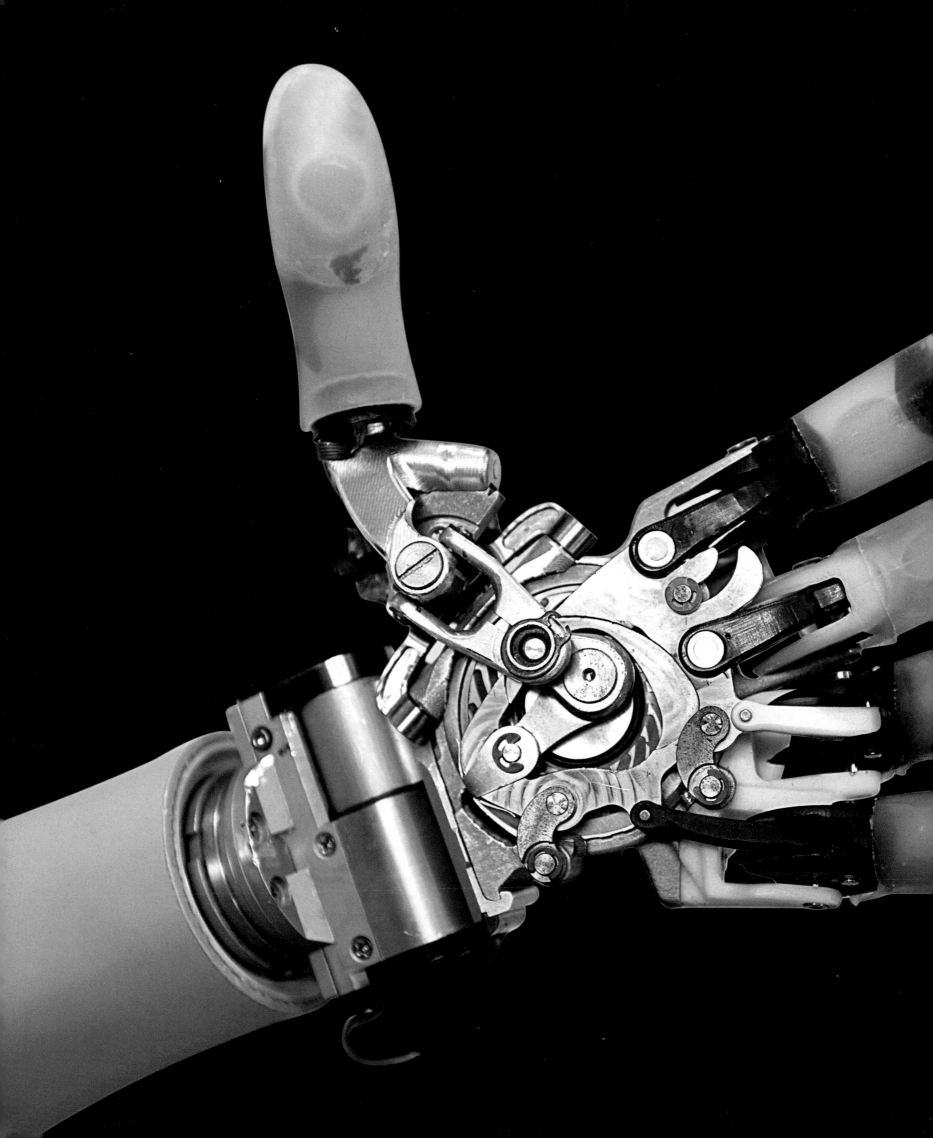

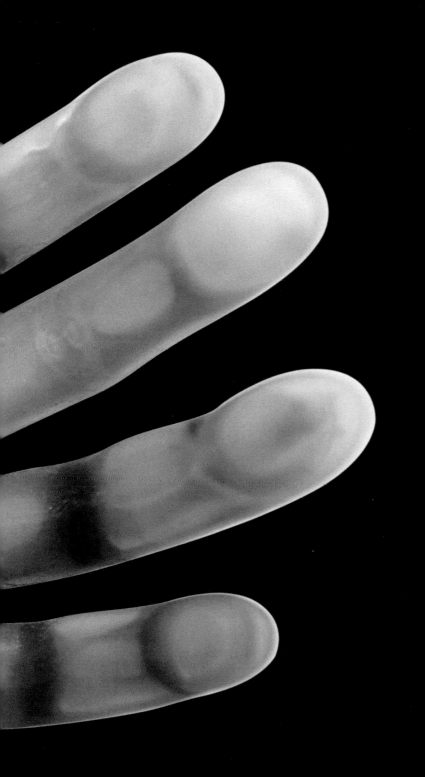

5
PROMISES
OLD AND NEW
1960 – PRESENT

PROMISES OLD AND NEW
1960–PRESENT

1960	**1970**	**1980**

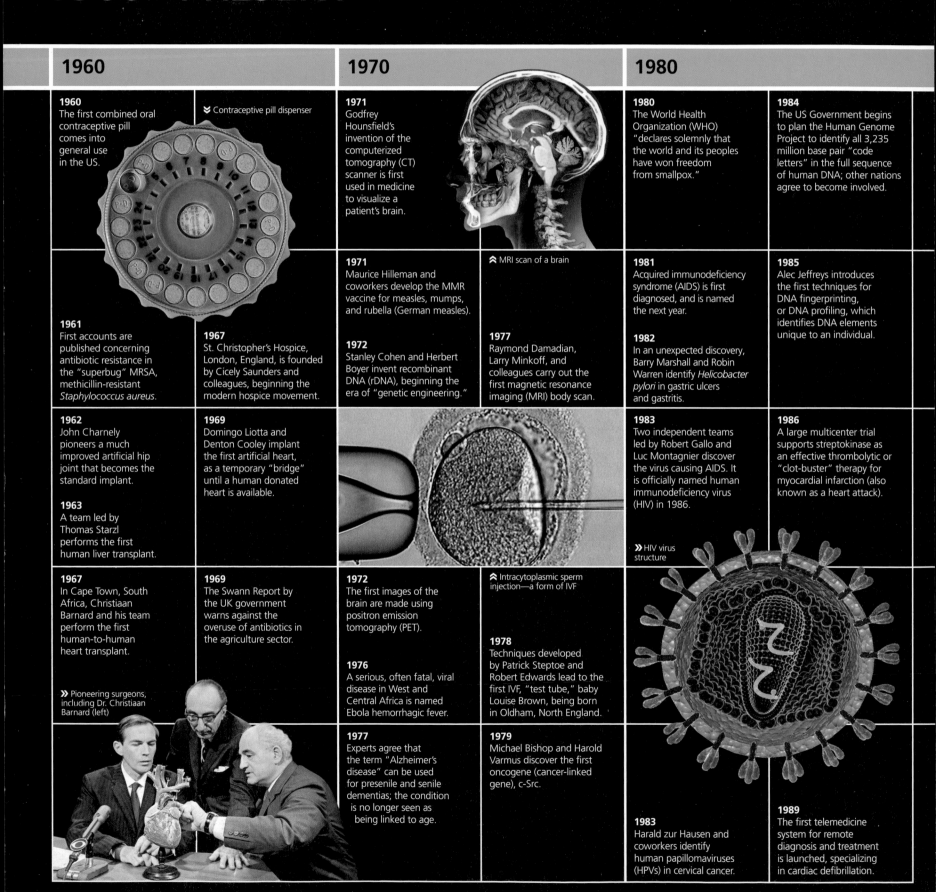

1960
The first combined oral contraceptive pill comes into general use in the US.

⌄ Contraceptive pill dispenser

1961
First accounts are published concerning antibiotic resistance in the "superbug" MRSA, methicillin-resistant *Staphylococcus aureus*.

1962
John Charnley pioneers a much improved artificial hip joint that becomes the standard implant.

1963
A team led by Thomas Starzl performs the first human liver transplant.

1967
In Cape Town, South Africa, Christiaan Barnard and his team perform the first human-to-human heart transplant.

❯❯ Pioneering surgeons, including Dr. Christiaan Barnard (left)

1967
St. Christopher's Hospice, London, England, is founded by Cicely Saunders and colleagues, beginning the modern hospice movement.

1969
Domingo Liotta and Denton Cooley implant the first artificial heart, as a temporary "bridge" until a human donated heart is available.

1969
The Swann Report by the UK government warns against the overuse of antibiotics in the agriculture sector.

1971
Godfrey Hounsfield's invention of the computerized tomography (CT) scanner is first used in medicine to visualize a patient's brain.

1971
Maurice Hilleman and coworkers develop the MMR vaccine for measles, mumps, and rubella (German measles).

1972
Stanley Cohen and Herbert Boyer invent recombinant DNA (rDNA), beginning the era of "genetic engineering."

1972
The first images of the brain are made using positron emission tomography (PET).

1976
A serious, often fatal, viral disease in West and Central Africa is named Ebola hemorrhagic fever.

1977
Experts agree that the term "Alzheimer's disease" can be used for presenile and senile dementias; the condition is no longer seen as being linked to age.

⌃ MRI scan of a brain

1977
Raymond Damadian, Larry Minkoff, and colleagues carry out the first magnetic resonance imaging (MRI) body scan.

⌃ Intracytoplasmic sperm injection—a form of IVF

1978
Techniques developed by Patrick Steptoe and Robert Edwards lead to the first IVF, "test tube," baby Louise Brown, being born in Oldham, North England.

1979
Michael Bishop and Harold Varmus discover the first oncogene (cancer-linked gene), c-Src.

1980
The World Health Organization (WHO) "declares solemnly that the world and its peoples have won freedom from smallpox."

1981
Acquired immunodeficiency syndrome (AIDS) is first diagnosed, and is named the next year.

1982
In an unexpected discovery, Barry Marshall and Robin Warren identify *Helicobacter pylori* in gastric ulcers and gastritis.

1983
Two independent teams led by Robert Gallo and Luc Montagnier discover the virus causing AIDS. It is officially named human immunodeficiency virus (HIV) in 1986.

❯❯ HIV virus structure

1983
Harald zur Hausen and coworkers identify human papillomaviruses (HPVs) in cervical cancer.

1984
The US Government begins to plan the Human Genome Project to identify all 3,235 million base pair "code letters" in the full sequence of human DNA; other nations agree to become involved.

1985
Alec Jeffreys introduces the first techniques for DNA fingerprinting, or DNA profiling, which identifies DNA elements unique to an individual.

1986
A large multicenter trial supports streptokinase as an effective thrombolytic or "clot-buster" therapy for myocardial infarction (also known as a heart attack).

1989
The first telemedicine system for remote diagnosis and treatment is launched, specializing in cardiac defibrillation.

The late 20th century raised great hopes that the new gene and stem cell therapies would revolutionize many forms of treatment, but the early 21st century moderated that optimism—while some techniques achieved success, others showed slow, patchy progress. Smallpox was eradicated but other infections continue to be stubbornly resistant.

Other fields race ahead, including preventive medicine, vaccination, and the battle against cancers. Technology also brings great progress, for example, noninvasive imaging and "bionic" implants. Advances in medicine have helped raise global life expectancy from 31 years in 1900 to 70-plus today, with more of those years in good health.

1990

2000

2010

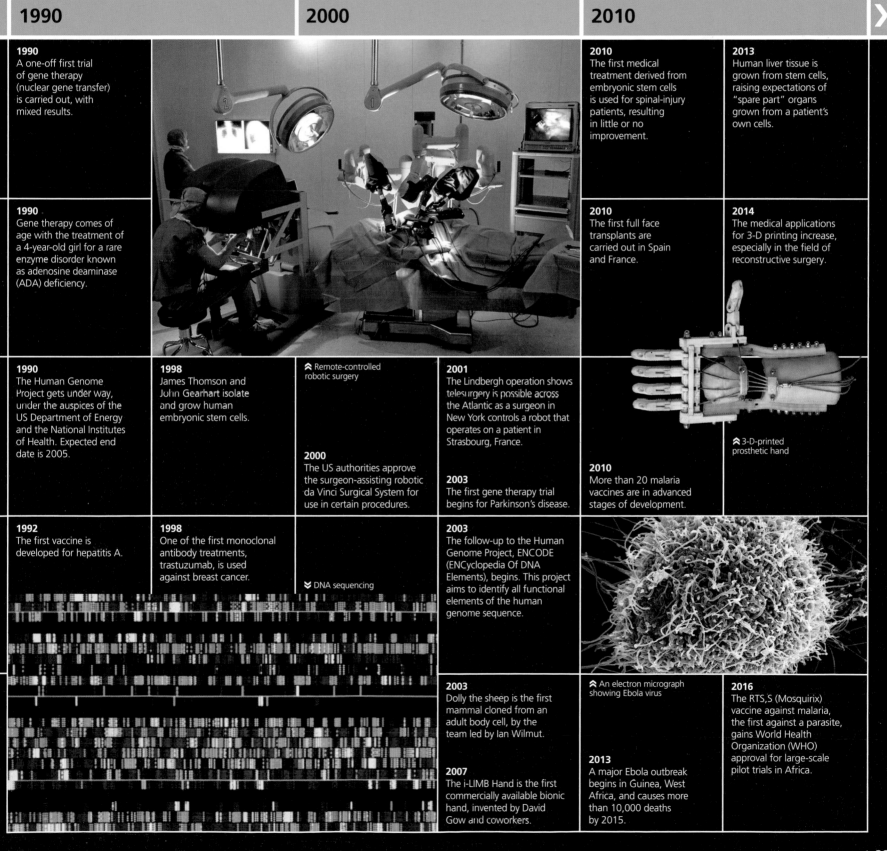

1990
A one-off first trial of gene therapy (nuclear gene transfer) is carried out, with mixed results.

1990
Gene therapy comes of age with the treatment of a 4-year-old girl for a rare enzyme disorder known as adenosine deaminase (ADA) deficiency.

1990
The Human Genome Project gets under way, under the auspices of the US Department of Energy and the National Institutes of Health. Expected end date is 2005.

1998
James Thomson and John Gearhart isolate and grow human embryonic stem cells.

1998
One of the first monoclonal antibody treatments, trastuzumab, is used against breast cancer.

≫ Remote-controlled robotic surgery

2000
The US authorities approve the surgeon-assisting robotic da Vinci Surgical System for use in certain procedures.

1992
The first vaccine is developed for hepatitis A.

≫ DNA sequencing

2001
The Lindbergh operation shows telesurgery is possible across the Atlantic as a surgeon in New York controls a robot that operates on a patient in Strasbourg, France.

2003
The first gene therapy trial begins for Parkinson's disease.

2003
The follow-up to the Human Genome Project, ENCODE (ENCyclopedia Of DNA Elements), begins. This project aims to identify all functional elements of the human genome sequence.

2003
Dolly the sheep is the first mammal cloned from an adult body cell, by the team led by Ian Wilmut.

2007
The i-LIMB Hand is the first commercially available bionic hand, invented by David Gow and coworkers.

2010
The first medical treatment derived from embryonic stem cells is used for spinal-injury patients, resulting in little or no improvement.

2010
The first full face transplants are carried out in Spain and France.

2010
More than 20 malaria vaccines are in advanced stages of development.

≫ An electron micrograph showing Ebola virus

2013
A major Ebola outbreak begins in Guinea, West Africa, and causes more than 10,000 deaths by 2015.

2013
Human liver tissue is grown from stem cells, raising expectations of "spare part" organs grown from a patient's own cells.

2014
The medical applications for 3-D printing increase, especially in the field of reconstructive surgery.

≫ 3-D-printed prosthetic hand

2016
The RTS,S (Mosquirix) vaccine against malaria, the first against a parasite, gains World Health Organization (WHO) approval for large-scale pilot trials in Africa.

The **Contraceptive Pill**

An oral contraceptive pill that prevented pregnancy first became available in 1960 in the US. Because it offered women the chance to control when they became pregnant, it rapidly ushered in enormous social, public health, and economic changes.

Historically, women had used a variety of methods to prevent conception. Some, such as the pessaries used by ancient Egyptian women made of cotton soaked in date juice, honey, and acacia, may have been effective. More dubious methods included the medieval advice of wearing outlandish wreaths of herbs, desiccated cat's livers, or weasel testicles to prevent pregnancy.

Condoms were first mentioned by Italian physicist Gabriele Fallopio in a 1564 work on syphilis as a means of preventing the disease and were in common use as contraceptives by the 17th century. By the late 19th century the most common form of contraceptive was the pessary, a hollow rubber hemisphere covering the outer cervix to prevent sperm entry; later designs included the diaphragm and cervical cap. All these methods, however, were cumbersome, costly, or simply did not work well.

17.1 The percentage of American women aged 15–44 using the pill as a contraceptive method in 2010.

Developing the pill

In the 1920s, research on hormones active in the female reproductive cycle (see p.205) offered hope that a chemical means of impeding pregnancy could be found. In 1921 Austrian physiologist Ludwig Haberlandt transplanted the ovary of a pregnant rat into one that was not pregnant, and the second rat stopped ovulating. This led to the discovery of progesterone, one of the main hormones involved in pregnancy, in 1934.

In 1942 American chemist Russell Marker found a way to extract progesterone from Mexican yams, and in 1951 Australian-born American Chemist Carl Djerassi developed a synthetic form that was powerful enough to mimic the body's natural production of progesterone. The final advances were made in the US by chemist Gregory Pincus and gynecologist John Rock, whose research found that progesterone halted ovulation in rabbits. They began tests using Djerassi's synthetic progesterone and a similar compound developed by Frank Colton at the Searle pharmaceutical company.

Consequences of contraception

In 1956 full-scale human trials of an oral combined (progesterone-estrogen) contraceptive began in Puerto Rico, involving more than 200 women. The trials had a 100 percent success rate in preventing pregnancy. In 1960 the Food and Drug Administration (FDA)—the US drugs regulator—approved the pill for contraceptive use, and by 1961 the UK and Germany had followed suit. By 1965 6.5 million American women were using the pill. Women could now control the timing of their pregnancies and avoid the risky, and often illegal, process of abortion. This led to a rise in female participation in the labor force, which in the US rose from 26.2 million in 1965 to 73 million by 2014. Some, though, were uneasy that the "sexual

revolution" the pill unleashed—in which intercourse was unlikely to lead to pregnancy—would have a corrosive effect on public morality. In 1968 Pope Paul VI banned the pill's use by Catholics.

Side effects, which had been glossed over in the initial enthusiasm of the pill's development, proved problematic too. Some of the Puerto Rican test subjects had experienced nausea and dizziness; subsequently oral contraceptive use was linked to thrombosis (blood clots) and coronary embolisms (heart vessel blockages). A US Senate hearing in

CONTRACEPTIVE PILL DISPENSER, 1960s

CONCEPT

HOW THE PILL WORKS

The contraceptive pill contains the female hormones estrogen and progesterone. Estrogen prevents the pituitary gland from secreting another hormone which normally triggers ovulation. The progesterone thins the lining of the womb as well, which makes it difficult for an egg to implant there, and it thickens the mucus around the cervix, making it harder for sperm to pass through.

> **"...** reproductive health includes **contraception** and **family planning** and access to legal, safe abortion."
>
> HILLARY CLINTON, US SECRETARY OF STATE, AT THE G8 FOREIGN MINISTERS' MEETING, 2010

1970 debated the problem and decided not to ban the pill but to enforce the addition of a health warning on its packaging.

In 1982 a pill with a lower dosage of progesterone became available, which reduced the side effects. In the 1990s emergency contraception pills with high progesterone dosages (or "morning-after pills") became available that could be taken days after intercourse. With research focusing on a male chemical contraceptive, using testosterone to retard sperm formation, the 21st century looks set to be an era of ever-wider reproductive choice.

▷ **Feminist movement**
Effective contraception emboldened feminist movements to campaign for equality of rights. In 1975 the UK passed a Sexual Discrimination Act, barring unequal treatment of men and women; in the US, however, an Equal Rights Amendment, put forward in 1972, was never ratified.

▽ **The summer of love**
The pill, with its sexually liberating effect, played a role in enabling counterculture movements, such as the "Summer of Love" in 1967, during which tens of thousands of hippies descended on San Francisco, California.

AMERICAN NURSE AND ACTIVIST Born 1879 Died 1966

Margaret Sanger

"No woman can call herself **free** until she can **choose** whether she will or will not be a **mother.**"

MARGARET SANGER, FROM *WOMAN AND THE NEW RACE*, 1922

Liberal thinker Margaret Sanger is credited with having given women the right to choose and use modern methods of "birth control"—a term she invented. A tireless activist, health worker, and campaigner, she argued against the Comstock Laws—US legislation that limited access to contraceptives and information regarding them.

Early years

Sanger was greatly affected by her upbringing as one of 11 offspring surviving from her mother's 18 pregnancies over a period

▷ **Activist on trial**
This picture of Sanger was taken at the time of her 1916 trial, after she opened the first US birth control center in New York City. Although she was found guilty and the clinic closed, her campaign gained huge momentum.

of 23 years. Her mother's early death at the age of 49 set Sanger on a course toward the nursing profession. She interrupted her training to marry, have three children of her own, and share her developing radical views about society and politics with her husband William Sanger. After moving to New York City, she began to join left-wing groups, attend socialist rallies, and also nurse in the city's poorer areas. Here she encountered women from working families who had no knowledge or means to obtain contraception and who had suffered at the hands of "backstreet," or illegal, abortionists, often with serious, long-term harm.

Sanger felt that women should be in charge of their bodies, their health, and their ability to enjoy sex without it leading to motherhood. But, in the early 1900s, the Comstock laws—anti-obscenity legislation originally passed in 1873—bracketed contraception with indecency and prostitution. It was therefore a federal offense to disseminate

△ **Radical women's journal**
Margaret Sanger published the first issue of *The Woman Rebel* in March 1914. The journal embraced the view that every woman should be the "absolute mistress of her own body."

▷ Birth control in court

Crowds gather outside a courtroom in New York City in 1929 to attend hearings regarding a raid on a birth control clinic. Clinics that dispensed contraceptive information faced constant harassment under the Comstock legislation.

birth control information or items through the mail or across state lines. Individual US states strengthened the legislation and Connecticut made it a punishable offense to use birth control, even by a married couple in private.

Contraceptive methods at the time were less varied and effective than today. Condoms were crudely made and under the control of the male partner. Women had "feminine hygiene" products such as foams, creams, douches, and suppositories, but these were embarrassing to use, often unreliable, and sometimes harmful.

Decriminalizing contraception

Around 1911 Sanger began her prolific writing career in the *New York Call*, a daily newspaper aligned with the Socialist Party of America. Her pamphlets described sexuality and associated matters in an open manner. She specialized in providing birth control information and contraceptives.

In 1915 she was charged with sending diaphragms by mail. The next year, she opened the US's first birth control clinic in Brownsville, Brooklyn. She was promptly arrested and sent to jail. The clinic was shut down but the case brought much publicity and an increase in support.

A new ruling in 1918 allowed doctors to provide contraceptive information for medical reasons—the first relaxation of the Comstock

> # "**Contraceptives** can put an **end** to the horrors of **abortion** and **infanticide.**"
>
> MARGARET SANGER, FROM *WOMAN AND THE NEW RACE*, 1922

laws. In 1921 Sanger and her colleagues founded the American Birth Control League. She toured, lectured, and wrote in order to reach a wider audience. In 1923 she and her associates set up the Clinical Research Bureau to provide women with birth control for "therapeutic reasons." She invited prosecution again in 1936 when she mail-ordered contraceptives. The case resulted in further easing of the

Comstock laws. The next year, in a landmark decision, the American Medical Association agreed that birth control should be a standard service provided by physicians.

The finish line

In the 1940s Sanger took her efforts to the wider world, becoming a founding member of the International Committee on Planned Parenthood. She had envisioned a contraceptive pill that was cheap, effective, easy to use, and under the woman's control, and she played a major role in the development of this pill—the oral contraceptive (see pp.224–25)—in the 1950s. The year before she died, Sanger also saw the final Comstock restrictions removed when the case of Griswold v. Connecticut ruled that private use of birth control was a legal right of all US citizens.

◁ Lobbying for birth control

Two women sit in an office of the National Committee on Federal Legislation for Birth Control. Sanger and her co-workers set up the Committee in 1929 to lobby against laws that categorized contraception as obscene and immoral.

Known since antiquity, different cancers affect varied parts of the body, have diverse causes, and some tend to affect people of a certain age, gender, or occupation. This is probably why they were not grouped as a single disease until the 18th century.

Despite their diversity, all cancers have the same underlying basis: out-of-control cells. Instead of following the usual path of growing, doing their job, and dying as part of the body's natural cell turnover, mutant cells fail to perform normally, multiply too fast, accumulate into a tumor or growth, and may also spread to where they should not be. Self-contained benign tumors are noncancerous. In malignant (or cancerous) tumors, cells detach and disperse to other parts of the body, forming secondary tumors, in a process called metastasis. Some cancers do not form physical tumors but cells still multiply out of control, as in certain leukemias that affect the blood.

Cancer through the ages

Evidence of cancerous growths dates back more than 3,000 years to ancient Egypt. Abnormal bone shapes in excavated mummies are suggestive of tumors, and the Edwin Smith and George Ebers papyri (see pp.20–21) describe probable cancers, especially of the breast. Even older accounts from the Sumerian civilization report "ulcers that spread," while in India, the *Susruta Samhita* (see pp.30–31)

8.5 MILLION Estimated number of cancer deaths in the world every year. Cancers are the second biggest killers in the US, and are responsible for two in five deaths in the UK.

describes growths on the skin, in the rectum, and in urinary passages.

Ancient Greek physician Hippocrates (see pp.36–37) noted that formations of new blood vessels in and around a tumor resembled the limbs of a crab, or *karkinos* in Greek. Galen (see pp.40–41) used *onkos* to refer to a nonspreading swelling, mass or tumor; oncology, the branch of medicine dedicated to cancer, is derived from this term.

Great Islamic physicians (see pp.48–51) such as al-Razi and Ibn Sina described various growths of the eye, nose, tongue, stomach, liver, kidney, bladder, testes, and breast.

Theories of cause

Ancient Greeks attributed the causes of cancers to imbalance in the body's four humors (see pp.34–35). In the 17th century some form of contagion was favored; later cancers were thought to have a parasitic origin. In 1761 a more scientific basis for cancers was established when Italian anatomist Giovanni Morgagni began linking features of his patients' illnesses to abnormal findings in autopsies after their deaths.

In 1838 German physiologist and microscopist Johannes Müller proposed the blastema theory: according to this, cancers arose from budding elements (blastema) sprinkled between normal tissues. His student, German pathologist Rudolph Virchow (see pp.152–53) proposed some form of tissue

Cancers

One of the five major causes of death worldwide, cancer has always been regarded as a mystery, and its causes the subject of many theories. However, in the last half-century there has been huge progress in the understanding and treatment of cancers.

CONCEPT

HOW CELLS TURN CANCEROUS

Genes are the instructions for how a cell grows, functions, divides, and dies in a preprogrammed way. They are carried on threadlike chromosomes, found in all cells. Proto-oncogenes are specific genes in a cell responsible for regulating normal processes such as cell division, mending damaged genes, and self-destruction of cells with faulty genes. Exposure to carcinogens, such as ultraviolet light, certain chemicals, and viruses can alter or mutate proto-oncogenes, and although they can usually be repaired naturally, this sometimes fails. Progressive damage changes proto-oncogenes into oncogenes, which may then cause the cell to function abnormally, and eventually to become cancerous. When faulty oncogenes are inherited, a cell may turn cancerous sooner.

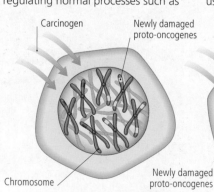

Carcinogen
Newly damaged proto-oncogenes
Chromosome
Newly damaged proto-oncogenes

CARCINOGENS PENETRATE CELL

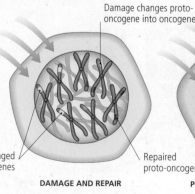

Damage changes proto-oncogene into oncogene
Repaired proto-oncogene

DAMAGE AND REPAIR

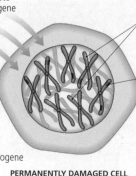

Oncogenes accumulate
Fewer repaired proto-oncogenes

PERMANENTLY DAMAGED CELL

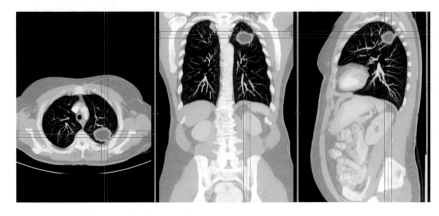

▷ CT scan showing lung cancer
Like many cancers, lung cancer can begin with no or only few, mild symptoms. It may already have dispersed to other body parts, or metastasized, before it can be detected in a scan as a growth (seen here in blue).

irritation, while another suspected cause that held sway until the 1910s was trauma or physical damage.

In the mid-20th century, studies suggested a link between smoking tobacco and a rise in lung cancers. A chemical trigger for cancer had already been demonstrated as early as 1775 by Percivall Pott (see p.230), but in the 1940s and 1950s a wealth of incriminating evidence showed that chemicals in tobacco smoke were carcinogens (cancer-causing). The tobacco industry fought long and hard against the accumulating

medical evidence, but by the 1960s smoking was established as a major cause of cancers.

As well as carcinogenic chemicals, factors such as exposure to radiation and ultraviolet light, viruses, and genetic tendencies can cause cancer by mutating or changing healthy genes into oncogenes that interfere with the growth and multiplication of cells (see panel). The 1989 Nobel Prize in Physiology or Medicine was awarded to US researchers J. Michael Bishop and Harold Varmus for revealing the role of oncogenes. Their research showed that oncogenes did not originate in viruses, as was previously

thought, but are human genes mutated and carried by viruses. Normally, a gene family known as the tumor suppressor gene slows down cell division, repairs faulty genetic material, and programs cell death, but in the case of cancer cells it does not.

A number of viruses are now known to induce cancerous changes. They include hepatitis B or C viruses, which are linked to liver cancer, and HIV, which is linked with cancers such as Kaposi's sarcoma and a type of lymphoma.

Staging and screening

In 1977 the American Joint Committee on Cancer (AJCC) published its first guidelines to TNM staging—a method of »

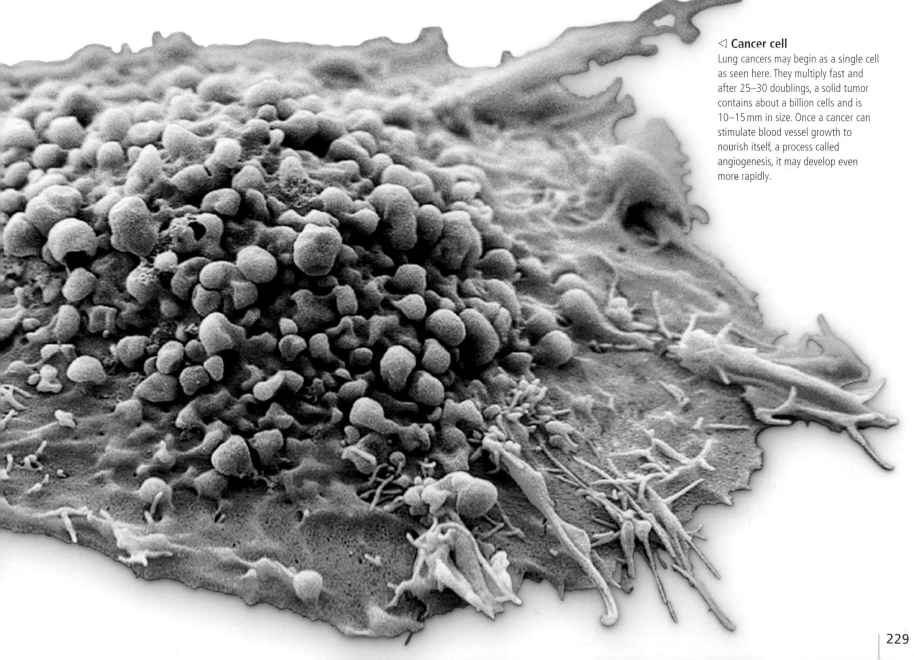

◁ Cancer cell
Lung cancers may begin as a single cell as seen here. They multiply fast and after 25–30 doublings, a solid tumor contains about a billion cells and is 10–15 mm in size. Once a cancer can stimulate blood vessel growth to nourish itself, a process called angiogenesis, it may develop even more rapidly.

▷ **Early chemotherapy**
Methotrexate was an early anticancer agent, first recommended for leukemias in the 1950s. It is an antifolate that interferes with the way cells use folic acid, a B vitamin used in the making of DNA.

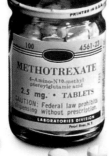

» assessing how far a cancer has developed and spread. T is the size of the primary tumor; N shows involvement of nearby lymph nodes, or glands, since cancers often spread through the lymph system (see p.290–91); while M refers to metastases.

Two types of cancer screening are used. One tests individuals for inherited genes that increase the risk of cancer developing, for example, breast cancer genes BRCA1 and BRCA2 (BReast CAncer types 1 and 2), both discovered in the early 1990s.

The second type of screening is to detect cancers in apparently healthy people for improved outcomes. The first was the Pap, or smear, test for cervical cancer (see

pp.204–05), developed in the 1920s by Greek doctor George Papanicolaou. It came into general use in the 1960s, and was followed in the 1970s by mammography—X-ray to detect lumps in the breasts. US advice also includes annual colorectal cancer screening from 50 on.

Surgery and pathology

In the 18th century Scottish surgeon John Hunter was one of the first to draw distinctions between tumors that might be safely removed and those that might not. US surgeon William Halsted introduced radical mastectomy for breast cancer in 1882 to improve survival rates. Much progress in surgery came with advances such as anesthetics and antisepsis (see pp.128–29).

The use of microscopes enabled pathological studies of abnormal cells and tissues to help establish whether surgery had eliminated a growth completely. Further advances included X-rays (see pp.172–73), scans, and endoscopy to locate tumors. Advances in treatment included highly targeted liquid nitrogen to freeze cancer cells and lasers to burn them.

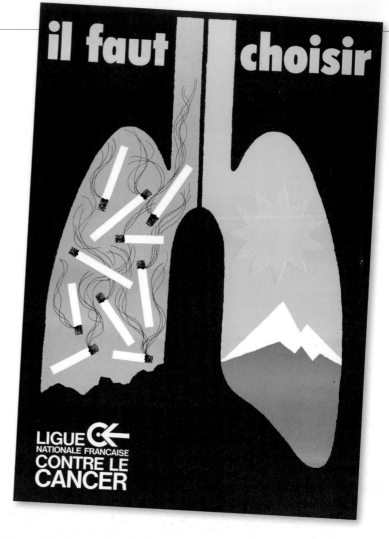

il faut choisir

LIGUE NATIONALE FRANÇAISE CONTRE LE CANCER

Radio- and chemotherapy

Radiotherapy treatment for cancer began soon after the discovery of X-rays in 1896 as researchers realized that X-rays could be effective against rapidly dividing cells. In the 1920s French doctor Claudius Regaud found that successive smaller doses of X-rays worked as well as one large dose, but had fewer harmful side effects. Technological advances allowed for better control in the power and direction of radiation. Internal radiotherapy, developed in the 1900s, involved inserting pellets of a radioactive material next to the tumor.

Paul Ehrlich, a German immunologist, pioneered chemotherapy —use of synthetic chemicals to fight disease—in 1910. One of the first chemotherapeutic anticancer agents was chlormethine (mustine), used in the 1940s. The toxic effects of mustard gas on fast-dividing healthy cells—such as those in bone marrow that produce blood cells—had been noticed in World War I. Later research into chemicals

△ **Potent carcinogens**
Tobacco smoke carries more than 70 chemicals linked to cancers—not only of the lungs and airways, but also the gullet, stomach, bowel, pancreas, liver, kidney, blood, ovary, and breast. The cost of smoking to world health is immense, estimated at $500 billion yearly.

related to mustard gas showed they could suppress tumor growth. Hundreds of other chemotherapeutic agents have since been discovered, sometimes combined into "cocktails."

Anticancer immunotherapy harnesses the body's own immune system to seek and destroy cancer cells with, for example, specially made antibodies that are called monoclonal antibodies or MCAs. Among the first was trastuzumab, which was first used against breast cancer in 1998.

While chemotherapy tends to attack most kinds of fast-dividing cells and causes side effects, targeted therapies aim more precisely at cancer cells, which reduces collateral damage. One

15 PERCENT of female cancer deaths worldwide are from breast cancer.

25 PERCENT of male cancer deaths worldwide are from lung cancer.

BRITISH SURGEON (1714–88)

PERCIVALL POTT

With more than 40 years of service at St. Bartholomew's Hospital in London, Pott was the most renowned surgeon of his day. He was accepted into the Company of Barber-Surgeons in 1736; in 1745 he became assistant surgeon, and full surgeon in 1749. He wrote on many diseases and injuries; some still bear his name.

In 1775 he described how scrotum cancer was more common among chimney sweeps, especially boys who worked for many years and developed cancer around puberty. This was one of the first medical associations between a carcinogenic substance (soot), an occupation (chimney sweep), and cancer. His work led to laws to improve working conditions for chimney sweeps.

targeted approach, known as angiogenesis inhibition, prevents the growth of new blood vessels, which a tumor needs to survive.

Hormonal therapy dates back to 1896 when British surgeon Thomas Beatson tried ovary removal in patients with breast cancer, with some success. This therapy also led to development of drugs such as tamoxifen, which stops the female sex hormone estrogen from instructing breast cancer cells to grow. US-based researcher Charles Huggins received the 1966 Nobel Prize in Physiology or Medicine for his work on hormonal treatment for prostate cancer. Other Nobel recipients for cancer research include Leland Hartwell, Tim Hunt, and Paul Nurse for their work on key regulators of the cell cycle, and German virus expert Harald zur Hausen for discovery of the human papilloma viruses that cause cervical cancer (see pp.244–45).

> " Cancer is an uneven **swelling, rough, unseemly, darkish, painful**… and if operated on, it becomes worse and **spreads by erosion…**"
>
> PAUL OF AEGINA (625–690 CE), *MEDICAL COMPENDIUM IN SEVEN BOOKS*

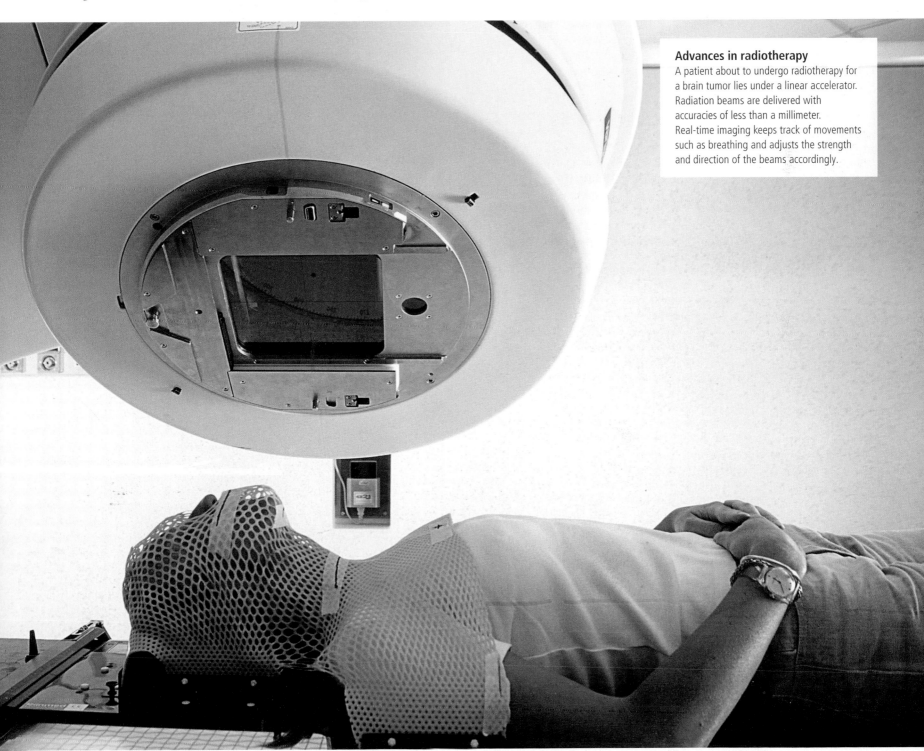

Advances in radiotherapy
A patient about to undergo radiotherapy for a brain tumor lies under a linear accelerator. Radiation beams are delivered with accuracies of less than a millimeter. Real-time imaging keeps track of movements such as breathing and adjusts the strength and direction of the beams accordingly.

Advanced Imaging

In 1895 the use of X-rays (see pp.172–73) provided the first noninvasive internal views of the human body. The following century saw much progress in imaging technologies, and the foremost of these was magnetic resonance imaging (MRI), introduced in 1980.

Medical imaging often developed alongside, and perhaps relied on, other areas of research such as in biology, physics, electronics, and computing. The concept of MRI stemmed from nuclear magnetic resonance (NMR), in which, under the influence of a magnetic field, atomic nuclei absorb and release electromagnetic radiation (see p.217). MRI exploits this behavior of nuclei, particularly of hydrogen—a common substance in the body. During an MRI scan, nuclei in the body are exposed to intense magnetic fields, which causes them to align in the same direction. They are then subjected to a strong pulse of radio frequency energy, after which the nuclei return to their original alignment. Now, they emit micropulses of their own and, using complex computer analysis, are turned into visual cross sections, and then assembled into 3-D images.

MRI has advantages over other kinds of imaging, such as a lack of potentially harmful radiation, finer resolution for detail, and the ability to differentiate between various kinds of soft and hard tissue. It is particularly used in neurology to visualize the brain and nerves. Functional magnetic resonance imaging (fMRI)—an offshoot of MRI—reveals real-time energy used by parts of the brain, indicating which areas are "thinking" hardest. MRI is now a springboard for generating new imaging techniques to view the body in even greater detail.

> "Maybe we could build a **scanner** that would… **hunt down cancer**… I had hope."
>
> RAYMOND DAMADIAN, INVENTOR OF THE MRI MACHINE, 2011

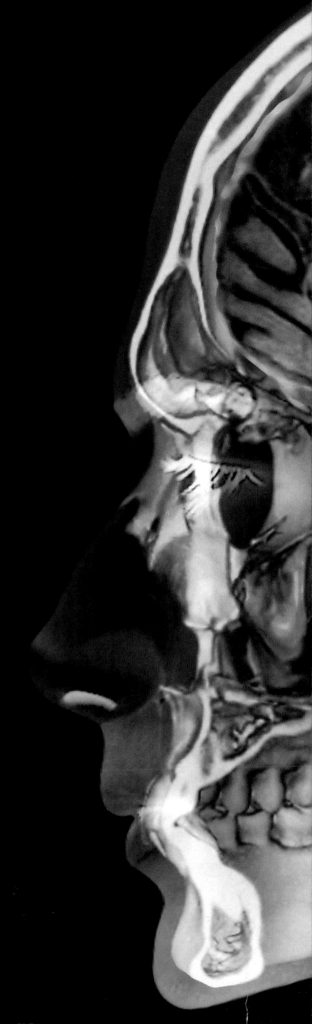

▷ Brain scan
This MRI scan of a 35-year-old patient's head reveals fine details, including nerve fibers, blood vessels, connective tissue, and the fluid-filled cavity under the wrinkled cortex. The face and neck bones have been imaged by computerized tomography (CT).

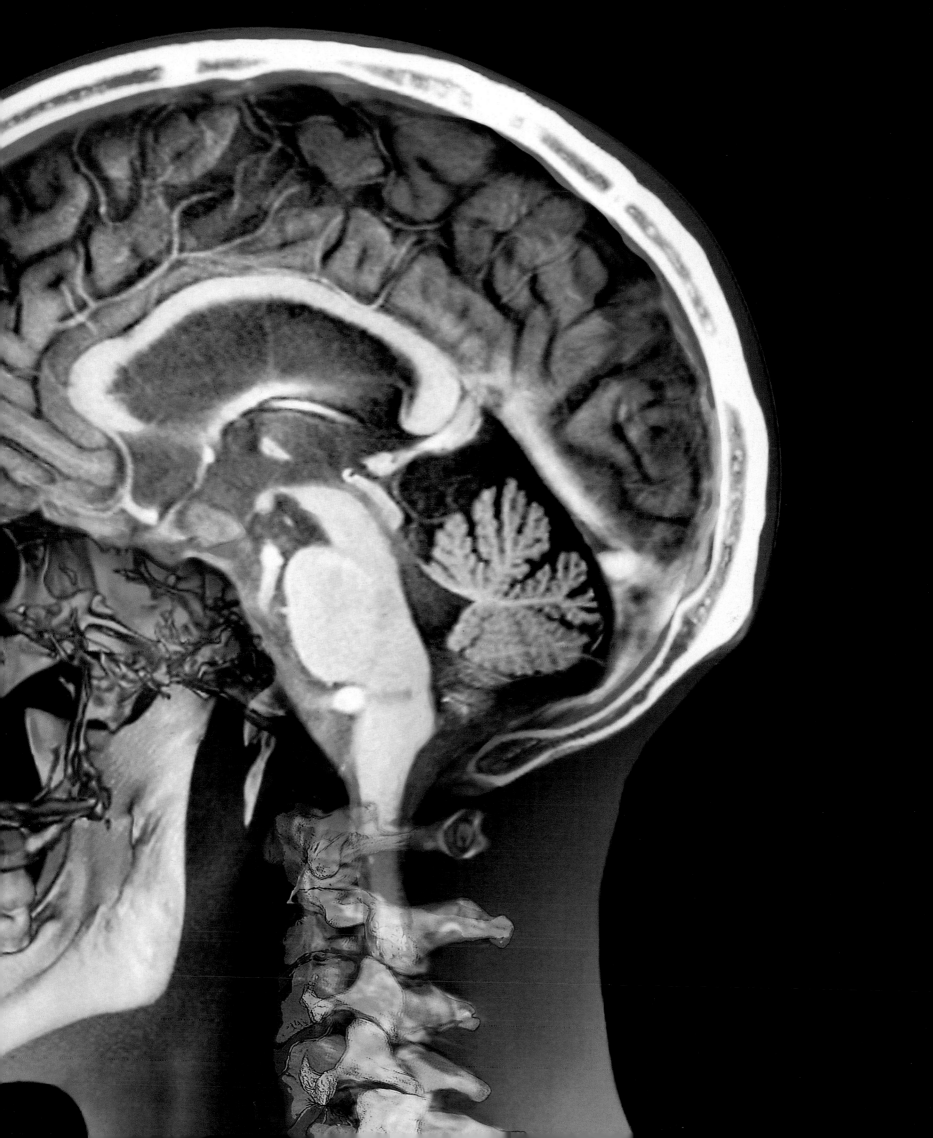

The **First Heart Transplant**

Before the 1960s severe coronary artery disease and congestive heart failure spelled a death sentence for patients. Then in 1967 South African surgeon Christiaan Barnard transplanted a heart from a human donor to a critically ill patient, heralding a new era in transplant surgery.

The first slow steps toward successful transplantation began in the 1890s, when skin grafts using the patient's own tissue were successfully carried out; but those using donor skin, for example from corpses, had little success. An attempt to transplant a donor pancreas in 1894 also failed. Doctors did not as yet understand the role of the immune system in accepting donor organs.

Another prerequisite for the complex surgery needed to remove diseased organs and replace them with healthy donated ones, was the ability to perform vascular suturing, that is, sew back torn or severed blood vessels. This procedure was established by French surgeon Alexis Carrel from 1901 to 1910.

First transplants

The first transplantation attempts were carried out on dogs, starting with their kidneys. The earliest successful canine heart transplant was carried out by Norman Lumway and Richard Lower at Stanford, California, in 1959. They used a technique of topical hypothermia, in which the donor heart is frozen outside the body, preserving its functions for several hours while surgery is carried out. In 1954 the science of transplantation took a major step forward with the first successful human kidney transplant—between identical twins. Rejection—when the recipient's immune system attacks a donor organ because it recognizes it as foreign tissue—was a danger, but in this case it was low as donor and recipient were genetically identical, and the kidney recipient lived for nine years. In general the survival rate was much lower because

▷ **Pioneering surgery**
Christiaan Barnard (left) with Michael DeBakey (centre) and Adrian Kantrowitz. Kantrowitz performed the first paediatric heart surgery, on a 2-day-old baby, just three days after Barnard's first operation, while DeBakey developed an early form of artificial heart.

rejection was common. The only way to prevent it was by massive irradiation with X-rays to suppress the recipient's immune system. In 1959 the first immunosuppressant drugs—which dampen down the

33 YEARS Longest period survived by a heart transplant patient— John McCafferty, who died on February 9, 2016.

body's immune system—were developed by British surgeon Roy Calne, and survival rates soon improved. These drugs, as well as refinements to the heart-lung machines that could take over the function of those organs during surgery, made heart transplants feasible. However, ethical concerns delayed surgeons from performing the operation for several years.

The big leap

On December 3, 1967, Christiaan Barnard, a South African surgeon at Groote Schuur Hospital in Cape Town, successfully transplanted a heart from a donor—a 24-year-old woman who had died in a traffic accident—into the body of 54-year-old Louis Washkansky, who was suffering from a terminal heart disease. The operation took nearly

---

IN PRACTICE

HEART TRANSPLANT

Transplant surgery can now be performed in a range of different ways, and can also take into account the condition of the old and donor heart.

In heterotopic transplantation, the patient's old heart is left in place, giving it a chance to recover or take over if the new heart fails. On the other hand, in orthotopic heart surgery, the patient's chest is opened up and the blood vessels (such as the aorta and pulmonary artery) are dissected before the old heart is

removed. The new heart—kept alive by hypothermia during the procedure—is then sutured into place.

In 2006 "beating heart" surgery was introduced, where the donor heart is not cooled but connected to a machine that allows it to pump blood, increasing the amount of time it can be kept viable.

HETEROTOPIC HEART TRANSPLANT

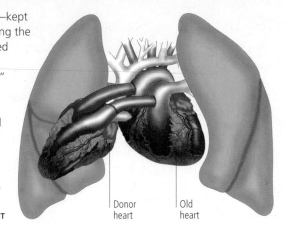

Donor heart | Old heart

---

five hours: the surgeons first removed Washkansky's diseased heart, and then carefully sutured the blood vessels in his chest to the donor organ. The new heart failed to beat at first, but it was shocked into life with a defibrillator. The operation had worked, but 19 days later Washkansky succumbed to pneumonia, aggravated by an immune system that had been suppressed to stop organ rejection.

In January 1968 Barnard operated on his second patient, Philip Blaiberg, who lived for 594 days. Surgeons in other countries began

conducting heart transplants and by 1971, 180 operations had been performed. However, survival rates remained disappointing. There was a high rate of rejection and the immunosuppressant drugs had severe side effects. In 1976 Belgian immunologist J. F. Borel discovered the immunosuppressant qualities of cyclosporine. It had far fewer toxic side effects than previous antirejection drugs and was licensed for use in transplant surgery in 1983. When tried in heart surgery, survival rates for patients improved and the number

of transplant operations rapidly increased, reaching around 3,500 per year by the early 21st century.

Some heart transplant patients have now lived for over 30 years and the 10-year survival rate has reached 65 to 70 percent. A continuing problem is transplant coronary artery disease—the excessive narrowing of arteries where blood vessels are sewn back together during surgery. Resolving this, the most common cause of death after heart transplantation, is one of the most serious challenges faced by heart surgeons today.

"It is infinitely **better to transplant** a heart **than** to **bury** it to be devoured by worms."

CHRISTIAAN BARNARD, SOUTH AFRICAN SURGEON, TO *TIME* MAGAZINE, 1969

Left inflow

Left outflow

▷ **Artificial heart**
The first artificial heart—designed to keep a patient alive while waiting for a donor heart—was implanted in 1982. This AbioCor artificial heart, first implanted in 2001, has an internal battery that lasts 4 hours, and has a lifespan of 18 months.

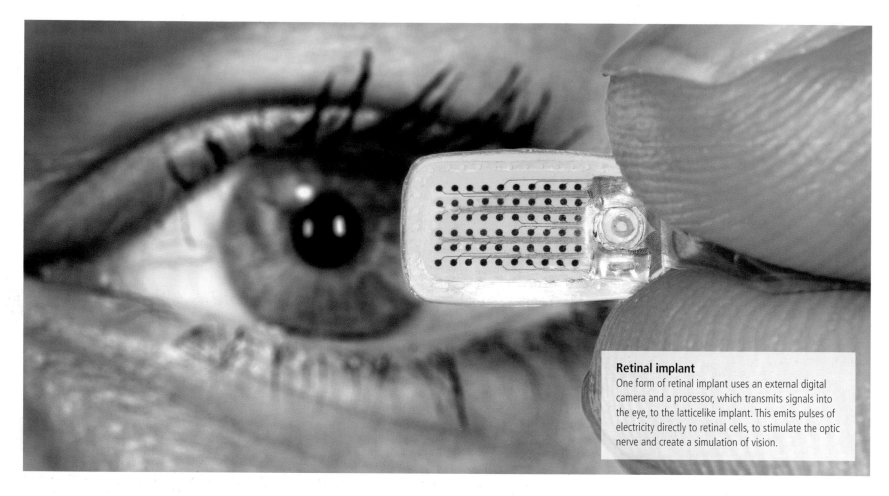

Retinal implant
One form of retinal implant uses an external digital camera and a processor, which transmits signals into the eye, to the latticelike implant. This emits pulses of electricity directly to retinal cells, to stimulate the optic nerve and create a simulation of vision.

Implants and Prostheses

Since ancient times, physicians have used artificial body parts to replace damaged limbs, eyes, teeth, and other organs. As anatomical understanding grew and new materials were developed, the functionality of protheses increased. Researchers are now working on implants that enable prostheses to be controlled by nerve signals.

▷ **Early eyeball**
Dating to 4,800 years ago, this artificial eyeball found in Iran is the oldest known eye prosthesis. Made of tar and animal fat, it has gold wire inserted in it to give a realistic impression of the natural capillaries of the eye.

Artificial limbs were among the first prostheses to be fitted, mainly to replace legs or arms amputated after war injuries or accidents. Protheses are mentioned in the *Rig Veda*, a Hindu religious text from the 2nd millennium bce. The oldest surviving artificial leg, made of bronze and iron with a wooden core, was found in Capua in Italy and dates from 300 BCE. Early artificial legs were "peg-legs" used to replace limbs amputated

below the knee. A projecting pole was strapped to the stump, which, although ungainly, allowed the wearer some mobility. It remained the most common leg prosthesis throughout the medieval period.

In 1575 Ambroise Paré (see pp.78–79) produced *Les Oeuvres*, the first work on limb prosthetics. Based on his career as a French military surgeon, Paré described more sophisticated artificial legs that incorporated adjustable

leather harnesses and knee-locking mechanisms. Over the years incremental improvements were made in comfort and functionality, but the problem of being unable to flex the leg persisted in those who had undergone below-knee amputations. Finally, in 1805, British prosthetist James Potts created a leg that could be articulated at the knee, ankle, and toe joints (later known as the "Anglesey leg" after the Marquess of Anglesey, who lost a leg at the Battle of Waterloo). A lighter, aluminum leg was invented by the Desoutter brothers in 1913 after aviator Marcel Desoutter lost his leg in a plane crash.

In spite of a pressing need for artificial limbs after two world wars that resulted in a huge number of amputees, the next major advance in prosthetic legs

350,000 The number of wearers of artificial glass eyes in the United Kingdom in 1939.

was not made until the 1980s, when the American prosthetist John Sabolich invented the Sabolich socket. This new interface for artificial legs spread the user's weight more evenly across surviving limbs and muscles, providing enhanced comfort to the user.

The 1990s saw the introduction of microprocessor controlled limbs, which, by converting muscle movements into electrical signals, allowed a near-normal gait to be achieved. The development of advanced materials, such as carbon fiber, permitted the manufacture of much lighter and more durable prostheses.

Arms and hands
Mentions of artificial arms go as far back as the 3rd century BCE, to the time of the Second Punic War (218–201 BCE), when Marcus Sergius, a Roman general, was fitted with an iron hand. Hand prostheses were mostly made of rigid metal, but as early as the

▷ **Neural stentrode**
The paper-clip sized "stentrode" (stent electrode) is intended to be inserted in a blood vessel adjacent to the brain's motor cortex or movement center. It will interpret electrical activity associated with movement and transmit corresponding radio signals to control robotic limbs.

16th century German knight Goetz von Berlichingen was known to wear an artificial right hand with movable joints that could close around an object. By 1812 Berlin dentist Peter Baliff had developed a prosthesis for below-elbow amputees that used shoulder muscles to allow fingers to flex or extend. Further developments in hand and arm protheses followed a similar path to that of artificial legs.

False eyes
The first eye prostheses were produced in ancient Egypt and were generally worn outside the eye socket. It was not until the development of artificial glass eyes by Venetian glassblowers around 1561 that it became practical to fit ocular prosthetics inside the socket. Essentially unchanged, these remained in use until the 1930s, when restrictions on their export from Germany—which had become the leading manufacturer—led to the development of plastic and acrylic artificial eyes. In all cases, however, the artificial eye was an esthetic device that improved the appearance of the wearer, but did not bring back vision. Finally, in the early 21st century, a number of projects to develop "bionic eyes," or prosthetic implants to replace damaged retinas were undertaken. From 2007 onward, US and European trials of a retinal implant developed in California showed that partial sight could be restored to blind patients, allowing them to perceive images and movement.

FUTURE OF PROSTHETICS

The DEKA arm was funded by the US Department of Defense to provide a better type of prosthetic for wounded military veterans. Developed by American inventor Dean Kamen, it was first licensed for use in 2014.

This battery-powered arm is a myoelectric prosthetic, directed by the patient's residual nerve impulses to closely mimic human function. It can be controlled either by foot movements or by electrodes placed against the nerves remaining in the shoulder or arm. These nerve impulses are decoded by a microprocessor to direct movement in the artificial hand.

The arm has multiple movable joints, which permits greater control of the fingers, offering six different types of grip and allowing objects as varied as a grape, a zipper, and a drill to be grasped. Here Fred Downs, a former soldier who underwent amputation after being injured in action, demonstrates a DEKA arm.

Technological leaps
The use of advanced technology today is enabling prostheses to be more sophisticated, durable, and cheaper. It is redefining the experience of using artificial limbs, especially for children who no longer have to worry about outgrowing their artificial limbs because they are made of plastic and are relatively inexpensive. The first prosthetic hands created by a 3D printer were manufactured in 2012 and used nerve impulses to control movement of the hands.

Experiments are underway to allow direct control of robotic arms by the brain and nervous system. Recent decades have seen the development of a wider range of implants and prostheses including devices as diverse as artificial hearts (developed in 2005) fitted with batteries that can be charged through the skin, synthetic tracheas (developed in 2011), and a bionic spine (first announced in 2016). Research that may in the future result in artificial livers, lungs, and ovaries is also underway.

> " A Martian might be taken aback by the energy with which **we knock limbs off** and the tenderness with which **we replace them.**"
>
> FROM *THE LANCET*, MAY 1944

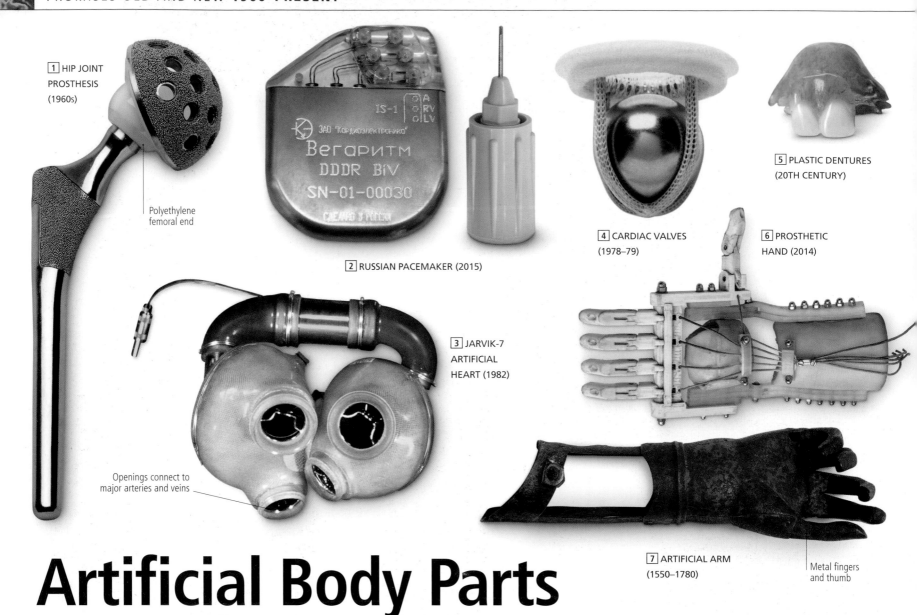

1 HIP JOINT
PROSTHESIS
(1960s)

Polyethylene
femoral end

2 RUSSIAN PACEMAKER (2015)

3 JARVIK-7
ARTIFICIAL
HEART (1982)

Openings connect to
major arteries and veins

4 CARDIAC VALVES
(1978-79)

5 PLASTIC DENTURES
(20TH CENTURY)

6 PROSTHETIC
HAND (2014)

7 ARTIFICIAL ARM
(1550-1780)

Metal fingers
and thumb

Artificial Body Parts

One of the oldest known prostheses—an ancient Egyptian wood-and-leather toe—dates back to c.1000 BCE. Today's bionic body parts have advanced composites and powered joints that respond to signals from muscles or even direct from the brain.

1 **Hip joint prosthesis** This low-friction model used polyethylene ends to reduce wear, an improvement on the first hip replacements, which used prostheses made of glass and metal. 2 **Russian pacemaker** This model is "three-chambered," with an extra lead to harmonize the beating of the right and left ventricles; electric pacemakers to regulate the heartbeat were first implanted in 1958. 3 **Jarvik-7 artificial heart** The first artificial heart to be implanted in a human, this device required a 397-lb (180-kg) power unit to function. Modern models have portable external batteries. 4 **Cardiac valves** These Starr-Edwards valves were used for mitral valve replacements, a procedure developed in 1960. 5 **Plastic dentures** In the 20th century plastic dentures replaced those made of ivory, porcelain, or sometimes the teeth of dead soldiers. 6 **Prosthetic hand** The medical applications of 3-D printing, as used here, are increasing, especially in the area of reconstructive surgery. 7 **Artificial arm** This iron hand included a forearm and was designed for below-the-elbow

amputees; it allowed no real use of the hand for manipulation. 8 **Prosthetic eye** During World War II, artificial eyes made of acrylic were substituted for those made of glass. 9 **Bionic arm** This powered arm was made using 3-D printing, which may soon allow mass-production of such prostheses at a fraction of the previous cost. 10 **Modern artificial leg** Made of light material, this leg includes customized joints to spread weight and microprocessor controls that allow gait to be adapted to walking speed. 11 **Prosthetic leg** This leg has lockable knee and ankle joints, and a perforated leather corset that helps the wearer stay cool. 12 **Prosthetic arm** Made of aluminum, this arm has more flexible joints than its wooden predecessors. 13 **Knee arthroplasty replacement prosthesis** The plastic total condylar knee—with single tibial unit and a central stem—invented in the 1970s closely mimicked the action of a natural knee joint. 14 **Electric prosthetic foot** This was the first artifical foot that was powered using the patient's nerve impulses.

Aluminium
casing

Attachment
strap

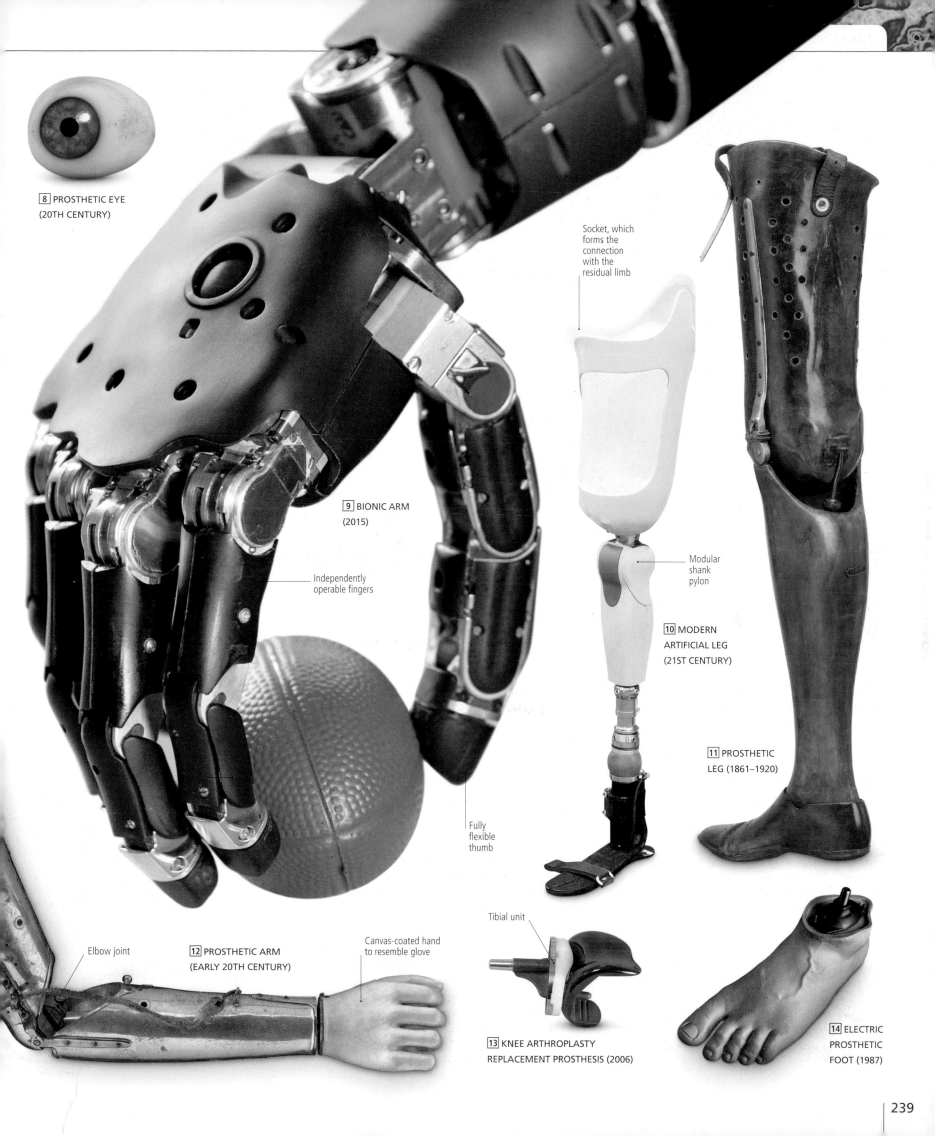

8 PROSTHETIC EYE
(20TH CENTURY)

Socket, which
forms the
connection
with the
residual limb

9 BIONIC ARM
(2015)

Independently
operable fingers

Modular
shank
pylon

10 MODERN
ARTIFICIAL LEG
(21ST CENTURY)

11 PROSTHETIC
LEG (1861–1920)

Fully
flexible
thumb

Elbow joint

12 PROSTHETIC ARM
(EARLY 20TH CENTURY)

Canvas-coated hand
to resemble glove

Tibial unit

13 KNEE ARTHROPLASTY
REPLACEMENT PROSTHESIS (2006)

14 ELECTRIC
PROSTHETIC
FOOT (1987)

In Vitro Fertilization

The world's first baby created in vitro (by fertilizing a mother's egg outside the body), Louise Brown was born in Oldham, North England, in 1978. A ground-breaking and controversial event, it divided communities, as well as religious and political leaders. Since then, more than 3 million in vitro fertilization (IVF) babies have been born.

The first IVF birth, pioneered by British physiologist Robert Edwards and British obstetrician Patrick Steptoe at their clinic in Cambridge, UK, would never have taken place without the work of scientists and physicians before them. In 1884 the first case of artificial insemination by a donor was recorded in the US when American physician William Pancoast took drastic, albeit unethical, action to help an infertile couple. While the lady was under anesthesia, unbeknown to her, he injected her with sperm taken from a medical student. The lady gave birth to a boy, but Pancoast's actions only came to the medical world's attention after his death.

It was not until 1934 that conception outside the body began to seem feasible. Harvard scientist Gregory Pincus conducted IVF experiments on rabbits and proposed that a similar procedure could work for humans. Much of the scientific community condemned his work but American infertility specialist John Rock was inspired to try IVF in humans. Together with his colleague and laboratory assistant Miriam Menkin, he fertilized an egg in a test tube. The process was repeated by others, including Robert Edwards who, along with Steptoe, was the first to implant a fertilized egg into a woman. Ten years of their research and experimentation led to the world's first test tube baby for John and Leslie Brown, and a Nobel Prize for Edwards.

> **" ... when this life-giving treatment was first considered, it was massively controversial."**
>
> ROBERT WINSTON, BRITISH SCIENTIST AND FERTILITY EXPERT

▷ **Infertility**
Intracytoplasmic sperm injection (ICSI), like all forms of IVF, involves the fertilization of the egg by the sperm outside the body. However, rather than the egg being placed in a dish with lots of sperm, in ICSI, the egg is directly injected with a single sperm.

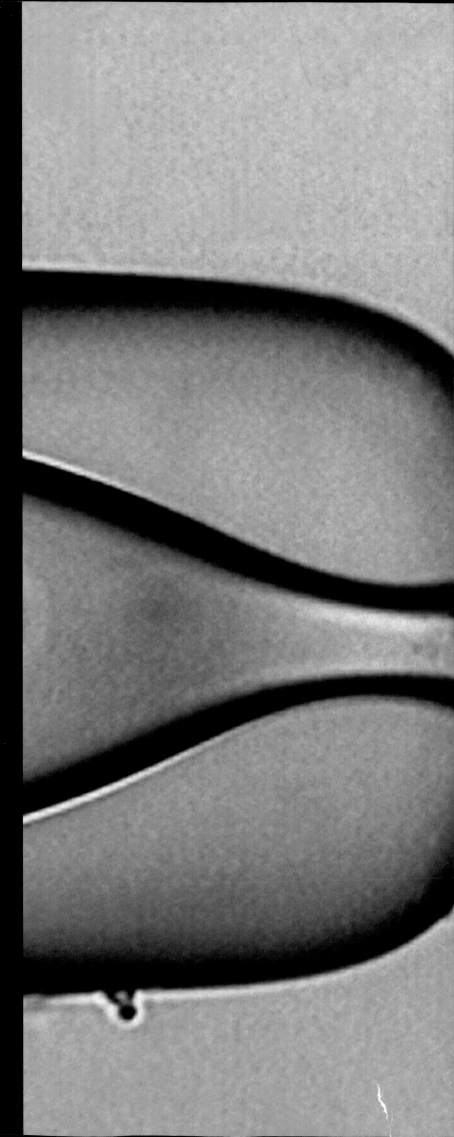

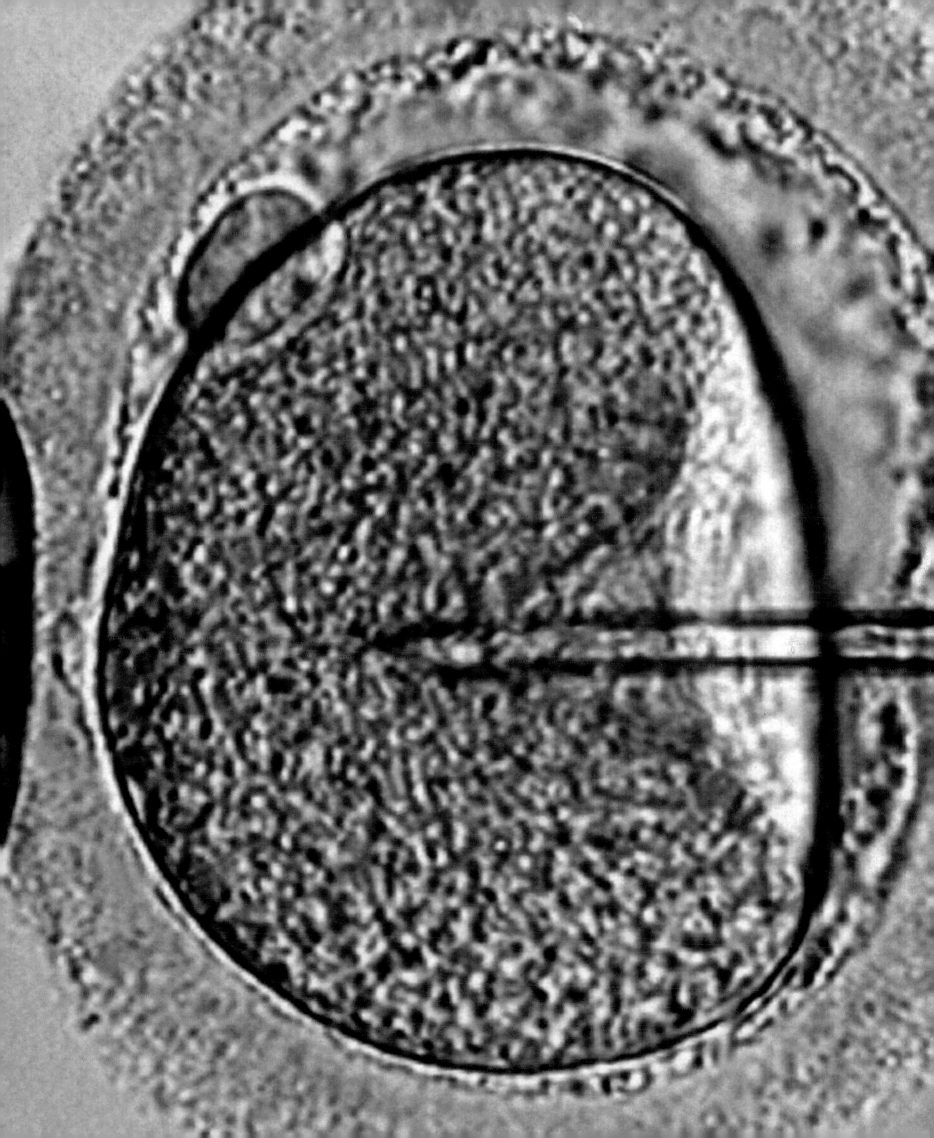

HIV and AIDS

In 1982 doctors in the US recognized a new illness, AIDS, which operated by suppressing the patient's immune system, rendering it susceptible to other, opportunistic infections. More than 40 million people worldwide have died of AIDS-related illnesses. A cure, or even a vaccine, remains elusive.

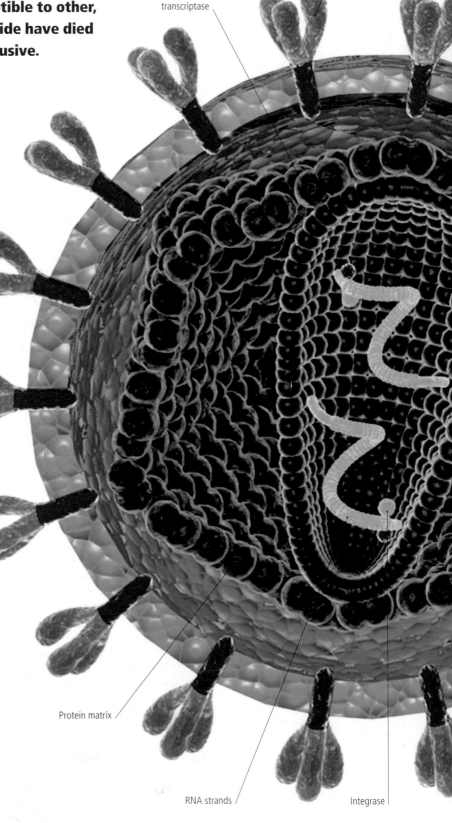

Glycoprotein gp120

Reverse transcriptase

Protein matrix

RNA strands

Integrase

In the late 1970s physicians in California noted an increasing occurrence of Kaposi's sarcoma, a rare type of cancer, and of *Pneumoncystis carinii*, a form of pneumonia previously seen only in patients whose immune system had been compromised (for example by chemotherapy). By 1981 the US Centers for Disease Control and Prevention (CDC) diagnosed these patients with a new disease, which came to be called Acquired Immune Deficiency Syndrome, or AIDS.

Pattern of transmission

Initially, clusters of victims were identified among the male homosexual community, intravenous drug users, and users of blood products such as hemophiliacs or patients undergoing transfusions. This seemed to indicate a cause associated with the transfer of blood or other human body fluids.

In 1983–84 two research teams identified a viral agent believed to be responsible. The French team, led by Luc Montagnier, named it lymphadenopathy-associated virus (LAV); Robert Gallo's US team called it human T-lymphotropic virus III (HTLV-III). It was realized that these were the same microorganism, which, in 1986, was named human immunodeficiency virus (HIV).

The search for a cure became urgent as, from the first isolated cases, AIDS turned into an epidemic. By 1989 there were 100,000 AIDS cases in the US and a further 142,000 worldwide, rising to 30 million by 1993. The early cases were predominantly in the homosexual community, but the balance tipped as rates among intravenous drug users in the US and Europe rose. Millions of cases were also reported in sub-Saharan Africa, where the disease appeared to be spreading primarily as a result of

▽ **Tip of the iceberg**
Public health campaigns used television advertisements such as this one from 1987 to emphasize the severity of AIDS, which had claimed many lives and would continue to do so if ignored.

THERE IS NOW A DEADLY VIRUS.

"HIV/AIDS is the greatest danger we have faced for many, many centuries."

NELSON MANDELA, SOUTH AFRICAN POLITICIAN, LAUNCHING AN
ANTI-AIDS AWARENESS STRATEGY, 2002

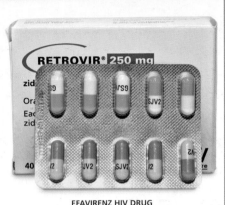
heterosexual intercourse and mother-to-baby transmission at birth, or through breastfeeding.

How HIV works

A test for HIV antibodies was developed in 1984. Researchers found that HIV was a retrovirus, a type of virus that had its genetic material encoded in an RNA (ribonucleic acid) strand. The virus penetrates a host cell and binds its RNA to the host's DNA (deoxyribonucleic acid), rendering the virus safe from the host's immune system. HIV targets CD4 T-helpers—a type of cell that assists with the body's general immune response. The virus replicates and begins to kill the host's CD4 cells; at this stage the patient is termed HIV positive. When the CD4 cell numbers fall below a certain level, and the sufferer's immune system collapses, the patient is said to have AIDS. Untreated, the average time from initial infection to death is two to three years. By 1986 a drug therapy had been found in AZT (azidothymidine), which prevents the viral RNA from becoming incorporated into the host cell's DNA. The advent of the more potent Highly Active Antiviral Therapy (HAART) in 1995 then enabled HIV-positive patients to stay AIDS-free for up to seven years. Today, antiretroviral drugs (see panel, above) can control the virus and extend life expectancy considerably.

HIV: past and future

Finding a definitive cure or vaccine for HIV/AIDS is complex, partly because there are many and varied subtypes of the virus. Research has also focused on finding its origin. In 1989 a similar virus SIV (simian immunovirus) was identified in chimpanzees in West Africa. It became clear that at some point the virus had crossed over to humans, possibly through the hunting of chimpanzees for bushmeat. It had then spread through prostitution in the growing urban sprawls of West Africa and by the reuse of infected needles in health programs that were crippled by underinvestment and the effects of regional civil wars. In the 1970s doctors had noted a disease in Uganda called "slim," characterized by the severe wasting away of patients—a sign of end-stage AIDS. By 2015 some 70 percent of HIV-positive patients were from sub-Saharan Africa.

Analysis of tissue samples from 1959 and 1960 from the Belgian Congo, which became Zaire (now DR Congo), indicated that they had HIV/AIDS and the virus may have crossed over to humans around 1920.

Today AIDS remains a serious healthcare and economic challenge for African countries, where many adults are rendered economically inactive by the disease. Research indicates that an estimated 40 million children will have been orphaned by AIDS by 2020.

36.9 MILLION **The number of people worldwide with HIV in 2014.**

59 PERCENT **People with HIV who are not receiving treatment.**

△ HIV structure and replication

HIV uses its reverse transcriptase to bind its genetic material, RNA, to the host cell's DNA as "pro-viral DNA." The infected cell makes proteins which are cut up by HIV's enzyme protease to produce new copies of the virus.

Capsid –
protein coat

▷ **Spreading awareness**
Public awareness campaigns are crucial in controlling the spread of HIV/AIDS. Here, Nigerian healthcare workers show posters explaining how a reduction in the number of sexual partners reduces the chances of contracting the disease.

New Discoveries for Old Diseases

In the late 20th century, several chronic diseases, such as stomach ulcers and cancers of the skin, cervix, and bladder, were found to be caused by microorganisms. These discoveries opened up the prospect of treatment and control through vaccination.

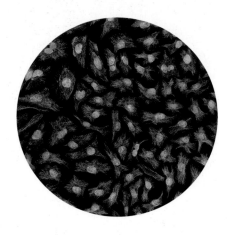

△ **HeLa cells for research**
These cells are a part of a line of cells that were first extracted in 1950 from the tumor of a woman with cervical cancer. The cell line was later found to be infected with HPV-18 and played a key role in the identification of a vaccine against HPV.

Scientists had long been puzzled by a range of diseases that were characterized by inflammation or cancerous tumors. Although the progression of these diseases was understood, the causes remained unclear. Factors such as pollution, lifestyle, heredity, and aging were blamed for conditions such as stomach ulcers and several forms of cancer.

Linking bacteria and cancer
Ulcers and abrasions in the lining of the stomach or duodenum (the first section of the small intestine) had long been understood to be associated with an excess of acid, and patients were commonly advised to consume a bland diet and reduce stress in their lives. An alternative link was found in 1979, when Australian pathologist Robin Warren found the curved bacteria *Helicobacter pylori* in the stomach of a patient suffering from dyspepsia (indigestion), a milder condition of the upper gut. Warren conducted further investigations with his colleague Barry Marshall and found a correlation between the presence of *Helicobacter* and duodenal ulcers. The pair announced their findings in 1982, but the medical community was slow to accept them, and the treatment of ulcers with antibiotics was not approved until 1996. It is now known that *Helicobacter pylori* causes 80 percent of all gastric ulcers, and has a role in the development of stomach cancer too.

Viruses and vaccinations
Cervical cancer is the fourth most common cancer among women, causing more than 250,000 deaths worldwide each year. In developing countries, where screening programs are unaffordable, it is the most common form of cancer among women. In 1974 German virologist Harald zur Hausen first suggested a possible link between cervical cancer and Human Papilloma Virus (HPV), which is part of a family of viruses that cause infections including genital warts. In 1986 zur Hausen identified two subtypes, HPV-16 and HPV-18, as the agents that cause most cervical cancers. This discovery led to the development of a vaccine for the condition, potentially saving thousands of lives.

Inflammatory diseases
Recent research has indicated that a range of other inflammatory diseases and cancers may be linked to infectious agents. For example, the small flatworm that causes schistosomiasis—a waterborne disease endemic in parts of the Middle East and East Africa that affects the bladder, kidney, and liver—has been linked to bladder cancer since the 1970s. Meanwhile

Chlamydia pneumoniae, identified in 1986 as a cause of respiratory diseases, has been shown in laboratory tests conducted with animals to play a role in athero-sclerosis (hardening of the arteries).

Recent studies have also suggested that Parkinson's disease—a progressive condition affecting the nervous system—may be linked to inflammation of the brain caused by infections such as influenza or Japanese encephalitis. Further research is likely to add to the number of diseases understood to be caused by infectious agents.

AUSTRALIAN PATHOLOGISTS
BARRY MARSHALL (1951–) AND ROBIN WARREN (1937–)

Marshall (left) and Warren (right) conducted biopsies on 100 patients as part of their research to discover the bacterial causation of stomach and duodenal ulcers. In 1984, as part of their study, Marshall even drank a culture of *Helicobacter* bacteria to infect himself, contracting a case of acute gastritis (a mild inflammation of the gut) as a result. The Australian scientists suffered almost a decade of scorn from gastroenterologists before their theory was accepted. They were awarded the 2005 Nobel Prize in Physiology or Medicine.

△ **Protection against HPV**
First approved in 2006, the HPV vaccine is part of routine vaccination programs for young girls in nearly 60 countries today. However, as HPV takes many years to cause cervical cancer, the efficacy of the vaccine is yet to be fully established.

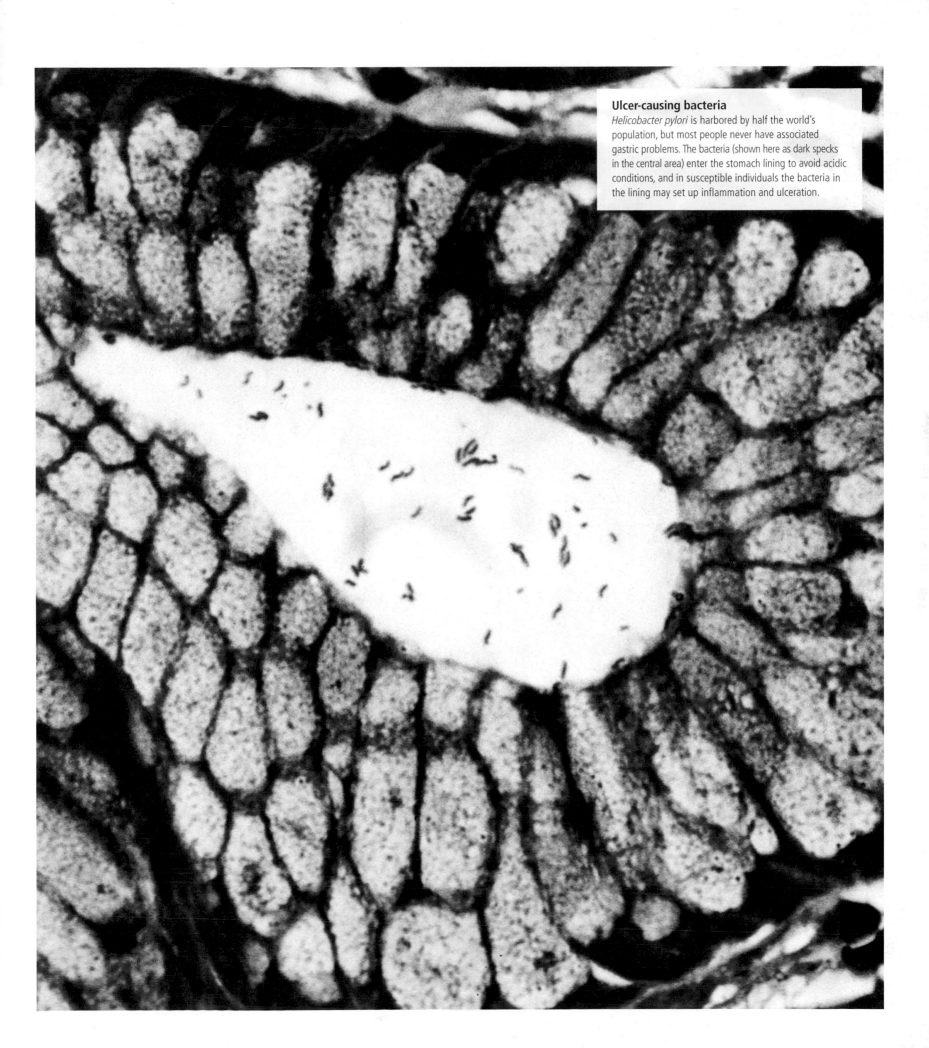

Ulcer-causing bacteria
Helicobacter pylori is harbored by half the world's population, but most people never have associated gastric problems. The bacteria (shown here as dark specks in the central area) enter the stomach lining to avoid acidic conditions, and in susceptible individuals the bacteria in the lining may set up inflammation and ulceration.

Genetic Revolution

The Human Genome Project began in 1990 with the aim of mapping and understanding all the genes found in humans (collectively known as the genome). In 2003 the project announced that it had identified all 3 billion bases, or "code letters," of human DNA (see pp.212–13). The same year the ENCODE (Encyclopedia of DNA Elements) project began, setting out to determine what all the genes and DNA instructions do.

Following the discovery of the structure of DNA in 1953, the genetic code in the 1960s, and the principles of how genes operate in the 1970s, researchers turned their attention to locating all of the human genome and understanding how genes work. It was found that the amount of the genome that instructs how to make proteins, called protein-coding DNA, was less than 2 percent, while the rest was considered to be only "junk DNA." Since 2010 it has emerged that the initial estimate, made in the 1990s, of more than 100,000 genes was incorrect and the human genome consists of 20,000 genes (about the same number as a 1-mm (0.4-in) roundworm *Caenorhabditis*—the first animal whose genome was sequenced). It has also been found that much of what was dubbed "junk DNA" actually contains instructions for thousands of noncoding ribonucleic acids (ncRNAs) that are involved in control of the genes.

These advances in understanding the genome have huge implications for medicine. With faster, cheaper DNA profiling, parts of individual DNA could be sequenced for personalized medicine. Drugs that target the gene regulation system may also be developed in the future.

"To identify all functional elements in the **human genome.**"

GOAL OF THE ENCODE PROJECT, LAUNCHED BY THE US NATIONAL HUMAN GENOME RESEARCH, 2003

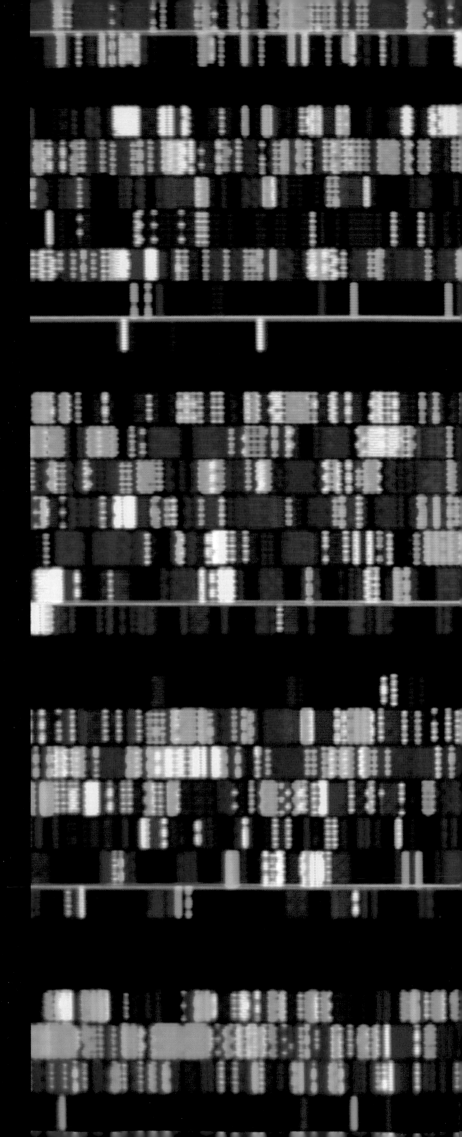

▷ DNA sequencing
Computers visualize DNA sequences in terms of their color-coded bases or "letters"— A, T, G, and C. Tiny variations among the 3.2 billion base pairs, differing on an average by 0.1 percent between individuals, are compared to find the genetic basis of health and disease.

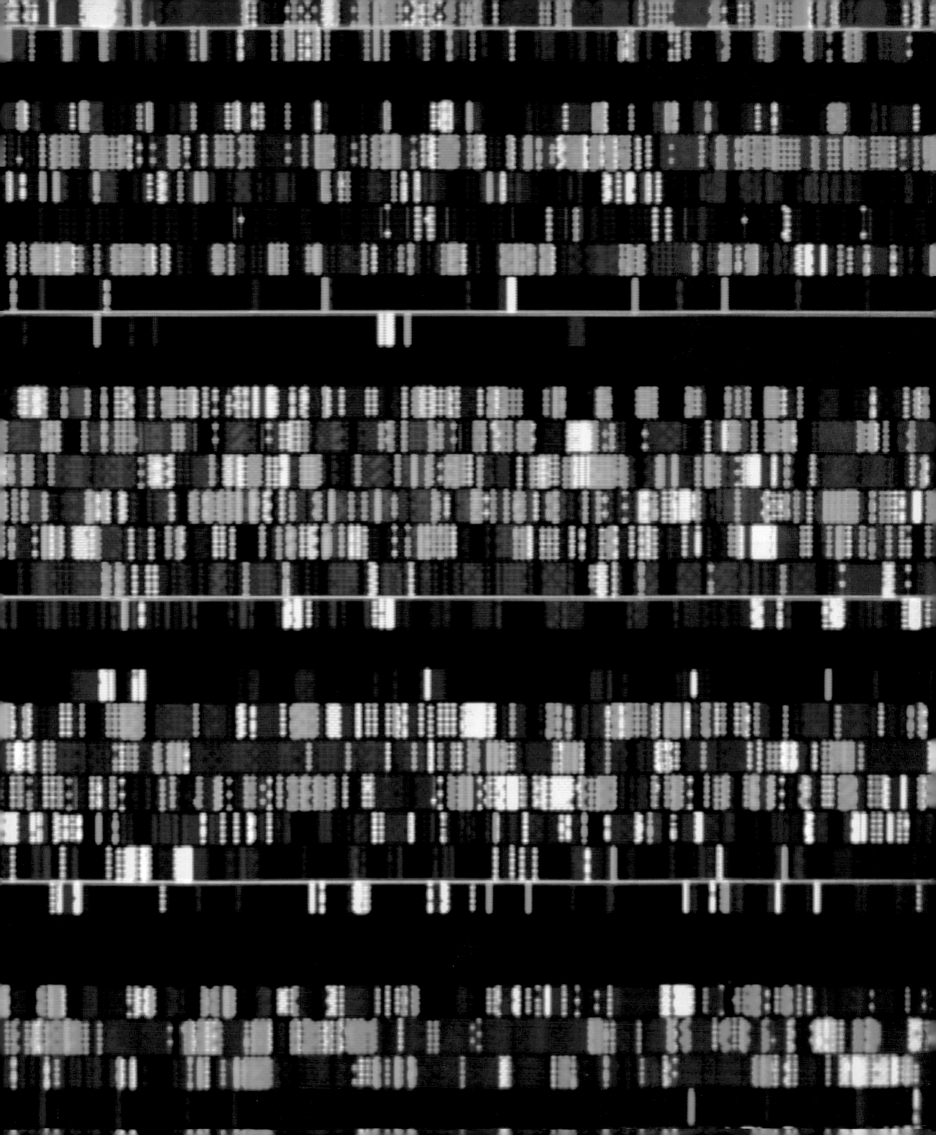

Genetic Testing

Research into the molecular structure of DNA (deoxyribonucleic acid) during the 1980s led to a greater understanding of genetic disorders. Consequently, a new field of medical genetics has emerged that offers the possibilities of predictive and personalized medicine.

One of the key discoveries that made the new field of genetic medicine possible was that each of us has a unique DNA "fingerprint," recognizable by differences in the arrangement of genes on the chromosome. Genetic profiling has proved invaluable in forensic investigations and for the purpose of establishing a child's parentage, but perhaps most importantly genetic testing has

revolutionized many different areas of medicine. In particular, it has enabled the detection of mutant or damaged genes and the specific genes associated with many genetic diseases. Research, especially that conducted by the international Human Genome Project from 1990 to 2003, led to the development of a range of techniques to identify the approximately 20,000 protein-making genes found in human DNA.

Techniques of analysis

The procedure for obtaining a sample of a patient's DNA is simple and painless. Genetic testing can be performed on almost any body tissue, such as blood, skin, or hair, but the usual method is to take cells from the inside of the patient's mouth using a swab or a mouthwash of saline solution.

◁ **Biochemical analysis**
Samples containing a person's DNA are pipetted into a multiwell plate prior to analysis. Chemical analysis can identify individual genes. This makes it possible to detect genetic disorders, which may indicate a risk of genetic disease.

Genetic tests can be carried out on unborn children by taking a sample of the amniotic fluid around the fetus. Sample cells are then sent to a laboratory, where the DNA is isolated and the genetic structure analyzed. Sophisticated biochemical analysis using machines, such as DNA sequencers combined with computer programs that can annotate the results, has become routine and provides an accessible and accurate method of testing for

> " Gene therapy is **ethical** because it can be supported by the fundamental moral principle of beneficence: it would **relieve human suffering.** "
>
> WILLIAM FRENCH ANDERSON, AMERICAN SCIENTIST, FROM *GENETICS AND HUMAN MALLEABILITY*, 1990

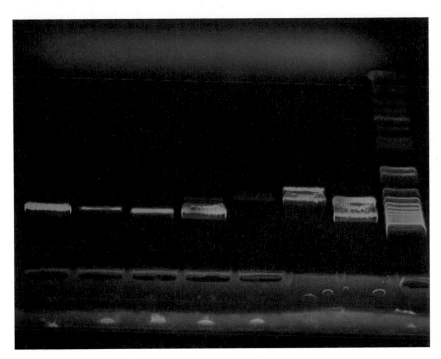

◁ **DNA electrophoresis under UV light**
Gel electrophoresis is a technique used to analyze the molecular composition of DNA samples. These are placed into a gel, which has an electrical current passed through it that separates the DNA molecules based on their size and charge.

Diagnosis and treatment

Initially, genetic testing focused on identifying the defective genes associated with inherited genetic disorders and mutated or damaged genes. More recently, particular genetic characteristics have been identified that may indicate a patient's predisposition to diseases, including certain types of cancers and heart conditions, and even how well the patient might respond to different medicines. Today there are several thousand specific genetic tests that are used in many different ways. For example, genetic testing can be used as a diagnostic tool, when a patient displays symptoms that may relate to a genetic disorder. By checking for the presence of specific mutations or damage at the gene level, genetic tests can confirm or rule out the diagnosis of a particular genetic condition. This form of testing is especially useful in children who may possibly have inherited a genetic condition that can be treated if it is caught early enough, and an increasing number of newborns are receiving genetic screening for this reason. If there is a significant risk of a child being born with a genetic disorder, such as Down syndrome, this can be confirmed or ruled out by prenatal genetic testing, generally of fluid taken by amniocentesis.

Genetic testing is not, however restricted to diagnosis of existing conditions. Expanding areas of genetic medicine are predictive and presymptomatic testing, which are typically used for patients with a family history of a particular condition, or belong to an ethnic group with a high risk of a genetic disorder, such as sickle-cell anemia among people of African-American descent. The exciting, specialized field of pharmacogenetic testing has also emerged, which involves studying genes to determine the effects of a patient's genetic make-up on the efficacy of different drugs. Results can be used to tailor a patient's treatment to the optimum dose of the safest, most effective medication—so-called "personalized meds."

3.2 BILLION The approximate number of base pairs in the human genome.

genetic disorders. As more people become aware of genetic disorders, the demand for genetic screening has grown. Today many companies offer home testing kits to collect samples of DNA to send off for genetic analysis.

Diseases and disorders

Many diseases—including cystic fibrosis, sickle-cell anemia, and hemophilia—are present at birth and are caused by a defective gene inherited from one or both parents. However, other diseases—such as cancers—occur when genes mutate and become damaged. Every time DNA is copied, there is a chance that an error can be made, which leads to mutations. Damage to DNA is more likely to occur with age, and it can also be triggered by environmental factors, such as radiation, sunlight, and tobacco smoke, as well as diet, alcohol, and possibly stress.

IN PRACTICE

GENE THERAPY

Advances in medical genetics have enabled scientists not only to identify the genes responsible for genetic disorders but also to develop means of treating them. By introducing nucleic acid polymers (large biomolecules) into a patient's cells, their DNA can be modified, as damaged or mutant genes (genes in a specific position on the chromosome) are replaced with healthy versions. Clinical trials of gene therapy have been conducted since 1990, leading to the approval of a growing number of drugs for clinical use in the 21st century.

Many copies are made

Normal version of gene is cut out

Normal gene is prepared for insertion

Healthy chromosome

Normal gene is inserted into cells of a person with a genetic disorder

Mental Health and Talking Therapies

Attitudes to mental disorders changed radically through the 19th century as more was discovered about their physical and psychological causes. Neurology and psychiatry were established as branches of medical science, and psychological treatments began to emerge.

For much of history, mental disorders were considered to be incurable. "Madness," or "mania," was generally regarded as a congenital abnormality, and "melancholia" (depression) as a personality disorder caused by an imbalance of the humors (see pp.34–35). Rather than being given treatment, the mentally ill were often simply isolated from society (see pp.164–65).

A new outlook

In the 19th century a more enlightened approach to mental health arose as a result of a greater understanding of the brain and its functions. New branches of medicine developed, including neurology, which viewed mental disorders as having physical or anatomical causes that could be treated and cured. Toward the end of the century the idea emerged that mental disorders might have a psychological cause,

and so would need a psychological treatment. This shift in thinking was inspired by the work of several neurologists in Europe, including Jean-Martin Charcot (see pp.160–61) from France, who described the physical characteristics of the human brain and studied the effects of hypnosis. Charcot's research then influenced Austrian

1 in 4 people worldwide will suffer from mental or neurological disorders at some point in their lives.

physician Josef Breuer and his colleague, neurologist Sigmund Freud (see pp.182–83). Freud used hypnosis on patients to treat them for what are now termed affective disorders—which include depression, bipolar disorder, and mania—and a variety of anxiety disorders such as phobias, panic attacks, and obsessive–compulsive

disorder (OCD). Freud developed the psychodynamic theory, in which he divided the mind into the "conscious" and the "unconscious," and believed that many mental disorders were caused by conflict between the two areas. He thought that hypnosis would help him unlock the contents of the patient's unconscious mind to help resolve this conflict. Freud also developed a method of psychoanalysis—a "talking therapy," based on Breuer's successful treatment of his patient Anna O—which became the model for many types of psychotherapy in the 20th century.

Freud's pioneering work attracted many followers, including the Swiss psychotherapist Carl Jung and the Austrian doctor and psychologist Alfred Adler. Different versions of the "talking therapy" began to emerge, incorporating ideas from various branches of psychology, but psychoanalysis remained the main form of psychotherapy until after World War II.

Advancements in therapy

In the 1950s some psychologists questioned the validity of Freud's psychodynamic theories and even the very notion of psychoanalysis as an effective therapy. Although the idea that psychological problems should be treated by psychological means rather than drugs or surgery was well established, there were different approaches to treatment,

△ **Scientific analysis**
Using modern imaging techniques, neuroscientists are now able to distinguish the differences in the patterns of neural activity in depressed (top) and healthy (bottom) brains.

based on development of cognitive and behavioral psychology. Rather than delving into the unconscious, new therapies found more practical ways of changing the behavior or thinking that affect the patient's mental well-being, thereby helping them deal with their problems.

Several cognitive and behavioral therapies were developed toward the second half of the 20th century, all of which came together in the cognitive-behavioral therapy (CBT) pioneered by the American psychoanalyst Aaron Beck. Under the guidance of a therapist, patients learn to understand distressing thought patterns, and find strategies to modify the way in which they react. More recently there has been a movement of "positive psychology," placing an emphasis on mental health rather than mental disorder.

BERTHA PAPPENHEIM (ANNA O)

Known by the pseudonym Anna O, Bertha Pappenheim—a German social worker—first experienced symptoms of hysteria, including headaches and hallucinations, while she was caring for her ailing father.

She was treated by Josef Breuer, who encouraged her to talk freely and express her thoughts and feelings. Pappenheim dubbed this therapy a "talking cure."

Her case study was later published in *Studies on Hysteria* that Breuer wrote with Freud in 1895.

" The word **'happiness…'** would lose its meaning if… not balanced by **'sadness'**."

CARL JUNG, SWISS PSYCHOTHERAPIST

Complex mechanical aids, such as heart-lung bypass machines, have been used in operations since the 1950s, but the surgery itself was carried out by humans. At the end of the 20th century major changes began to take place with the development of sophisticated robots—electro-mechanical machines that can be programmed to carry out a wide variety of functions and are able to manipulate objects.

Early robots were most suited to straightforward tasks that required precision, and the first robot used in surgery was Arthrobot, during a hip replacement procedure in Vancouver, Canada, in 1983. After this advances followed rapidly: in 1985 a PUMA 560 robot was used to insert a needle during a brain biopsy, in 1988 a robot carried out prostatic surgery at Imperial College London,

and in 1992 the Robodoc was employed to mill out bone tissue in the femur in order to create a smooth surface for a hip replacement. By 1999 robots had become advanced enough to assist in a heart bypass operation at Ohio State University.

There are many advantages to using robots to perform surgery. They can be capable of greater precision, flexibility, and control than a human; they make remote surgery possible, when the surgeon is not physically present at the operation (see Lindbergh operation, opposite); and they reduce physical stress on surgical staff, who can sit down during lengthy operations.

Minimally invasive surgery

One of the principal reasons for developing robots for surgery was to assist in minimally invasive, or

"keyhole," surgery, for example laparoscopy (see pp.188–89). Developed in the 20th century, these procedures usually involve making a small incision and inserting a miniature imager and light source so that the surgeon can examine the area, as well as small surgical tools such as biopsy forceps. The introduction of computer technology in the mid-1980s meant magnified images could be shown on a monitor so the surgeon could see inside the body clearly and guide the tools to the correct place.

In 2000 the da Vinci Surgical System was developed. With this system, instead of surgeons operating the tools manually, they direct the surgery from a computer console, which then transmits instructions to a robot that carries out the actions. Another way of controlling robots is telemanipulation—a process in which the surgeon wears a glove that transmits motion to the robot. Recent innovations have meant that a surgeon can now program a computer in advance to carry out the entire procedure—a method first employed in Italy in 2006.

Doctor interacts with patient

Mobile robot monitors patient

▷ **InTouch robots**
Telehealth increasingly uses robots for remote consultation. A screen, camera, speaker, and microphone allow for two-way communication, and with the latest machines devices such as a digital stethoscope or ultrasound machine can be connected to the robot for patient examination.

Robots and Telemedicine

Technological advances at the end of the 20th century allowed surgeons to use robots for basic surgical procedures. Physicians also benefited from telemedicine, which involves carrying out a range of consultative procedures using telephones, videoconferences, and the Internet without the need for the patient to be in the same room, or even country.

400,000 The number of robotic surgeries performed in the US in 2012.

Robotic laparoscopic surgery reduces the level of trauma for patients because smaller incisions are made, there is less blood loss, and less risk of infection than traditional open procedures, so recovery time is quicker. As a result, the scope of robotically assisted operations has increased to include bladder reconstructions (2007) and kidney transplants (2009).

Telemedicine
Technological advances have also made telemedicine possible—that is, the remote diagnosis and treatment of patients by means of telecommunications technology. An early pioneer of telemedicine was the Australian Royal Flying Doctor Service, set up in 1928 to provide

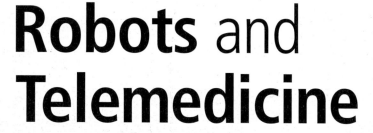

◁ **Teleconsultation**
A telemedical consultation is carried out involving doctors from two different hospitals. The monitor transmits information directly to the second doctor, giving access to additional medical expertise.

> "In my opinion, there is **no way back from robotic surgery.**"

PIER CRISTOFORO GIULIANOTTI, CHIEF OF MINIMALLY INVASIVE, GENERAL, AND ROBOTIC SURGERY AT THE UNIVERSITY OF ILLINOIS, CHICAGO, 2013

remote consultation by radio to faraway communities. The practice spread as technology became increasingly sophisticated, starting with the telephone, then video, and now via the Internet. It has become possible for patients to consult with doctors, and for diagnosis to be made from a distant location. Teleradiology, which enables the electronic exchange of images, such as X-rays and scans, is a key example of the growing use of technology to share information and records for more efficient patient care. The use of telemedicine permits the provision of care to patients in remote areas without the need for expensive medical infrastructure, particularly in developing countries where there are few doctors or specialists. Remote surgery is another growing innovation in telemedicine.

Further developments

Robotic surgery and telemedicine continue to make great strides as technology advances. The falling cost of electronic communications means that telemedicine can be adapted to new fields, for example telerehabilitation, in which patients can be monitored by physical therapists remotely. Meanwhile, in robotics, the development of nanobots (see pp.264–65)—tiny specialized robots less than a millimeter long—which can perform tasks such as clearing arteries, offer great potential.

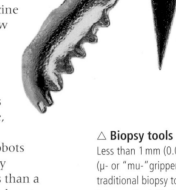

TRADITIONAL BIOPSY FORCEPS

MU-GRIPPER

△ **Biopsy tools**
Less than 1 mm (0.04 in) across, microgrippers (μ- or "mu-"grippers)—a fraction of the size of traditional biopsy tools—are released in clusters into a patient. Their star-shaped gripping arms collect tiny tissue samples and they are retrieved using a magnetic tool.

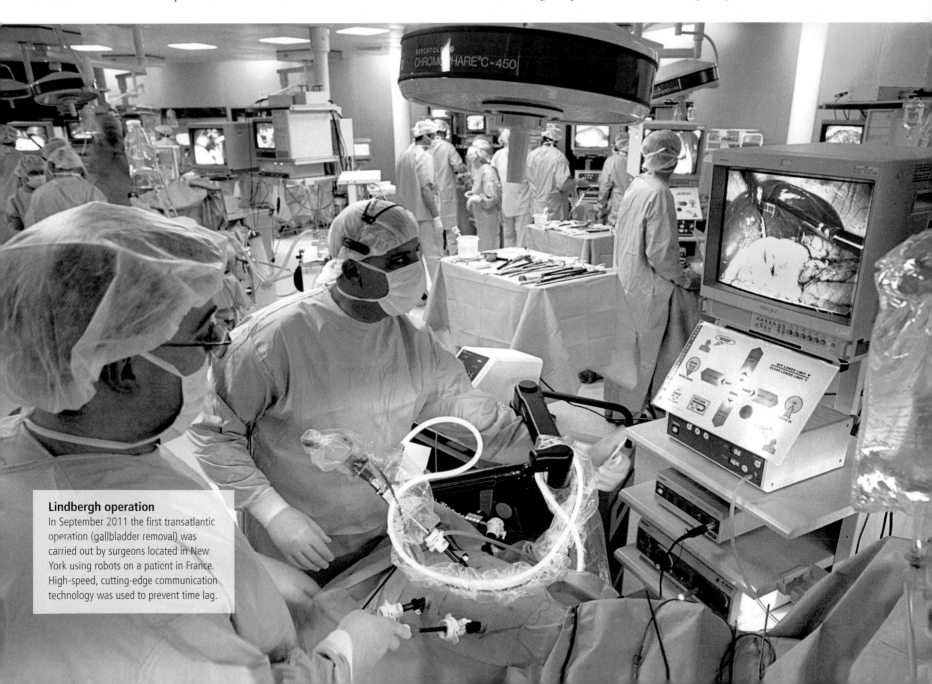

Lindbergh operation
In September 2011 the first transatlantic operation (gallbladder removal) was carried out by surgeons located in New York using robots on a patient in France. High-speed, cutting-edge communication technology was used to prevent time lag.

Robotic Surgery

Until the 1990s, robots were not considered adept enough to match the skills of a surgeon. However, advances in robotic technology since then have enabled the invention of sophisticated surgical robots. Today, robotically assisted surgery is performed regularly, aiding rather than replacing human surgeons.

Robotic surgical systems were developed in the late 1990s, mainly to aid the growing field of minimally invasive surgery (see pp.188–89). Among the first successful surgical robot systems were AESOP and ZEUS. They were superseded by Da Vinci Surgical System, approved by the regulatory body in the US, the Food and Drug Administration, in 2000.

Robotic surgical systems typically consist of two components: the robot itself and a separate console used by the human surgeon to control it. Mounted on a cart, the robot has several arms, one of which is equipped with an endoscopic camera. The other arms are designed to hold surgical tools such as scalpels, scissors, and cauterizing equipment. These arms are capable of a wide range of movements, and can be controlled by the surgeon with a high degree of precision. The robot responds to the surgeon's hand and foot movement, but can also improve them by detecting and removing any tremor in the surgeon, and scaling down the motion to enable extremely precise micro-movements.

" … snakelike arms through [a small] hole… would change [the] nature of surgery."

MICHAEL PALESE, DIRECTOR OF MINIMALLY INVASIVE UROLOGY AT MOUNT SINAI HOSPITAL, NEW YORK CITY, 2012

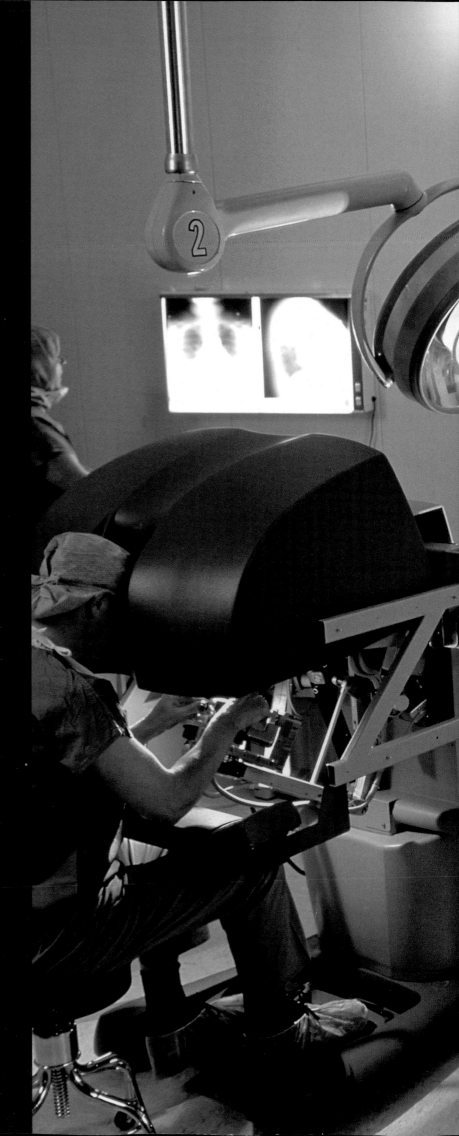

▷ Remote control
Sitting at a console, a surgeon performs minimally invasive surgery using a remotely controlled surgical robot. Three of the arms of the robot are equipped with surgical tools, which the surgeon manipulates via the console, while the fourth holds a camera to provide him with a 3-D image of the operation site.

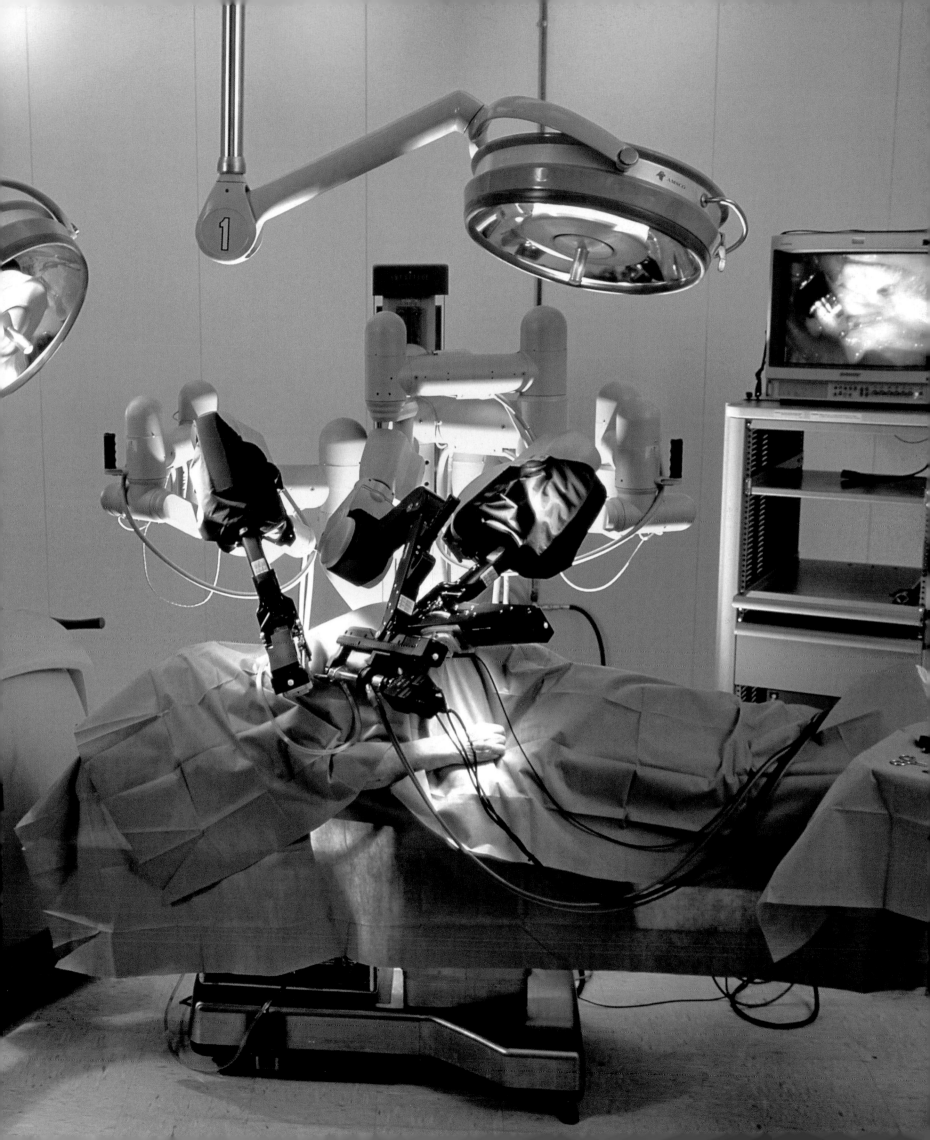

Emergency Medicine

Unexpected illness and injury occur in every lifetime. Emergency medicine has been in use for millennia, progressing from desperate measures on the battlefield to high-tech diagnosis and treatment in one of the newest areas of medical specialization.

Like so much else in the medical sciences, emergency medicine has taken great strides forward in wartime (see pp.192–95). Urgent care for the sick and wounded, such as tourniquets to stem blood loss, has been around since battles in ancient Rome, East and South Asia, and China. At the time of the First Crusade in the 11th century, groups such as the Knights of the Order of St. John of Jerusalem (the Knights Hospitaller) specialized in first aid for battle casualties as well as for pilgrims.

▷ **Pioneering service**
This 1869 ambulance operated out of Bellevue Hospital, New York City. Bellevue was the second US hospital to start an ambulance service in 1869. The Commercial Hospital, Cincinnati, Ohio, set up the first one in 1865.

Mobile medicine

During the Napoleonic wars of the late 18th and early 19th centuries, the French army chief surgeon Dominique Jean Larrey brought many innovations to emergency medical care. He is credited with introducing ambulances in the form of rapid horse-drawn carriages, which he adapted from "flying artillery" that was trained to move quickly around the battle area. His teams established mobile field hospitals with medically trained staff, and brought back wounded soldiers from the battlefield during the conflict, rather than waiting until fighting ceased, by which time many would have died.

Larrey formalized the concept of triage, the selection of patients based on the severity of their condition, which is still used today. When resources are stretched to the limit, casualties are assigned into three groups: those who will probably recover without medical care; those who will probably die even with medical care; and those who might be saved by medical care. The last group is given priority.

Modern advances

Larrey also described the critical choice between bringing limited emergency medical care to the patient or taking the patient to more highly equipped emergency facilities. This choice remains at the heart of emergency care, although recent advances in equipment and training have dramatically improved outcomes for onsite treatment.

By the time of the American Civil War (1861–65), almost all regiments had ambulances; railroad trains and steamships were also used in emergencies. In 1899 the first motorized ambulances, powered with electricity, were introduced in Chicago, Illinois. Gasoline-engined versions followed in 1905 and took over from horse-drawn carriages, especially during World War I.

10 MINUTES Ambulance response time in New York City, US.

180 CALLS HOURLY to the London Ambulance Service.

FRENCH SURGEON (1766–1842)

DOMINIQUE JEAN LARREY

Starting as an apprentice surgeon in Toulouse, France, Dominque Jean Larrey moved to the Hôtel-Dieu in Paris before beginning his long army career in 1792. France's many military campaigns needed talented surgeons, and Larrey was noted for his skill at rapid surgery and his compassion. He became Napoleon's surgeon-in-chief in 1797, and traveled extensively from Egypt to Russia with the army. He was taken prisoner at Waterloo, but was later released. After Napoleon's death, he again became chief army surgeon.

"There is a golden hour between life and death."

R. ADAMS COWLEY, US EMERGENCY MEDICINE AND TRAUMA SPECIALIST, 1957

In the early 1950s disasters such as the Richmond Hill rail crash in New York City, three air crashes in Elizabeth, New Jersey, and the Harrow and Wealdstone rail crash in London left hundreds dead or injured. These events shifted the role of the ambulance from a fast means of transport to a mobile "mini-hospital," equipped with compact ECG machines and heart monitors developed using newly invented transistors. Emergency medicine was established in hospitals in the same decade.

Portable cardiac defibrillators to restart or normalize the heartbeat were introduced in the 1960s; from the 1990s these have been installed for use in public places such as shopping malls.

Life-saving procedures such as cardio-pulmonary resuscitation (CPR) for a person whose breathing or heart has stopped, are assessed regularly, and guidelines are updated by authorities such as the Red Cross, the American Heart Association, and the UK Resuscitation Council. In the 1950s US physicians Peter Safar and James Elam developed the protocol A-B-C for CPR: first ensure Airway is open, then assist Breathing, then assist Circulation. In 2010 A-B-C was revised to C-A-B, to reflect the finding that chest compressions offer the best chance of saving a life.

Ambulance services and hospital emergency departments have also benefitted from new medications such as thrombolytic or "clot-buster" drugs that disperse thrombi (blood clots) in the body. Most effective if administered soon after the event, they are used in medical emergencies such myocardial infarction (heart attack), deep-vein thrombosis (a condition in which blood clots form in the veins), and ischemic stroke (in which the blood supply to part of the brain is reduced or blocked).

In the 1950s R. Adams Cowley introduced the idea of a "golden hour," to highlight that fact that treatment in the first hour after a medical emergency greatly improved chances of survival and recovery. With the introduction of smartphones and the Internet in the 1990s, along with a rise in the number of trained paramedics and first responders, the golden hour has become the "platinum 10 minutes," reflecting today's rapid assessment, treatment, and transportation.

△ **Air ambulance crew**
World War I saw the first casualty evacuations in specially equipped airplanes. Helicopters were used from World War II and many nations now have air ambulance services for the most rapid emergency response.

▷ **Help on two wheels**
Cycle response units can reach incidents faster than motorized ambulances in congested cities and large pedestrian spaces, such as airports.

Antibiotic Resistance and **Superbugs**

Within decades of the discovery of the miracle of antibiotics, some bacteria began to develop immunity to the drugs. By the late 20th century the spread of "superbugs" with resistance to multiple antibiotics threatened to return medicine to the days of untreatable infections.

The discovery of penicillin by Scottish bacteriologist Alexander Fleming in 1928 (see pp.198–99) was followed, over the next few decades, by more new antibiotics such as methicillin, tetracycline, and erythromycin (see pp.200–01). When penicillin-resistant strains of bacteria were discovered in 1940, there was not much concern; it was reasoned that infections could always be treated with another class of antibiotic. Gradually, however, fewer new antibiotics were discovered, revealing this to be dangerous complacency.

Resistant strains

Strains of diseases, such as tuberculosis, that were immune to one or more of the antibiotics used to treat them began to appear. Slowly, the realization dawned that overuse of antibiotics was a major factor in the emergence of resistant bacteria. This occurred when patients who were prescribed prescription antibiotics failed to complete the course of treatment, allowing bacteria to survive and acquire immunity. Self-medication, the uncontrolled use of cheap antibiotics in underdeveloped countries, and use of antibiotics to promote growth in livestock were also factors.

The problem of over-prescription demanded a change in medical practice. In the US the prescription rate for antibiotics to treat children dropped 25 percent between 2003 and 2010, although prescription rates for adults stayed the same.

The overuse of antibiotics in agriculture was the subject of a warning in the UK government's Swann Report in 1969. But it was only in 1985 that Sweden became the first country to forbid antibiotic growth promoters in livestock. An EU-wide ban was imposed in 2006.

Horizontal transfer

The mechanism of antibiotic resistance was understood as early as 1959, when scientists in Japan discovered the phenomenon of horizontal gene transfer. It was known that bacteria that mutated and acquired immunity against an antibiotic could transfer the resistance when they divided— a process known as vertical gene transfer. However, scientists had not suspected that bacteria could transfer genes to separate bacterial organisms or to different species of bacteria—the mechanism that became known as horizontal gene transfer. This explained why antibiotic resistance spread so fast.

The mutations either modified enzymes to make the bacteria insensitive to the antibiotic, or pumped out or even destroyed the antibiotic from the bacterial cell. Bacteria began acquiring immunity not just to one antibiotic, but to a range of them as medical staff used the next drug in their armory.

Superbugs

The first "superbug" to be identified was *Staphylococcus aureus*, a bacterium associated with throat infections that is harbored by 30 percent of the population. The bacteria acquired immunity to penicillin and then to methicillin, just three years after the latter drug was introduced in 1959. The resistant strain MRSA (methicillin-resistant *Staphylococcus aureus*) led, in 2005 alone, to 18,650 deaths in the US.

Further drug-resistant strains followed. In 2011 an estimated half a million new cases of multi-drug-resistant tuberculosis (MDR-TB) were reported worldwide. VRE

> " ...the ignorant man may easily **underdose** himself... and make [microbes] **resistant.**"
>
> ALEXANDER FLEMING, SCOTTISH BIOLOGIST, FROM HIS NOBEL PRIZE LECTURE, 1945

CONCEPT

ANTIBIOTIC RESISTANCE

When bacteria are exposed to an antibiotic, a very small number may have mutated and acquired a defense against it. This is more likely to occur where courses of drug treatment are not completed. The surviving bacteria, which are now drug-resistant, divide and transfer the resistant gene to their "offspring" in a process called vertical gene transfer. In environments such as hospitals bacteria are easily passed to other hosts. Once a particular defense against antibiotics has been acquired, genes for it can also be passed on to totally different species of bacteria through horizontal gene transfer.

NORMAL BACTERIUM   RESISTANT BACTERIUM

DEAD BACTERIUM

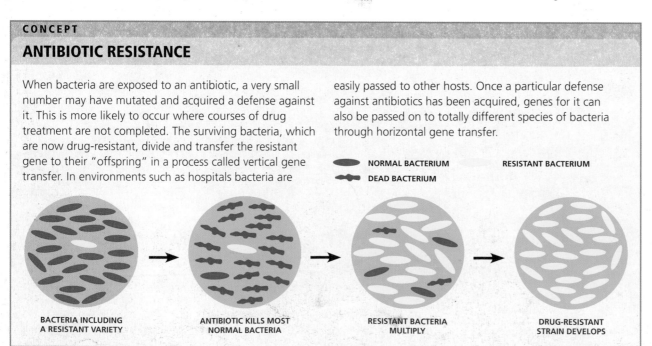

BACTERIA INCLUDING A RESISTANT VARIETY

ANTIBIOTIC KILLS MOST NORMAL BACTERIA

RESISTANT BACTERIA MULTIPLY

DRUG-RESISTANT STRAIN DEVELOPS

(vancomycin-resistant enterococci), ESBL (extended-spectrum beta-lactamase resistant bacteria), and *Clostridium difficile*, which flourishes when normal gut bacteria have been weakened by antibiotic use, claim thousands of lives each year.

In 2001 the first of a new class of antibiotics, the oxazolidinones, was used effectively to treat MRSA. In 2015 scientists identified a new antibiotic called teixobactin, found in a soil sample, with a unique mechanism that prevents bacteria from building cell walls. The fight against antibiotic resistance is not lost but is destined to continue.

▷ Antibiotics in livestock feed

In the US, 80 percent of all antibiotics sold continue to be used as growth promoters in livestock. In China, around 85 million pounds of antibiotics were added to agricultural feed in 2012.

▽ MRSA superbug

Methicillin-resistant *Staphylococcus aureus* (MRSA) accounts for around half of hospital-acquired infections, causing around 18,000 deaths in the US each year. It often infects patients who are already weakened by other infections, chronic diseases, or surgery.

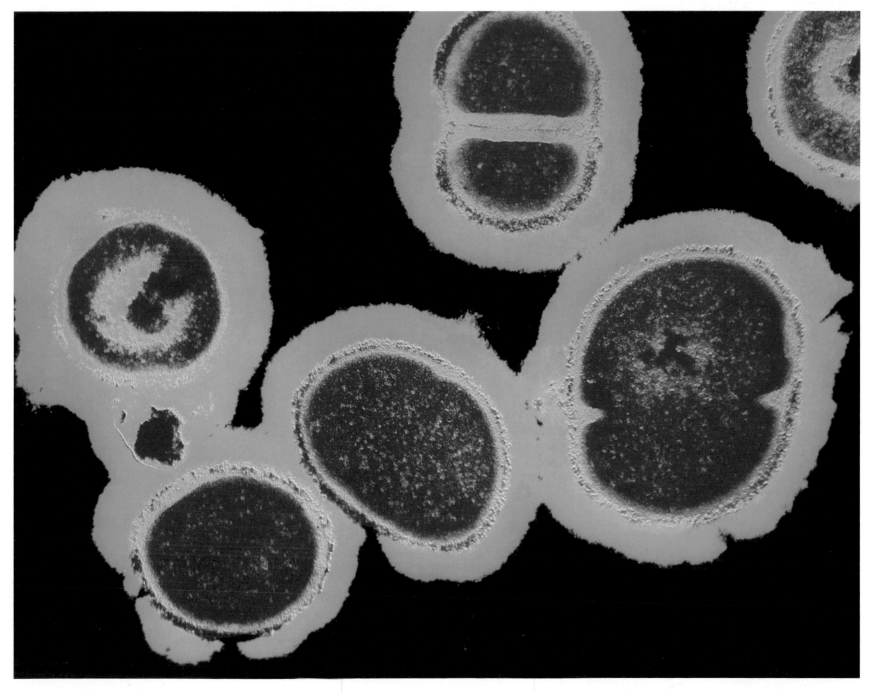

Alzheimer's Disease and Dementias

The radical increase in life expectancy during the 20th century saw a corresponding rise in diseases of old age, especially degenerative mental disorders. The most common of these are dementias, in particular Alzheimer's disease, which causes irreversible damage.

Since ancient times, medical practitioners have noted that mental capacity seems to decline with increasing age. Greek mathematician Pythagoras, writing in the 6th century BCE, defined "senium," or old age, as the period from the age of 63 when mental abilities regress to those of an infant. In the 1st century BCE the Roman physician Celsus used the term "dementia" to describe a persistent state of mental impairment.

However, until the 19th century, medical scholars had no clear idea of the causes of dementia, which came to describe the syndrome of severe memory loss that affected many older people. There was a tendency to accept it as a natural part of aging, rather than as a disease with a clinical foundation.

Over a period of 100 years the industrialized world had seen a marked increase in life expectancy: in Victorian England, for example, it had risen from around 35 years for women in 1800 to 48 years a

△ **Brain with Alzheimer's**
Comparing a cross-section of brain tissue from a healthy subject (top above) with that of the brain of an Alzheimer's patient (below) highlights the startling loss of tissue, as well as lesions and scarring, in the latter.

century later. The number of elderly patients rose accordingly and there was a surge of interest in their medical care.

In 1849 George Day, professor of medicine at the University of St. Andrews, Scotland, published *A Practical Treatise on the Domestic Management and Most Important Diseases of Advanced Life*, which gave one of the first full descriptions of the symptoms of dementia.

The loss of memory that characterized the disease troubled physicians, who began to seek a physical cause for this mental decline.

In 1894 Jean Noetzli, working in Zurich, examined the post-mortems of 70 patients with dementia and found degenerative changes and weight loss caused by the wasting away of the brain (atrophy) in almost all of them. Just under half had lesions in specific areas.

However, the key work in both identifying the external signs of dementias and the internal changes occurring in the brain was carried out by German neuropathologist

▷ **Diseased brain cells**
In a patient with Alzheimer's disease, the treelike structures formed by the brain's nerve cells are obstructed by tangled clusters of proteins that block the synapses and impede the passing of electrical impulses between the neurons.

Alois Alzheimer. In 1901 he treated Auguste Deter, a female patient at the Frankfurt Asylum who was suffering from severe short-term memory loss. After Deter died in 1906, Alzheimer examined her brain and found an abundance of abnormal structures. Insoluble deposits of a protein

35 MILLION The number of people who suffer from Alzheimer's disease worldwide. This number is expected to double by the year 2030.

called beta-amyloid had formed plaques between neurons (nerve cells), which inhibited the electrical and chemical signals between them that coordinate thought and memory (among numerous other processes). He also found buildups of twisted protein threads called "neurofibrillary tangles" clustered around the neurons (see panel right).

Alzheimer's precise description of the physical and mental symptoms of his patient's syndrome prompted his mentor, Emil Kraepelin—a German psychiatrist who believed that most mental diseases had a biological basis—to name the disorder after Alzheimer in the 1910 edition of his *Handbook of Psychiatry*.

> **" All in all we have to face a peculiar disease process..."**
>
> ALOIS ALZHEIMER LECTURE, 1907

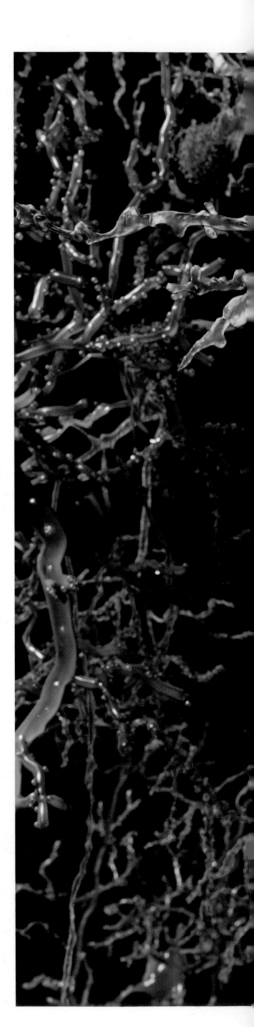

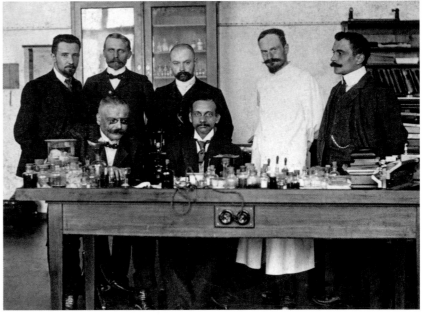

△ **Leading psychiatrists**
Alois Alzheimer (seated left) is pictured with leading psychiatrists at the University of Munich in 1905. The following year his study of the brain of his deceased patient Auguste Deter led him to discover the most common form of dementia.

Types and care

Gradually, neurologists began to distinguish between different types of dementia. Alzheimer's disease is the most common, affecting about two-thirds of sufferers, and is characterized by progressive memory loss, a decline in language and problem-solving ability, mood changes, and depression. Over time those affected become less able to cope with everyday life, leading to total dependence on caregivers in the final stages of the illness.

Vascular dementia makes up about 25 percent of cases, and is caused by multiple strokes in the brain, massive damage to blood vessels, and a mental decline that is far more rapid than in Alzheimer's.

Lewy body dementia, which is characterized by spherical objects (or Lewy bodies) in the brain, causes hallucinations and tremors.

Patients with frontotemporal dementia experience severe personality changes and language difficulties, but no loss of memory.

Although the processes by which malformed proteins cause changes in the brain are well understood, scientists cannot yet explain why only about 20 percent of 80-year-olds have dementia. There is as yet no cure. While progress has been made on drugs that inhibit the enzyme acetylcholinesterase, which may slow the process of damage to the neurons, the drugs cannot reverse it. As the world's aging population increases, an inevitable corresponding rise in the number of people with diseases such as Alzheimer's will make the need to find a cure even more urgent.

TANGLES AND PLAQUES

In Alzheimer's disease, insoluble beta-amyloid proteins are embedded in the membranes of neurons. Over time they form plaques that block communication between the neurons. In addition, strands of a protein called tau detach and cluster, forming neurofibrillary tangles which further inhibit neurons' activity, causing them to atrophy and die.

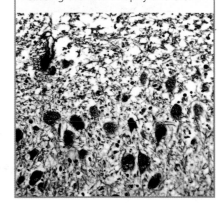

End-of-Life Care

Palliative care is the act of looking after someone at the end of their life, in the hospital or a hospice, or at home. This service, which traditionally combines spiritual and physical care, has a venerable history that dates back to early medieval times.

For centuries people in Europe died at home cared for by friends and family. However, as circumstances changed during the medieval period it was not always possible for them to spend the last months, or days, of their life at home. At the end of the 11th century, Pope Urban II called for Christians in Western Europe to travel to the Middle East and take possession of the Holy Land. This prompted thousands to undertake the long journey and fight to the death. In response to this the Knights Hospitaller (a Roman Catholic military order) established houses of rest and healing, where those traveling could seek shelter. The houses of rest also provided another important function—offering a place where the incurably ill and injured who were unable to get home could live out their last days.

Early hospices

The early efforts of the Knights Hospitaller were revived in the 17th century, when French Roman Catholic priest St. Vincent de Paul established the Sisters of Charity in Paris to care for the sick and dying. His work inspired religious organizations in other parts of Europe to do the same. The Irish Sisters of Charity, which were founded in 1815, were especially active, opening Harold's Cross Hospice in Dublin in 1879 and St. Joseph's Hospice in South London in 1902. Fifty years later a young nurse named Cicely Saunders began work at St. Joseph's as a full-time medical officer—an experience that would shape the future of end-of-life care.

Modern hospice movement

Saunders recognized that the care that was offered to the terminally ill in hospitals was inadequate, so she initiated a new movement in palliative care, with an emphasis on providing specific centers for patients with terminal illness. In 1967 she founded St. Christopher's Hospice in South London, where the terminally ill could spend their last days in peace. Her main aim was to combine the humanitarian principles of love and devotion with modern medical advances. The hospice provided the patient with drugs for pain management as well as emotional and spiritual support.

Saunders assembled highly skilled teams that represented a wide range of medical disciplines, from consultants and researchers to pharmacologists and nurses. The hospice even had a post-mortem room, with the aim of reaching a better understanding of the extent of each patient's symptoms and how to control the pain associated with them.

Central to Saunders' concept was that the hospice should be designed with comfort as well as practicality in mind, so that patients could maintain a sense of dignity and individualism to the end. Rather than replicating a typical hospital ward—usually a long, open space lined with about 30 beds—Saunders' design for St. Christopher's included single rooms, offering greater patient privacy (and also helped reduce infection). The designs included large windows to allow natural light to flood the interior, and clean modernist lines to help staff work efficiently. Artificial lighting was carefully considered too, with fluorescent and warmer, softer lighting used in different areas. Above all, St. Christopher's was designed to create a welcoming

3,200 The estimated number of hospice programs in the US.

◁ **Knights Hospitaller**
The history of palliative care can be traced back to the Knights Hospitaller, founded in Jerusalem in the 11th century to provide care for sick and dying pilgrims. Besides their charitable work, the Knights were also a military organization dedicated to joining battle in the Crusades and defending Christian pilgrims.

"You matter because **you are you**, and you matter to the **end of your life**. We will do **all we can** not only to help you **die peacefully**, but also to **live until you die.**"

CICELY SAUNDERS, FOUNDER OF THE MODERN HOSPICE MOVEMENT

space, which was neither hospital nor home, but an entirely new type of facility that set the standard for palliative care in future decades. The design served as the prototype for many hospices that followed.

Further developments

The next logical step for Saunders was to allow patients to die at home, should they choose to do so, with the same level of pain control and support that was provided in the

▽ **Palliative care nurses**

End-of-life care has improved immeasurably in the past 50 years. Specialized nurses are trained in pain and symptom management, and offer psychological and spiritual support. Some work in hospitals or hospices, while others are out in the community, visiting people in their homes.

hospice environment. In 1969 she helped launch the first palliative home-care nursing program, and took the care provided at St. Christopher's out into the community too.

Saunders' efforts in the area of palliative care had a major impact on professional medical care and the provision of services for the terminally ill, together with the recognition that all aspects of patient care need to be addressed, from day-to-day physical care and pain relief to emotional sustenance. By the late 1980s, palliative care was recognized as a certified medical speciality in the UK, and it had spread worldwide by the 1990s, although it was not officially certified in the US and many parts of Europe until the 2000s.

BRITISH NURSE, PHYSICIAN, AND WRITER (1918–2005)

CICELY SAUNDERS

The founder of St. Christopher's Hospice in London, Cicely Saunders is also considered the founder of the modern palliative care movement. After training as a nurse and medical social worker, her life's course was changed by her deep friendship with a young Polish man, David Tasma, who was dying of cancer, and the fact that she felt called by God to devote the rest of her life to the care of the terminally ill. Saunders was particularly focused on alleviating the pain and suffering associated with terminal illnesses, especially cancer, and subsequently became involved in researching pain-relieving drugs.

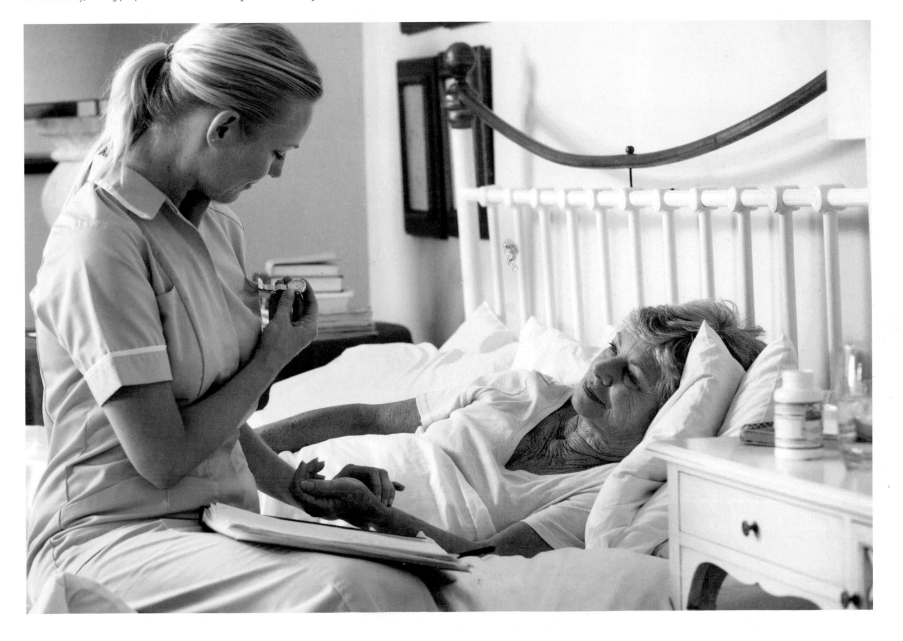

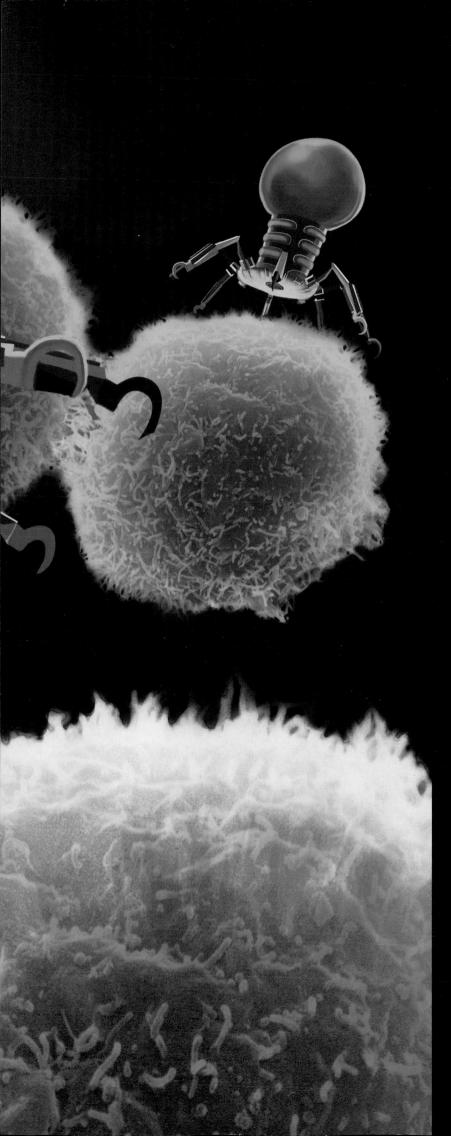

Nanomedicine

The science of diagnostics, therapeutics, and surgery conducted at a molecular scale—or nanomedicine—only became practically possible in the 1990s, the decade that saw the first use of targeted drug delivery mechanisms. Now, tiny robot surgical units, less than a millimeter long, are being tested, offering the potential of the ultimate noninvasive surgery.

The concept of nanomedicine originated in 1959 when American physicist Richard Feynman speculated that surgery might one day be carried out by small machines within the human body. Then in 1981 American engineer Eric Drexler described a theoretical cell repair machine that operated at a molecular level.

The first actual nanomedicines delivered targeted doses of drugs, some of which were "programmed" to attack cancer cells by reacting to their habitually acidic conditions or to transport small fragments of RNA or DNA for use in gene therapy. In 2006 dendrimers were developed—complex molecules with treelike branches that hold tiny drug doses, which are released on reaching the target cells. Semiconductor devices were also produced, such as quantum nanodots that emit light to provide illumination for endoscopics and are hundreds of times more powerful than conventional means. Mu-grippers, tiny biopsy tools for collecting tissue samples, were tested on animals in 2013. More recent development of corkscrew nanobots, which are manipulated through the body by a magnetic field to a blocked artery from which they then remove arterial plaque, are scheduled for clinical use in 2019.

" … it would be interesting in surgery if you could **swallow the surgeon.**"

RICHARD FEYNMAN, AMERICAN PHYSICIST, FROM THE LECTURE "THERE'S PLENTY OF ROOM AT THE BOTTOM," 1959

◁ Nanobots

The development of robots to conduct internal surgery on a tiny scale first became possible around 2013. In the future they may be able to repair DNA or map damaged vascular systems, rendering X-rays redundant.

Red Cross to the rescue
On September 19, 1985, a major earthquake registering 8.1 on the Richter scale hit Mexico City, killing between 5,000 and 10,000 people. The Red Cross organized humanitarian aid, rescuing bodies trapped in collapsed buildings and providing medical assistance.

Global Medical Bodies

Efforts to address healthcare crises at an international level began in the mid-19th century. By the end of the 20th century there was a proliferation of global medical bodies, some focusing on individual diseases and others on improving well-being and disaster relief.

International collaboration in health care began in 1851, when the first International Sanitary Conference met in Paris to discuss quarantine regulations in Europe. No agreement was reached, but at the 1892 conference a protocol concerning quarantining ships with cholera cases on board was signed.

Global governmental agencies

By the start of the 20th century it had become increasingly apparent that infectious diseases could not be confined by borders or by national policies. Prompted by a yellow fever outbreak that spread from Latin America to the US, the first international health agency—the Pan-American Sanitary Bureau (PASB)—was established in 1902. Shortly afterward, in 1907, 23 European countries founded the International Office of Public Hygiene, to prevent the spread of yellow fever, cholera, and plague. When the League of Nations was established in 1919, it formed a Health Committee that encompassed both the work of the PASB and that of the International Office of Public Hygiene. However, its effectiveness was limited due to lack of funding.

A major development in global health care came when the World Health Organization (WHO) was founded in 1948. As the healthcare arm of the United Nations, the WHO held considerable political sway. It expanded on existing work in monitoring communicable diseases and launched eradiction campaigns, its greatest successes to date being against smallpox (see pp.100–01) and polio (see pp.210–11). In recent years the WHO has strengthened its role in coordinating a swift response to disease outbreaks—for example Severe Acute Respiratory Syndrome (SARS) in 2003 and swine flu in 2009, both of which threatened to become worldwide pandemics. Increasingly, the WHO plays a key role in health education and the sponsorship of health programs throughout the developing world.

Nongovernmental bodies

In parallel with international governmental agencies, several nongovernmental organizations (NGOs) have developed. The first was the International Committee of the Red Cross (ICRC). Initially set up in 1863 to care for wounded soldiers (see panel, left), in 1929 the Red Cross extended its support to prisoners of war and civilians in war zones, and lobbied for the adoption of the Geneva Convention, which aims to protect people not taking part in armed conflict. Today, the ICRC (which includes the Red Crescent movement established in Islamic countries) consists of 190 national societies around the world, and is a neutral, independent presence in most areas of armed conflict.

Other such NGOs have been set up to coordinate medical responses to disasters or to manage national approaches to longer-term health-care challenges. These include the World Federation of Mental Health (1948), the International Planned Parenthood Federation (1952), and the World Medical Association (1947). Typical of a newer breed of NGO is Médecins Sans Frontières (MSF), set up by a group of French doctors in 1971 in response to humanitarian disasters such as the Biafran War in Nigeria. With a commitment to providing health care to all, MSF has projects in some of the world's most troubled countries, and works in areas—such as parts of Syria and Yemen—where few international medical bodies will send staff.

FOUNDER OF THE RED CROSS (1828–1910)

JEAN-HENRI DUNANT

Upon witnessing the bloody battle of Solferino in Italy in 1859, Swiss humanitarian and businessman Dunant argued that volunteers should be trained to help the wounded on the battlefield—an idea that led to the foundation of the International Committee of the Red Cross in 1863. In 1872 he lobbied for a court of arbitration to settle international disputes, but the idea was premature. Bankrupt and a recluse for the last 35 years of his life, Dunant was awarded the first ever Nobel Peace Prize in 1901.

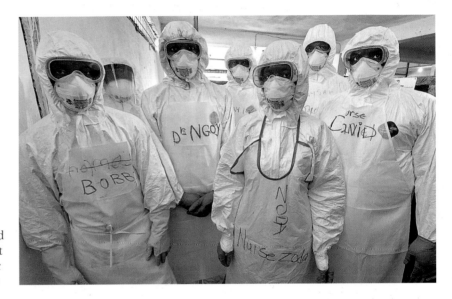

△ **Fighting Ebola**
In 2013 a outbreak of Ebola virus disease began in West Africa (see pp.268–69). Efforts by dedicated health workers helped control the epidemic, but the WHO was criticized for its delayed response.

▽ **Raising awareness**
From 2015 the Zika virus, carried by *Aedes aegypti* mosquitoes, began to spread rapidly in Brazil and beyond. In early 2016 the WHO declared this a "Public Health Emergency of International Concern" and coordinated multinational action.

Ebola Virus Disease

First discovered in Africa in 1976, Ebola virus disease has emerged as one of the deadliest diseases of the past 50 years. The first case originated near the Ebola River in what is now the Democratic Republic of Congo, and the virus has since infected more than 28,000 people and killed more than 11,000.

Despite the voracious nature of the Ebola virus disease, a cure has remained elusive. It is still not known why only some victims survive—the fatality rate can be as high as 70 percent, depending on age, location, and other variables. Transmitted from wild animals to humans, Ebola spreads in humans by close contact and transfer of bodily fluids from infected people. The incubation period of the virus ranges from 2 to 21 days, with the initial symptoms being sudden fever, muscle pain and weakness, headache, and sore throat, followed by nausea, vomiting, diarrhea, rash, kidney and liver problems, and both internal and external bleeding.

The worst outbreak of Ebola began in 2013, killing more than 10,000 people by 2015. It started in Guinea and spread across to Sierra Leone and Liberia, then by air to Nigeria and the US, and by land to Senegal and Mali. With no vaccines or accepted form of drug treatment, the World Health Organization (WHO) called an emergency meeting to handle the crisis, making an unprecedented decision to allow the use of experimental, unproven drugs. One such drug, ZMapp, was used with some success. The first person treated was American missionary Kent Bradley, who was working in Liberia, Africa. Bradley survived, as did many others who were given ZMapp, but the success was limited. Around 20–30 percent of those treated died but the odds of survival were better than not taking the drug, which is made up of three different antibodies. Research continues with the aim of developing both effective treatments and vaccines.

"This is a **virus** that is a **threat** to all **humanity.**"

GAYLE SMITH, SENIOR DIRECTOR, US NATIONAL SECURITY COUNCIL, 2014

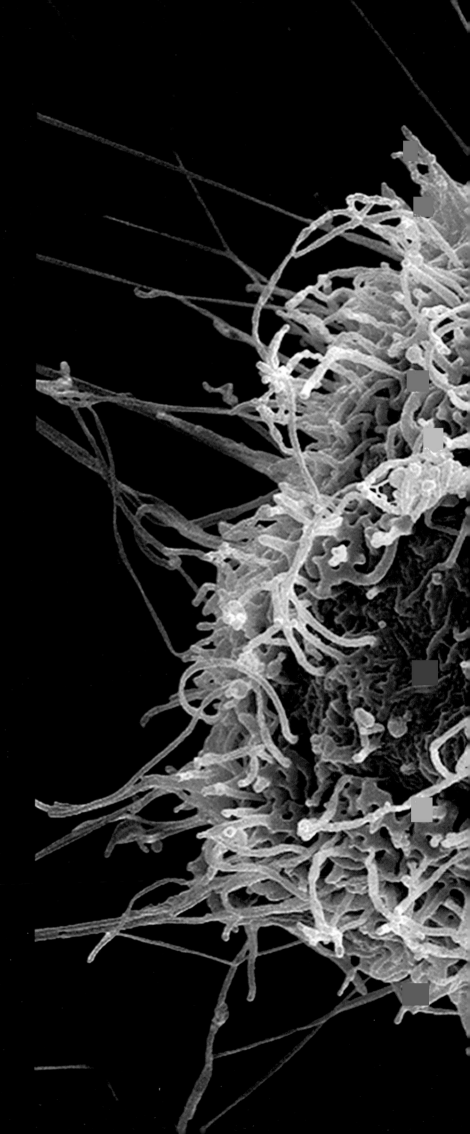

▷ Ebola virus in an infected cell
An electron micrograph shows Ebola virus particles (in blue) budding from an infected cell. Once it has infiltrated the cell membrane, the virus colonizes the cell to reproduce.

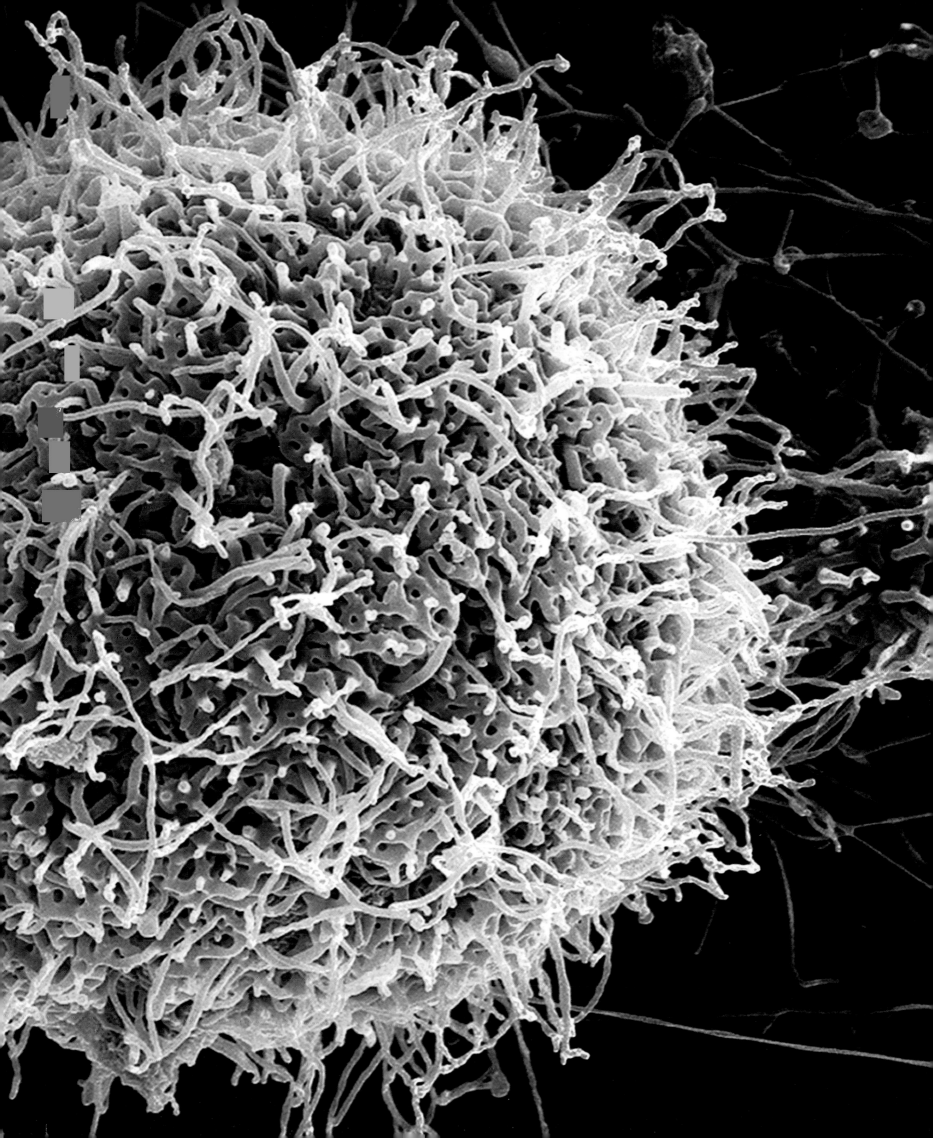

Stem Cell Therapy

Stem cells are unspecialized cells that are capable of renewing themselves and becoming another type of cell with a more specialized function, for example, a muscle, nerve, bone, or blood cell. Their ability to regenerate makes them potentially very useful in the treatment and prevention of many illnesses.

W hen a stem cell divides, each new cell can either remain a stem cell (to maintain their numbers) or become one of more than 200 different types of specialized cells. The cells multiply to replace those specialized cells that die as part of the body's natural cell turnover and tissue maintenance. For example, bone marrow stem cells produce more than two million new red and white blood cells every second, to replace a similar number that reach the end of their normal lives. Different types of body tissue replace cells at different rates—for instance, cells in the blood, skin, and digestive tract lining renew quickly, whereas nerve tissues are replaced very slowly, if at all.

300 BILLION The estimated number of new cells that stem cells produce in the body on a daily basis.

Stem cell types

There are many different types of stem cells, but broadly speaking they can be divided into two main groups: embryonic and adult.

Embryonic stem cells exist only at the earliest stage of development and are pluripotent, which means they have the potential to become almost any type of specialized cell. This makes them very valuable to medical science, as they provide a renewable resource for research and therapies. These cells are obtained from human embryos, usually those that were fertilized in vitro (IVF) and are no longer needed.

Adult (or somatic) stem cells appear during foetal development and remain in the body throughout a person's life. They are present in many mature organs and tissues, and are more specialized than embryonic stem cells. Typically, they produce specialized cells exclusively for the specific tissue or organ in which they live. For example, blood-forming (hematopoietic) stem cells in the bone marrow can create red or white cells or platelets, but cannot make liver or muscle cells. This ability to create multiple specialized cell types is referred to as multipotent, but there are also unipotent stem cells, which can develop into only one type of cell. Adult stem cells can be difficult to find in the body and grow in laboratory culture less easily than embryonic stem cells. However, they are still extremely useful for scientific research and therapies.

In 2006 there was a breakthrough in stem cell research, when Shinya Yamanaka, a Japanese researcher, reprogrammed mature, specialized skin cells to become pluripotent; these new cells were termed induced pluripotent stem cells (iPSCs). In 2012 Yamanaka and British researcher John Gurdon,

△ **Dolly, the cloned sheep**
Born in 1996, Dolly was the first mammal copied, or cloned, from a specialized adult stem cell—in this case, a mammary gland cell. In 2013 scientists produced human embryonic stem cells (hESCs) using a similar cloning technique.

who in 1962 had discovered that the specialization of mature cells is reversible, received the Nobel Prize in Physiology or Medicine for their groundbreaking findings.

Established therapies

Bone marrow transplant, or hematopoietic stem cell transplantation (HSCT), is well established to treat various blood disorders. First performed in 1956 in New York, its subjects were identical twins, one of whom had leukemia. Since the subjects were identical, the problem of rejection did not occur. Progress in immunosuppression, to prevent the immune system from rejecting cells, led to the first bone marrow transplant between nonidentical siblings in 1968, and between an unrelated but tissue-matched

CONCEPT

HOW STEM CELL THERAPY WORKS

All cells in the body contain the full set of human genes as their DNA (deoxyribonucleic acid). However, different types of cells have specific genes activated in them. For instance, in skin cells some genes are activated while others remain inactive. Similarly, each stem cell has its own particular set of genes activated. These instruct it to multiply and either replicate itself, producing copies that are identical to the original cell, or to become a specialized cell—a process known as "differentiation" in which the cell goes through various stages, becoming more specialized at each step. These instructions are triggered by various means, including natural signaling substances such as growth factors or cytokines. Stem cell research involves recreating these conditions to manipulate the cells along various pathways, for example, growing a new tissue to replace a diseased one.

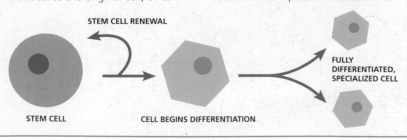

STEM CELL RENEWAL

STEM CELL

CELL BEGINS DIFFERENTIATION

FULLY DIFFERENTIATED, SPECIALIZED CELL

" **Stem cell research** is the **key to developing cures** for **degenerative conditions** like Parkinson's and motor neurone disease…"

STEPHEN HAWKING, BRITISH PHYSICIST AND COSMOLOGIST (AND MND SUFFERER), IN A PRESS RELEASE, 2006

donor and recipient in 1973. Today HSCT patients can be given their own hematopoietic stem cells that have been removed and stored. They also have a course of chemotherapy or radiotherapy to eradicate the diseased cells. In some cases, such as tissue matching problems, umbilical cord blood is used as an alternative.

Ongoing research

Stem cell therapy holds vast promise. Research involving iPSCs and adult stem cells is currently being carried out to treat a range of conditions, including certain cancers, diabetes, retinal problems, rheumatoid arthritis, repair of spinal cord and other tissues, and baldness. One of the primary aims of research is to develop the technology to take a few donated cells and grow them into new tissues or organs, such as a new heart, liver, or eye. Tissue or organ transplants using iPSCs should be safer than transplants from other people; since the cells would be taken from and put back into the same individual, they are less likely to provoke the patient's immune system into rejecting the new tissue or organ. Stem cell research is still in its infancy, but it already represents a medical revolution.

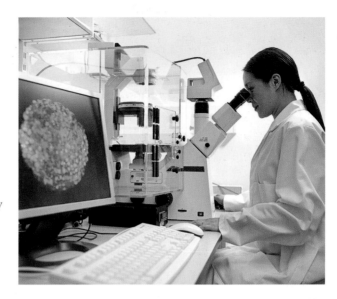

◁ **Research and development**
Research into stem cell therapy is one of the most exciting developments in medicine in recent years. However, there is some controversy surrounding the ethics of research using human embryos.

△ **Bone marrow stem cell**
Mesenchymal stem cells are multipotent adult stem cells present in the bone marrow (and possibly elsewhere) that can generate many kinds of tissue such as bone, cartilage, and fat.

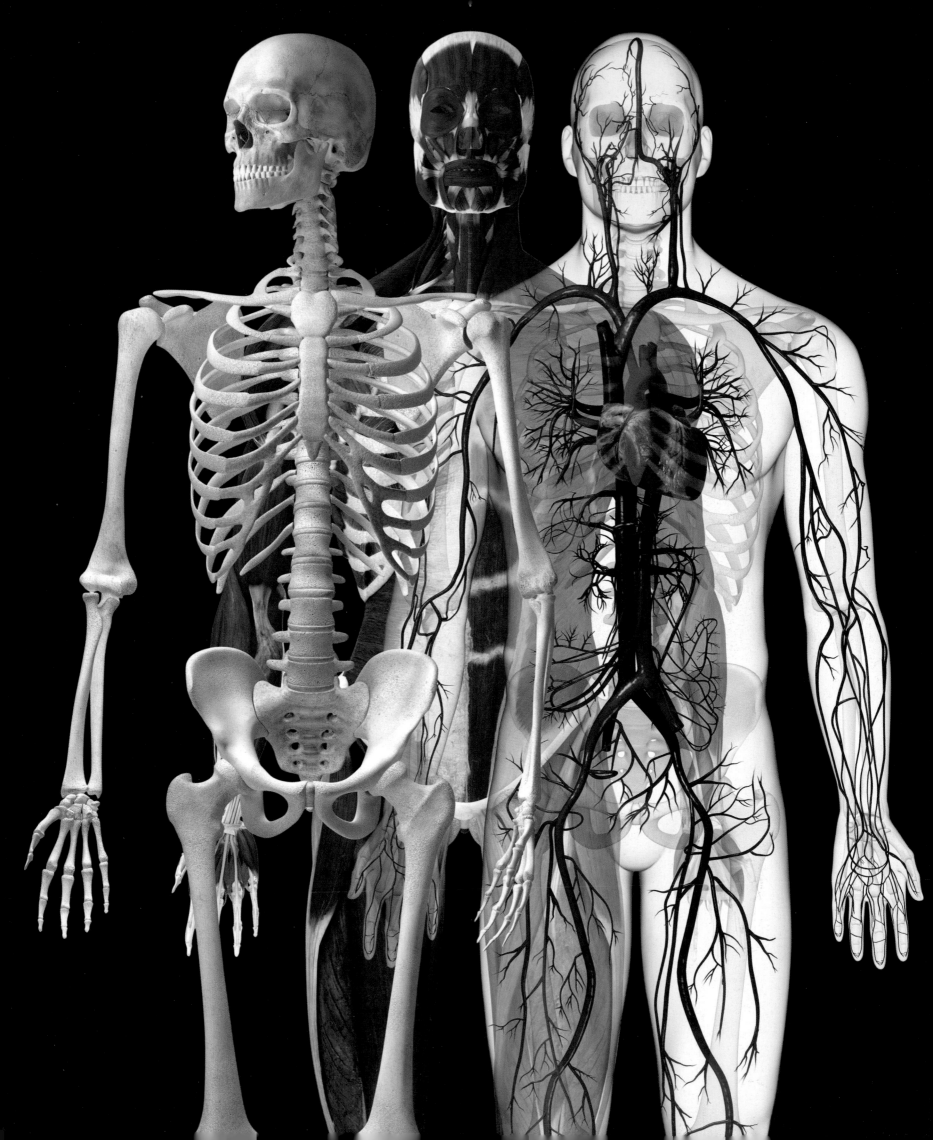

6
REFERENCE

Body Systems

The study of the human body has fascinated physicians for ages. Galen's writings were seen as definitive for centuries until 1543, when Andreas Vesalius first published images of dissected bodies. Once William Harvey had established the basic function of the heart and circulation, 19th-century scientists made advances in cardiovascular medicine. Doctors and scientists continued to piece together how all the different systems work together—for the most part, harmoniously. As a result, specialties developed and medical treatments advanced significantly. It helps to have an understanding of the basics of each body system, as well as to see how they are so interdependent, to appreciate the enormity of what has been achieved in the study of human biology and medicine, and what may continue to develop in the future.

SKELETAL

This is the solid movable framework that surrounds and protects delicate structures. Bones act as levers and anchor plates for the muscles to pull against, generating movement. They make blood cells and release minerals for use elsewhere in the body.
(See pp.276–77)

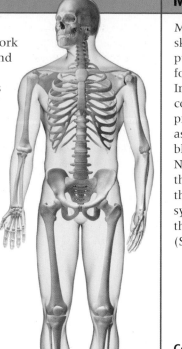

Components
Bones, cartilage, ligaments

MUSCULAR

Muscles work with the skeleton, providing pulling force for movement. Involuntary muscles control internal processes, such as digestion and blood distribution. Nerves control the muscles and the circulatory system supplies them oxygen.
(See pp.278–79)

Components
Skeletal muscle, tendons, smooth muscle in organs, cardiac muscle

ENDOCRINE

This system makes the chemical messengers, or hormones, that maintain an optimal environment in the body and govern long-term processes such as growth, puberty, and reproductive activity. Control of hormone production is linked to the nervous system, and hormones are circulated in the blood.
(See pp.286–87)

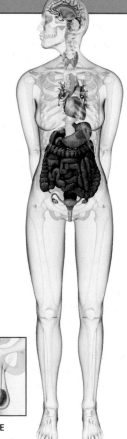

Components
Glands including hypothalamus, pituitary, thyroid, thymus, adrenal, pancreas; also heart, stomach, intestines; ovaries (female), testes (male)

MALE

DIGESTIVE

This system is responsible for chewing, storing, and processing food, passing nutrients into the liver via the circulatory system, and eliminating waste products from the body. Healthy digestion is dependent on healthy nervous and immune systems and is greatly affected by psychological state.
(See pp.288–89)

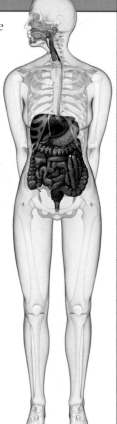

Components
Mouth, throat, esophagus (gullet), stomach, pancreas, liver, gallbladder, small and large intestines, rectum, anus

IMMUNE

Providing vital resistance to threats, such as infectious diseases, and malfunctions of internal processes, the immune system has an intricate relationship with many parts of the body. Lymph fluid delivers nutrients, collects waste matter, and carries immunity-providing white blood cells.
(See pp.290–91)

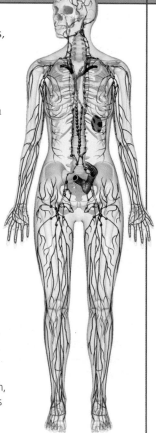

Components
White blood cells, antibodies, spleen, tonsils and adenoids, lymph, lymph vessels, lymph nodes (glands), lymph ducts

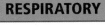

NERVOUS

The brain is the seat of consciousness and creativity, yet many of its functions are performed unconsciously. It controls the body's movements with the help of the spinal cord and nerves. It receives sensory information, controls the endocrine system, and maintains other systems.
(See pp.280–81)

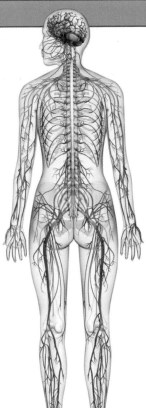

Components
Brain, spinal cord, nerves, sense organs

CARDIOVASCULAR

This system pumps blood around the body. It supplies the tissues with oxygen and nutrients and carries waste products from them so that they can be expelled. It also transports hormones and immune cells to the points where they are needed.
(See pp.282–83)

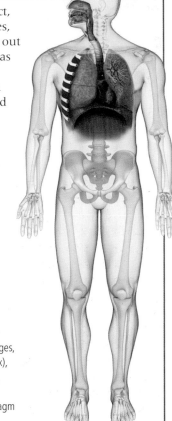

Components
Heart, arteries, veins, arterioles, venules, capillaries, blood (plasma and red and white cells)

RESPIRATORY

The respiratory tract, powered by muscles, carries air into and out of the lungs. This gas exchange allows oxygen to be taken up by the blood and carbon dioxide to be removed from it. A secondary function of this system is speech.
(See pp.284–85)

Components
Nasal and other air passages, throat (pharynx and larynx), trachea (windpipe), lungs, lung airways, respiratory muscles, including diaphragm

URINARY

The kidneys remove waste and excess fluid from the body for disposal as urine. They also help maintain fluid, salt, and mineral levels in the body. Urine production is controlled by hormones and influenced by internal factors such as blood flow and pressure, as well as external ones such as fluid intake and environmental temperature.
(See p.292)

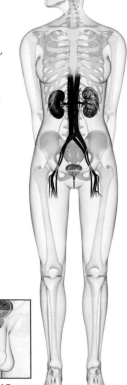

Components
Kidneys, ureters, bladder, urethra

MALE

REPRODUCTIVE

This is the one system that differs completely in males and females. It is controlled by the hormones and only functions for part of the human lifespan. Sperm production in men is continual while female egg production is cyclical. In males, sperm and urine are both expelled from the urethera.
(See pp.293–94)

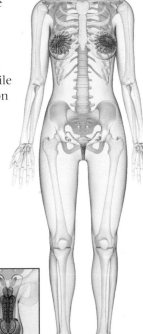

Components
Male: testes, epididymises, prostate and bulbourethral glands, seminal vesicles, spermatic ducts, urethra, penis
Female: Ovaries, fallopian tubes, uterus, vagina, and external genitalia

MALE

SENSORY

The skin, eyes, ears, tongue, and nose are the organs that provide information for the senses—touch, sight, hearing, taste, and smell—via nerve receptors. The skin, hair, and nails also form the body's outer protective covering against physical hazards and regulate body temperature. A layer of subcutaneous fat beneath the skin insulates the body, stores energy, and acts as a shock absorber.
(See pp.296–99)

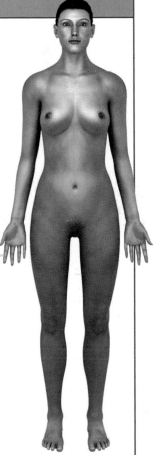

Components
Ears, eyes, mouth, nose, plus skin, hair, and nails

The **Skeletal System**

The human skeleton is a strong but flexible framework that supports the body and protects the internal organs. Bones act as levers and anchor plates for movement created by **the muscles. The skeletal system is also fully integrated with the rest of the body—blood cells develop in the bones, and essential minerals such as calcium are drawn from them.**

The skeleton

The average adult skeleton is made up of 206 bones, but there are always natural variations; for example, one person in 20 has an extra rib. A newborn infant has more than 300 bones, although they begin as soft cartilage, which turn to bone over time—a process called ossification—and many fuse as the child grows. Bone shape varies according to its function— flat bones have large surface areas for the attachment of muscles and long bones act as levers to enable movement.

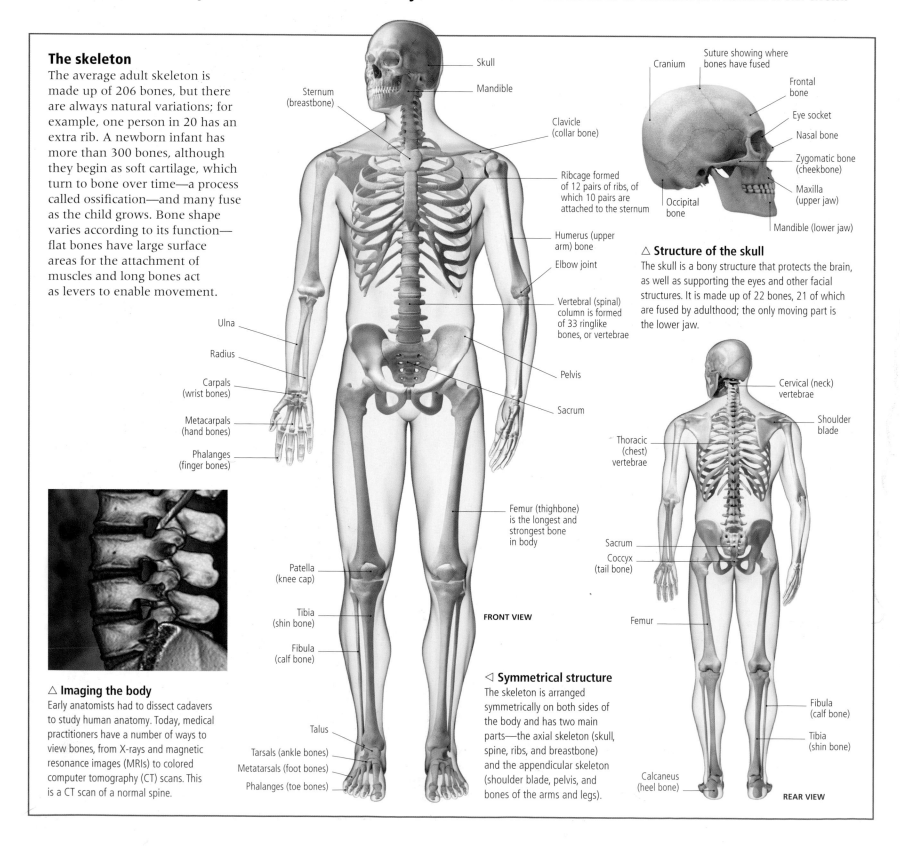

Skull
Mandible
Sternum (breastbone)
Clavicle (collar bone)
Ribcage formed of 12 pairs of ribs, of which 10 pairs are attached to the sternum
Humerus (upper arm) bone
Elbow joint
Vertebral (spinal) column is formed of 33 ringlike bones, or vertebrae
Pelvis
Sacrum
Ulna
Radius
Carpals (wrist bones)
Metacarpals (hand bones)
Phalanges (finger bones)
Femur (thighbone) is the longest and strongest bone in body
Patella (knee cap)
Tibia (shin bone)
Fibula (calf bone)
Talus
Tarsals (ankle bones)
Metatarsals (foot bones)
Phalanges (toe bones)
FRONT VIEW

Cranium
Suture showing where bones have fused
Frontal bone
Eye socket
Nasal bone
Zygomatic bone (cheekbone)
Maxilla (upper jaw)
Occipital bone
Mandible (lower jaw)

△ Structure of the skull
The skull is a bony structure that protects the brain, as well as supporting the eyes and other facial structures. It is made up of 22 bones, 21 of which are fused by adulthood; the only moving part is the lower jaw.

Cervical (neck) vertebrae
Shoulder blade
Thoracic (chest) vertebrae
Sacrum
Coccyx (tail bone)
Femur
Fibula (calf bone)
Tibia (shin bone)
Calcaneus (heel bone)
REAR VIEW

△ Imaging the body
Early anatomists had to dissect cadavers to study human anatomy. Today, medical practitioners have a number of ways to view bones, from X-rays and magnetic resonance images (MRIs) to colored computer tomography (CT) scans. This is a CT scan of a normal spine.

◁ Symmetrical structure
The skeleton is arranged symmetrically on both sides of the body and has two main parts—the axial skeleton (skull, spine, ribs, and breastbone) and the appendicular skeleton (shoulder blade, pelvis, and bones of the arms and legs).

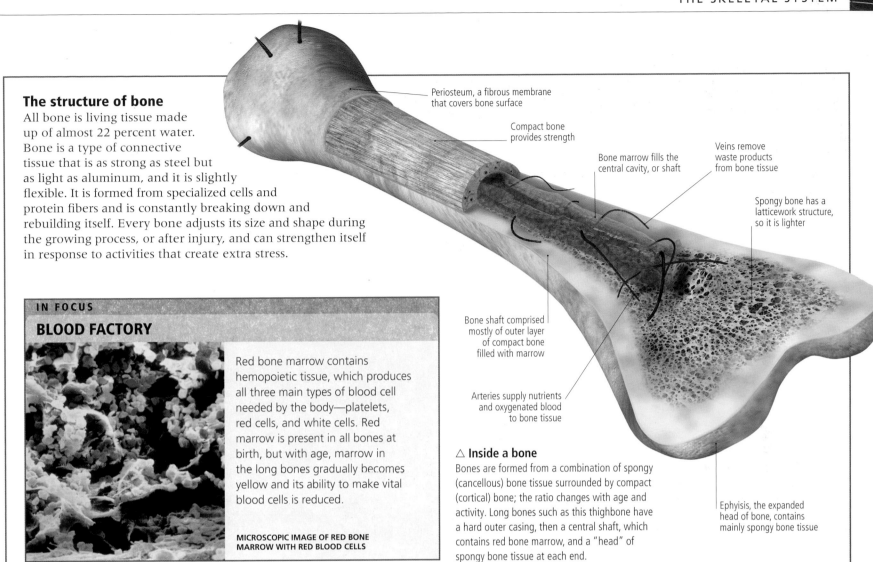

The structure of bone

All bone is living tissue made up of almost 22 percent water. Bone is a type of connective tissue that is as strong as steel but as light as aluminum, and it is slightly flexible. It is formed from specialized cells and protein fibers and is constantly breaking down and rebuilding itself. Every bone adjusts its size and shape during the growing process, or after injury, and can strengthen itself in response to activities that create extra stress.

Periosteum, a fibrous membrane that covers bone surface

Compact bone provides strength

Bone marrow fills the central cavity, or shaft

Veins remove waste products from bone tissue

Spongy bone has a latticework structure, so it is lighter

Bone shaft comprised mostly of outer layer of compact bone filled with marrow

Arteries supply nutrients and oxygenated blood to bone tissue

Ephyisis, the expanded head of bone, contains mainly spongy bone tissue

IN FOCUS

BLOOD FACTORY

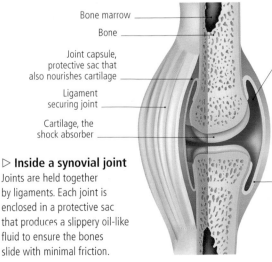

Red bone marrow contains hemopoietic tissue, which produces all three main types of blood cell needed by the body—platelets, red cells, and white cells. Red marrow is present in all bones at birth, but with age, marrow in the long bones gradually becomes yellow and its ability to make vital blood cells is reduced.

MICROSCOPIC IMAGE OF RED BONE MARROW WITH RED BLOOD CELLS

△ Inside a bone

Bones are formed from a combination of spongy (cancellous) bone tissue surrounded by compact (cortical) bone; the ratio changes with age and activity. Long bones such as this thighbone have a hard outer casing, then a central shaft, which contains red bone marrow, and a "head" of spongy bone tissue at each end.

All about joints

A joint is the point at which two bones meet. Some of these are fixed (those in the skull) or semimovable (symphysis pubis) and linked only by connective tissue, to allow for growth. By far the greatest number are freely moving, or synovial, joints, which are classified by the type of movement they allow—there are around 250 of them in the body.

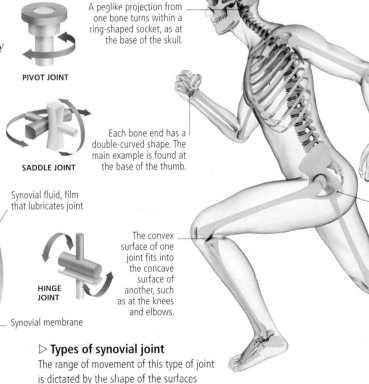

A peglike projection from one bone turns within a ring-shaped socket, as at the base of the skull.

PIVOT JOINT

Each bone end has a double-curved shape. The main example is found at the base of the thumb.

SADDLE JOINT

The convex surface of one joint fits into the concave surface of another, such as at the knees and elbows.

HINGE JOINT

ELLIPSOIDAL JOINT

An egg-shaped bone nestles in an ellipsoidal cavity, such as at the wrist, and movement can occur in most directions.

BALL-AND-SOCKET JOINT

The ball-shaped head of one bone fits into a cuplike cavity of another. The shoulder and hip are prime examples.

GLIDING JOINT

A joint where the surfaces of both bones are almost flat and they "glide" over each other, as in the small bones of the ankles and wrists.

Bone marrow

Bone

Joint capsule, protective sac that also nourishes cartilage

Ligament securing joint

Cartilage, the shock absorber

Synovial fluid, film that lubricates joint

Synovial membrane

▷ Inside a synovial joint

Joints are held together by ligaments. Each joint is enclosed in a protective sac that produces a slippery oil-like fluid to ensure the bones slide with minimal friction.

▷ Types of synovial joint

The range of movement of this type of joint is dictated by the shape of the surfaces that articulate and how they fit together.

The **Muscular System**

The skeleton forms the body's framework, but it is the muscles that provide the pulling forces that enable movement. Specialized muscles also power many internal **body processes, from heartbeats to the movement of food through the digestive system. Muscles rely on nerves to control them and blood to provide oxygen and energy.**

The muscles of the body

Muscles are the body's "flesh," they are the bulge and ripple under the skin, and are arranged in many crisscrossing layers right down to the bones. In a typical adult male, muscles make up two-fifths of the body weight. The job of the skeletal muscles is to contract and pull the bones to which they are anchored (see opposite).

IN FOCUS

TYPES OF MUSCLE TISSUE

The body has three types of muscle tissue. Skeletal muscles are joined to bones and cause movement. They are also known as voluntary or striated muscles, and have stripes or bands. Smooth muscles, sometimes called involuntary muscles because they cannot be consciously controlled, are found in the walls of the intestines, airways, and other organs. Cardiac muscles make up the walls of the heart.

SKELETAL MUSCLE

SMOOTH MUSCLE

CARDIAC MUSCLE

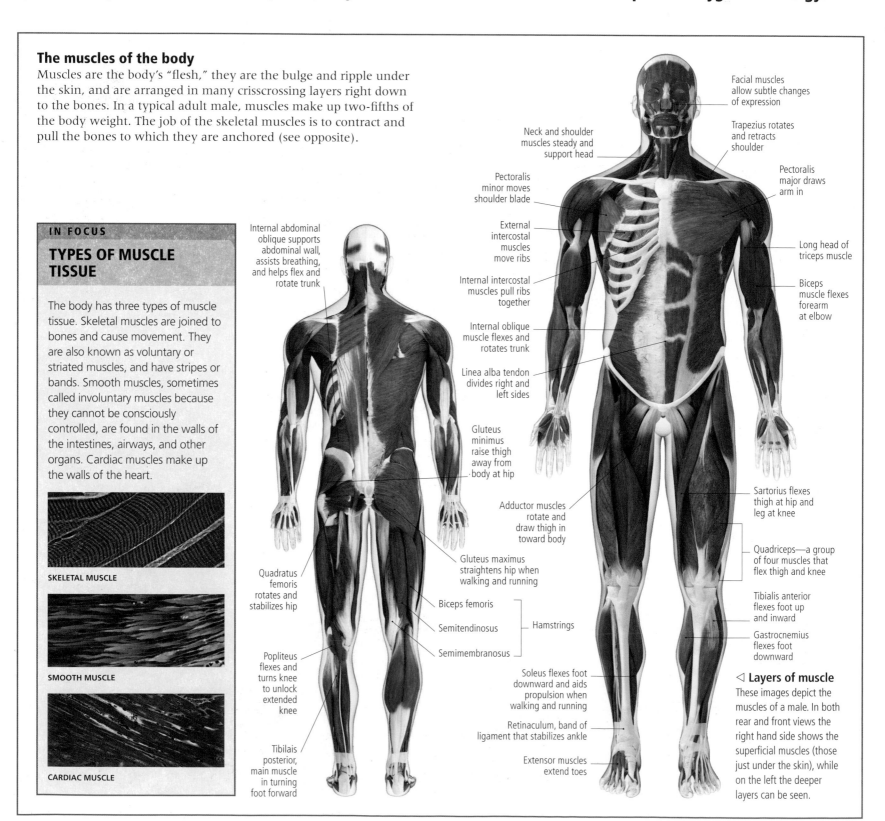

Internal abdominal oblique supports abdominal wall, assists breathing, and helps flex and rotate trunk

Quadratus femoris rotates and stabilizes hip

Popliteus flexes and turns knee to unlock extended knee

Tibialis posterior, main muscle in turning foot forward

Facial muscles allow subtle changes of expression

Trapezius rotates and retracts shoulder

Neck and shoulder muscles steady and support head

Pectoralis major draws arm in

Pectoralis minor moves shoulder blade

External intercostal muscles move ribs

Long head of triceps muscle

Internal intercostal muscles pull ribs together

Biceps muscle flexes forearm at elbow

Internal oblique muscle flexes and rotates trunk

Linea alba tendon divides right and left sides

Gluteus minimus raise thigh away from body at hip

Adductor muscles rotate and draw thigh in toward body

Sartorius flexes thigh at hip and leg at knee

Gluteus maximus straightens hip when walking and running

Quadriceps—a group of four muscles that flex thigh and knee

Biceps femoris

Semitendinosus   Hamstrings

Semimembranosus

Tibialis anterior flexes foot up and inward

Gastrocnemius flexes foot downward

Soleus flexes foot downward and aids propulsion when walking and running

Retinaculum, band of ligament that stabilizes ankle

Extensor muscles extend toes

◁ **Layers of muscle**
These images depict the muscles of a male. In both rear and front views the right hand side shows the superficial muscles (those just under the skin), while on the left the deeper layers can be seen.

Tendons

A typical muscle spans a joint and tapers at each end into bands of tough fibrous cords of connective tissue known as tendons, which anchor muscle to bone. Special fibers (Sharpey's fibers) within each tendon pass through the bone's outer covering (periosteum) and become embedded in it. The more stable muscle attachment, usually the end of the muscle nearest the center of the body, is known as the origin and moves little if at all during a contraction. The other end, called the insertion, moves more.

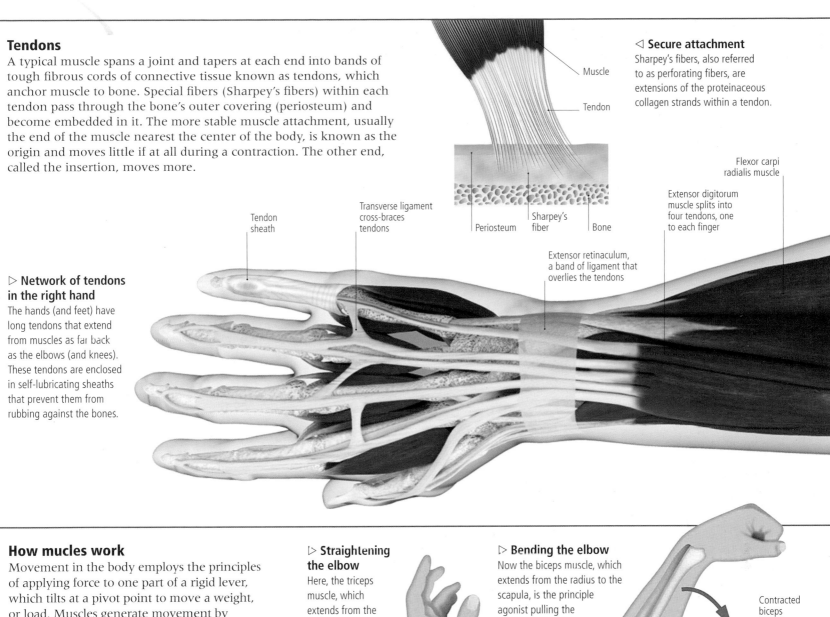

◁ Secure attachment
Sharpey's fibers, also referred to as perforating fibers, are extensions of the proteinaceous collagen strands within a tendon.

Muscle

Tendon

Periosteum

Sharpey's fiber

Bone

Flexor carpi radialis muscle

Extensor digitorum muscle splits into four tendons, one to each finger

Extensor retinaculum, a band of ligament that overlies the tendons

Tendon sheath

Transverse ligament cross-braces tendons

▷ Network of tendons in the right hand
The hands (and feet) have long tendons that extend from muscles as far back as the elbows (and knees). These tendons are enclosed in self-lubricating sheaths that prevent them from rubbing against the bones.

How mucles work

Movement in the body employs the principles of applying force to one part of a rigid lever, which tilts at a pivot point to move a weight, or load. Muscles generate movement by contracting, or pulling, which makes them shorter; they relax and lengthen passively as another muscle contracts. As a result, muscles are arranged in pairs that act in opposition to one another; agonists produce movement while antagonists relax.

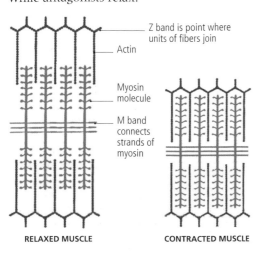

Z band is point where units of fibers join

Actin

Myosin molecule

M band connects strands of myosin

RELAXED MUSCLE

CONTRACTED MUSCLE

▷ Straightening the elbow
Here, the triceps muscle, which extends from the top of the humerus to the ulna, contracts to "pull" the arm straight.

◁ How a muscle fiber contracts
Skeletal muscle consists of elongated cells known as myofibers, which contain myofilaments made up of the proteins actin and myosin. As a muscle contracts, myosin filaments slide between actin filaments, shortening the muscle fiber.

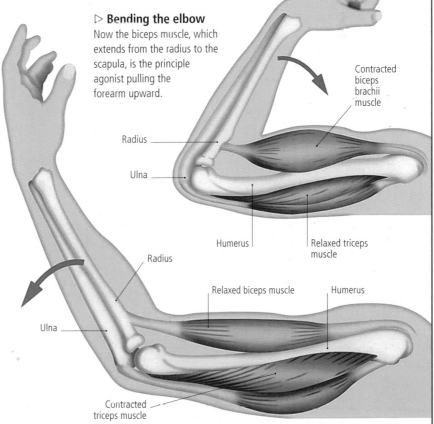

▷ Bending the elbow
Now the biceps muscle, which extends from the radius to the scapula, is the principle agonist pulling the forearm upward.

Radius

Ulna

Contracted biceps brachii muscle

Humerus

Relaxed triceps muscle

Radius

Relaxed biceps muscle

Humerus

Ulna

Contracted triceps muscle

The **Nervous System**

This is the body's central communication and control system. It has two main parts: the central nervous system (CNS) and the peripheral nervous system (PNS). A third component, the autonomic nervous system (ANS), controls involuntary functions of the body such as heart rate, breathing, and digestion.

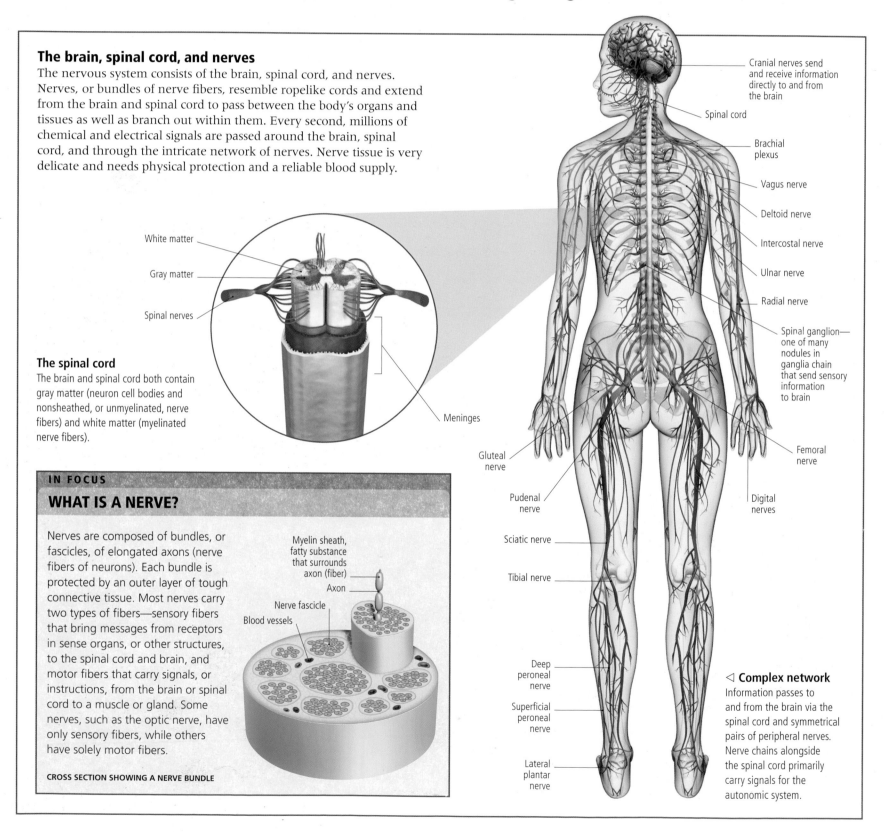

The brain, spinal cord, and nerves

The nervous system consists of the brain, spinal cord, and nerves. Nerves, or bundles of nerve fibers, resemble ropelike cords and extend from the brain and spinal cord to pass between the body's organs and tissues as well as branch out within them. Every second, millions of chemical and electrical signals are passed around the brain, spinal cord, and through the intricate network of nerves. Nerve tissue is very delicate and needs physical protection and a reliable blood supply.

White matter

Gray matter

Spinal nerves

The spinal cord

The brain and spinal cord both contain gray matter (neuron cell bodies and nonsheathed, or unmyelinated, nerve fibers) and white matter (myelinated nerve fibers).

Meninges

Cranial nerves send and receive information directly to and from the brain

Spinal cord

Brachial plexus

Vagus nerve

Deltoid nerve

Intercostal nerve

Ulnar nerve

Radial nerve

Spinal ganglion—one of many nodules in ganglia chain that send sensory information to brain

Femoral nerve

Gluteal nerve

Pudenal nerve

Digital nerves

Sciatic nerve

Tibial nerve

Deep peroneal nerve

Superficial peroneal nerve

Lateral plantar nerve

◁ **Complex network**
Information passes to and from the brain via the spinal cord and symmetrical pairs of peripheral nerves. Nerve chains alongside the spinal cord primarily carry signals for the autonomic system.

IN FOCUS

WHAT IS A NERVE?

Nerves are composed of bundles, or fascicles, of elongated axons (nerve fibers of neurons). Each bundle is protected by an outer layer of tough connective tissue. Most nerves carry two types of fibers—sensory fibers that bring messages from receptors in sense organs, or other structures, to the spinal cord and brain, and motor fibers that carry signals, or instructions, from the brain or spinal cord to a muscle or gland. Some nerves, such as the optic nerve, have only sensory fibers, while others have solely motor fibers.

Myelin sheath, fatty substance that surrounds axon (fiber)

Axon

Nerve fascicle

Blood vessels

CROSS SECTION SHOWING A NERVE BUNDLE

The anatomy of the brain

In many ways the human brain resembles a computer. The brain, in conjunction with the spinal cord, regulates both conscious and unconscious, or "autonomic," body processes and coordinates movement. This delicate and complex structure is surrounded and protected by the skull and several membranes. It also has a vast circulatory system: although the brain only accounts for 2 percent of the body's weight, it requires 20 percent of the blood to provide it with oxygen and nutrients.

IN FOCUS

THE PRIMITIVE BRAIN

Human behavior is not always rational. In times of stress or crisis, deep-seated instincts well up from within and take over awareness. Such events involve the primitive brain, which is based in a series of structures in the limbic system. The system influences subconscious, instinctive behavior similar to animal responses relating to survival and reproduction.

Midbrain

Cigulate gyrus, or limbic cortex

Olfactory bulb, or smell processors, are "wired" into the limbic system

Amygdala is concerned with emotions and drive

Pons, part of the brain stem

Hippocampus is linked with memory

STRUCTURES IN THE LIMBIC SYSTEM

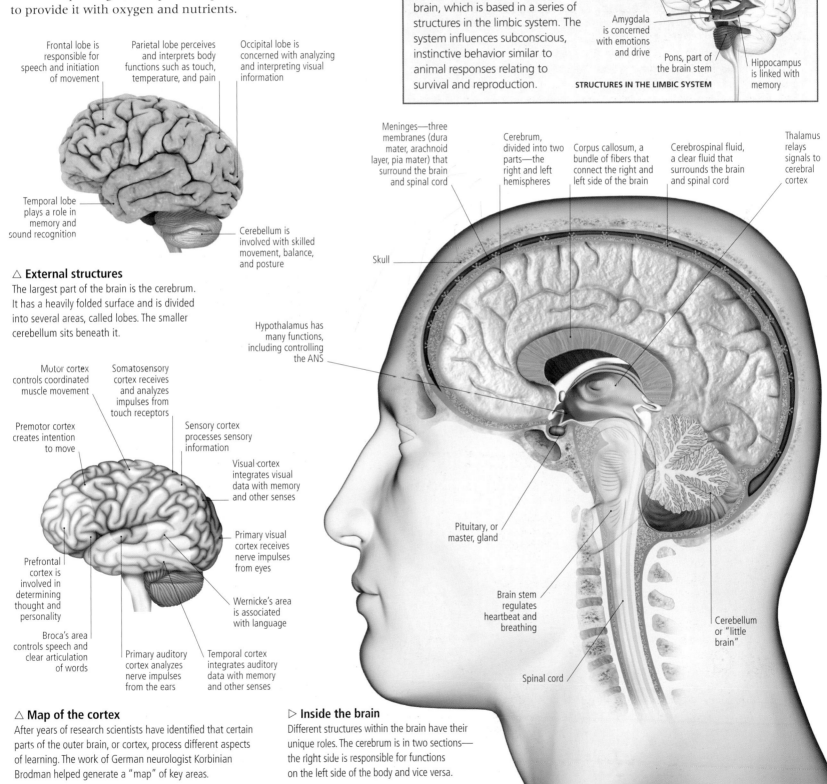

Frontal lobe is responsible for speech and initiation of movement

Parietal lobe perceives and interprets body functions such as touch, temperature, and pain

Occipital lobe is concerned with analyzing and interpreting visual information

Temporal lobe plays a role in memory and sound recognition

Cerebellum is involved with skilled movement, balance, and posture

△ **External structures**
The largest part of the brain is the cerebrum. It has a heavily folded surface and is divided into several areas, called lobes. The smaller cerebellum sits beneath it.

Motor cortex controls coordinated muscle movement

Somatosensory cortex receives and analyzes impulses from touch receptors

Premotor cortex creates intention to move

Sensory cortex processes sensory information

Visual cortex integrates visual data with memory and other senses

Prefrontal cortex is involved in determining thought and personality

Primary visual cortex receives nerve impulses from eyes

Broca's area controls speech and clear articulation of words

Wernicke's area is associated with language

Primary auditory cortex analyzes nerve impulses from the ears

Temporal cortex integrates auditory data with memory and other senses

Meninges—three membranes (dura mater, arachnoid layer, pia mater) that surround the brain and spinal cord

Cerebrum, divided into two parts—the right and left hemispheres

Corpus callosum, a bundle of fibers that connect the right and left side of the brain

Cerebrospinal fluid, a clear fluid that surrounds the brain and spinal cord

Thalamus relays signals to cerebral cortex

Skull

Hypothalamus has many functions, including controlling the ANS

Pituitary, or master, gland

Brain stem regulates heartbeat and breathing

Cerebellum or "little brain"

Spinal cord

△ **Map of the cortex**
After years of research scientists have identified that certain parts of the outer brain, or cortex, process different aspects of learning. The work of German neurologist Korbinian Brodman helped generate a "map" of key areas.

▷ **Inside the brain**
Different structures within the brain have their unique roles. The cerebrum is in two sections— the right side is responsible for functions on the left side of the body and vice versa.

The Cardiovascular System

This system delivers oxygen and other nutrients to virtually all of the body's cells and carries waste products away from them. While the ancient Egyptians had proposed theories about circulation, it was not until the 17th century that British physician William Harvey established that blood was pumped around the body by the heart (see pp.84–85).

Cardiovascular anatomy

Also known as the circulatory system, the cardiovascular system consists of the heart, blood vessels, and blood. The heart is a muscular pump that beats regularly to send life-giving blood to every part of the body. Blood travels to and from body tissues through blood vessels (see panel, below). The heart, or cardiac, muscle is supplied with its own blood by the coronary blood vessels.

IN FOCUS

BLOOD VESSELS

There are three types of blood vessel. Arteries carry blood away from the heart to the organs and tissues. Arteries subdivide to become thin-walled capillaries through which oxygen and nutrients are released and waste matter is collected. Capillaries join and enlarge to become veins, which carry blood back to the heart.

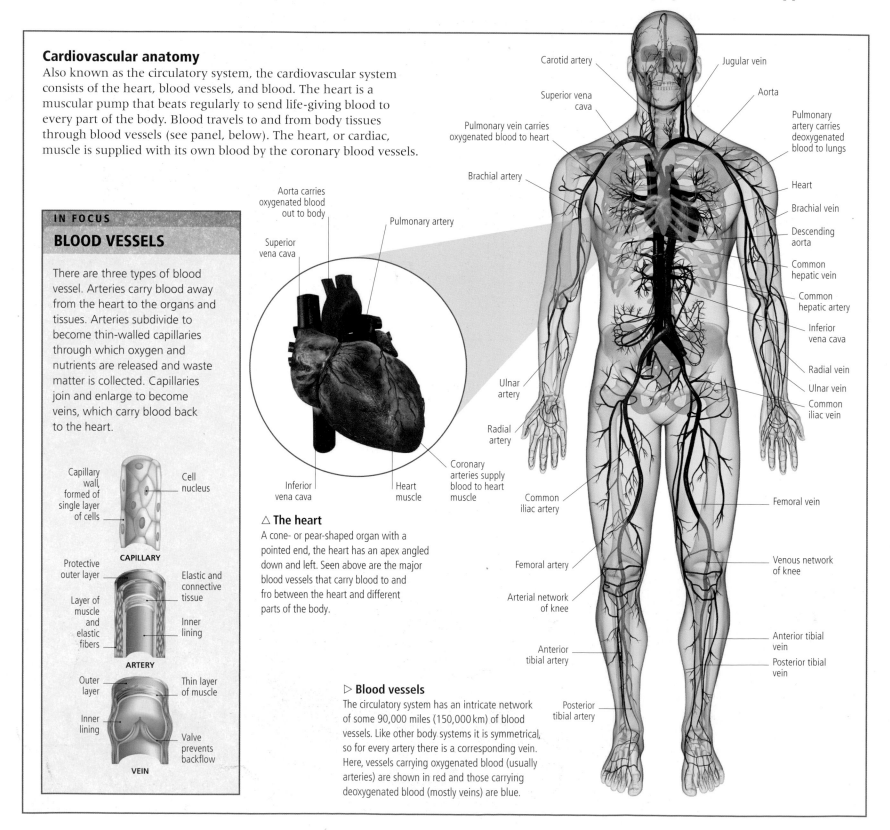

Capillary wall, formed of single layer of cells

Cell nucleus

CAPILLARY

Protective outer layer

Layer of muscle and elastic fibers

Elastic and connective tissue

Inner lining

ARTERY

Outer layer

Inner lining

Thin layer of muscle

Valve prevents backflow

VEIN

Aorta carries oxygenated blood out to body

Pulmonary artery

Superior vena cava

Inferior vena cava

Heart muscle

Coronary arteries supply blood to heart muscle

△ **The heart**
A cone- or pear-shaped organ with a pointed end, the heart has an apex angled down and left. Seen above are the major blood vessels that carry blood to and fro between the heart and different parts of the body.

▷ **Blood vessels**
The circulatory system has an intricate network of some 90,000 miles (150,000 km) of blood vessels. Like other body systems it is symmetrical, so for every artery there is a corresponding vein. Here, vessels carrying oxygenated blood (usually arteries) are shown in red and those carrying deoxygenated blood (mostly veins) are blue.

Carotid artery

Jugular vein

Superior vena cava

Aorta

Pulmonary vein carries oxygenated blood to heart

Pulmonary artery carries deoxygenated blood to lungs

Brachial artery

Heart

Brachial vein

Descending aorta

Common hepatic vein

Common hepatic artery

Inferior vena cava

Radial vein

Ulnar vein

Common iliac vein

Ulnar artery

Radial artery

Common iliac artery

Femoral vein

Femoral artery

Venous network of knee

Arterial network of knee

Anterior tibial artery

Anterior tibial vein

Posterior tibial vein

Posterior tibial artery

How blood circulates

The heart—a dynamic, untiring, double pump—forces blood through the network of blood vessels that make up the circulatory system. The heart beats more than three billion times in a person's lifetime. Each beat has two main phases: in the first (diastole), the heart relaxes and refills with blood; during the second stage (systole), it contracts, forcing blood out. This cycle takes, on average, less than a second. Each contraction creates a wave pressure along the arteries that can be felt where they lie close to the skin.

▷ Two circulation systems

The right side of the heart pumps blood to the lungs to "collect" oxygen and back into the left side in the pulmonary circulation. The left side of the heart pumps oxygenated blood out to the tissues in the systemic circulation.

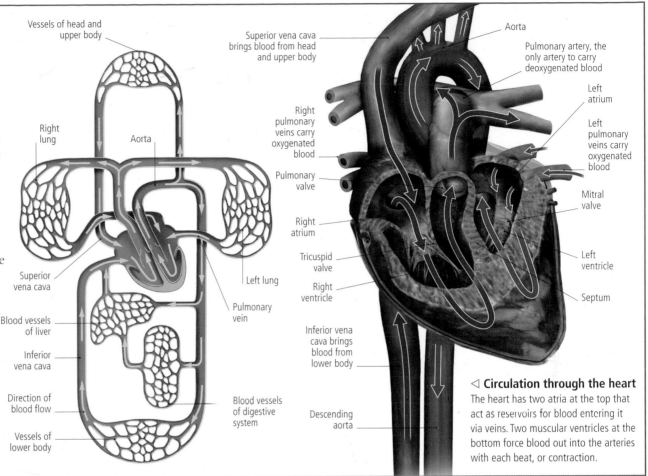

Vessels of head and upper body

Right lung

Aorta

Right atrium

Aorta

Superior vena cava

Blood vessels of liver

Inferior vena cava

Direction of blood flow

Vessels of lower body

Left lung

Pulmonary vein

Blood vessels of digestive system

Superior vena cava brings blood from head and upper body

Right pulmonary veins carry oxygenated blood

Pulmonary valve

Right atrium

Tricuspid valve

Right ventricle

Inferior vena cava brings blood from lower body

Descending aorta

Aorta

Pulmonary artery, the only artery to carry deoxygenated blood

Left atrium

Left pulmonary veins carry oxygenated blood

Mitral valve

Left ventricle

Septum

◁ Circulation through the heart

The heart has two atria at the top that act as reservoirs for blood entering it via veins. Two muscular ventricles at the bottom force blood out into the arteries with each beat, or contraction.

What is blood?

The average adult has about 8–11 pints (4–5 liters) of blood circulating in the body. Blood is made up of about 50–55 percent liquid (plasma), 1–2 percent white cells and platelets, and 45–50 percent red cells. Plasma is 90 percent water, but also contains glucose, waste products, proteins such as fibrinogen for clotting, and disease-fighting substances known as antibodies. Red cells contain a substance called hemoglobin (hem—an iron-rich protein—and a protein called globin), which carries oxygen.

▽ Blood make-up

In $1mm^3$ ($1/16000$ in^3) blood there are about 5 million red cells, 10,000 white cells, and 300,000 platelets floating in the plasma; the white cell count can double in hours if there is infection. Blood cells have a short lifespan, but are constantly replenished by bone marrow.

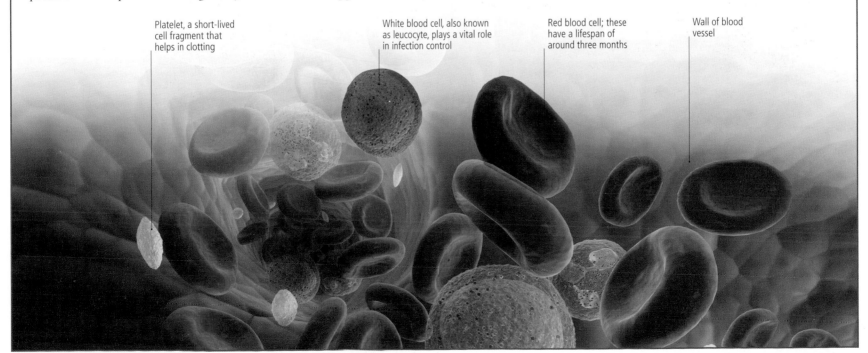

Platelet, a short-lived cell fragment that helps in clotting

White blood cell, also known as leucocyte, plays a vital role in infection control

Red blood cell; these have a lifespan of around three months

Wall of blood vessel

The Respiratory System

Oxygen is vital for life. The muscular and skeletal systems drive the movements of breathing so that the respiratory system can take oxygen from the air and transfer it to the **lungs where it can be absorbed into the bloodstream. The cardiovascular system then distributes oxygen to the body's cells via the blood vessels.**

Respiratory anatomy

This system is composed of the nasal cavity, mouth, pharynx (throat), trachea (windpipe) and main airways, and a pair of lungs. Air enters the body, mainly via the nostrils, into the nasal cavity, then the pharynx, the first part of which only carries air (the lower part conveying food and liquids too). From here air travels down the trachea, which splits into two tubes—the bronchi—one to each lung. Within the lungs these tubes divide and subdivide into smaller bronchi and finally into tiny bronchioles.

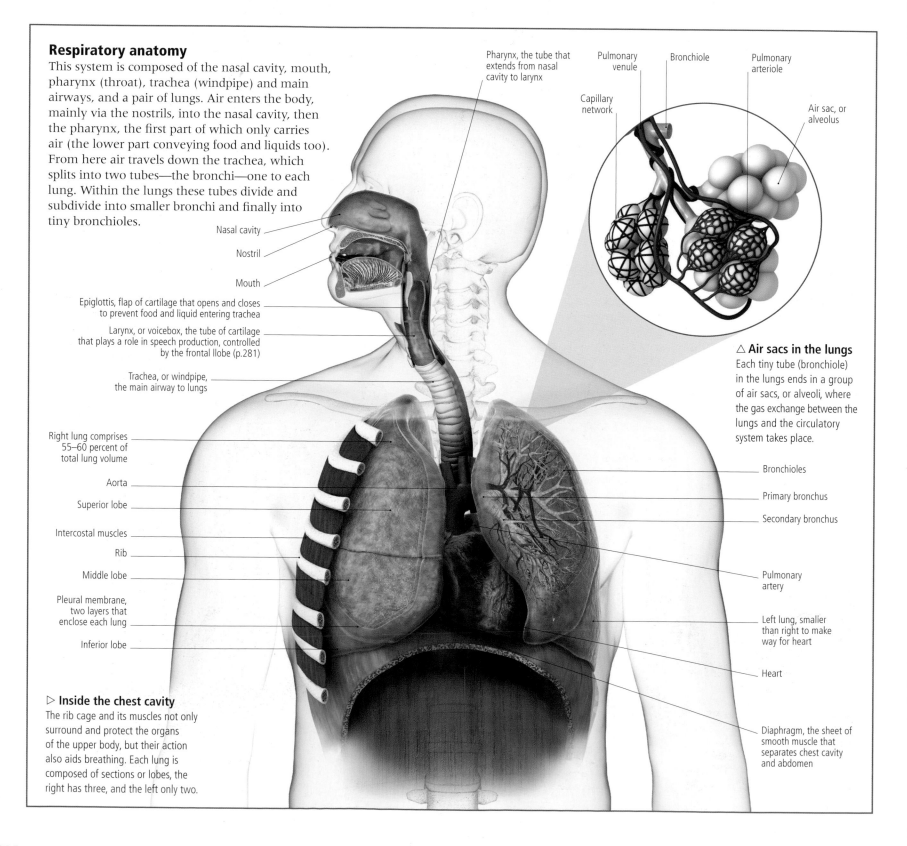

Pharynx, the tube that extends from nasal cavity to larynx

Pulmonary venule

Bronchiole

Pulmonary arteriole

Capillary network

Air sac, or alveolus

Nasal cavity

Nostril

Mouth

Epiglottis, flap of cartilage that opens and closes to prevent food and liquid entering trachea

Larynx, or voicebox, the tube of cartilage that plays a role in speech production, controlled by the frontal lobe (p.281)

Trachea, or windpipe, the main airway to lungs

Right lung comprises 55–60 percent of total lung volume

Aorta

Superior lobe

Intercostal muscles

Rib

Middle lobe

Pleural membrane, two layers that enclose each lung

Inferior lobe

△ Air sacs in the lungs
Each tiny tube (bronchiole) in the lungs ends in a group of air sacs, or alveoli, where the gas exchange between the lungs and the circulatory system takes place.

Bronchioles

Primary bronchus

Secondary bronchus

Pulmonary artery

Left lung, smaller than right to make way for heart

Heart

Diaphragm, the sheet of smooth muscle that separates chest cavity and abdomen

▷ Inside the chest cavity
The rib cage and its muscles not only surround and protect the organs of the upper body, but their action also aids breathing. Each lung is composed of sections or lobes, the right has three, and the left only two.

How breathing works

Chest movements during breathing draw fresh air into the lungs via the mouth and nose, and remove stale air. The physical movement of air is generated by creating differences in air pressure within the lungs and the atmospheric pressure around the body. These variations are produced by forceful expansion of the lungs through muscle contraction—so the pressure in the lungs is lower than that outside. Muscle relaxation then passively allows the lungs to return to their normal size—air is forced out as pressure is now higher in the lungs than outside the body.

Intercostal muscles contract so ribs swing up and out

Lungs expand as diaphragm and other muscles contract to increase capacity

Diaphragm contracts and moves down

▷ Breathing in and out

During breathing in (inspiration) the muscles of the chest and abdomen contract to pull the bones of the chest up and out to allow the lungs to expand. During breathing out, or expiration, the process is reversed to "force" stale air out.

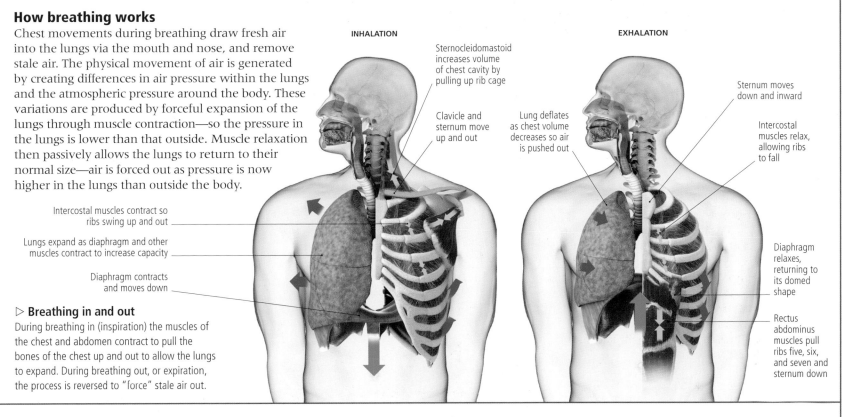

INHALATION

Sternocleidomastoid increases volume of chest cavity by pulling up rib cage

Clavicle and sternum move up and out

EXHALATION

Sternum moves down and inward

Lung deflates as chest volume decreases so air is pushed out

Intercostal muscles relax, allowing ribs to fall

Diaphragm relaxes, returning to its domed shape

Rectus abdominus muscles pull ribs five, six, and seven and sternum down

Gas exchange

The body cannot store oxygen and needs a constant supply. In addition, toxic carbon dioxide needs to be expelled. Gas exchange swaps oxygen and carbon dioxide in the lungs and body tissues; in both places the gases pass through thin cell walls by diffusion. Once inside body cells, oxygen is used to break down glucose (blood sugar) to free its energy in a chemical form; carbon dioxide is a by-product of this process.

▽ Blood flow between heart and lungs

The circulatory system returns blood depleted of oxygen, but that contains carbon dioxide from the tissues, to the heart. This blood is pumped to the lungs, to be replenished with oxygen, then returned to the heart for recirculation.

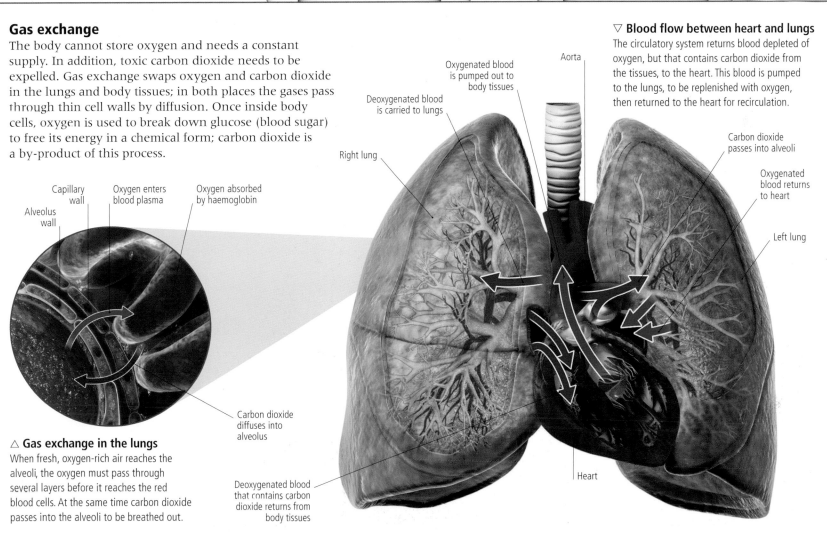

Capillary wall

Oxygen enters blood plasma

Oxygen absorbed by haemoglobin

Alveolus wall

Carbon dioxide diffuses into alveolus

Oxygenated blood is pumped out to body tissues

Aorta

Deoxygenated blood is carried to lungs

Right lung

Carbon dioxide passes into alveoli

Oxygenated blood returns to heart

Left lung

Heart

Deoxygenated blood that contains carbon dioxide returns from body tissues

△ Gas exchange in the lungs

When fresh, oxygen-rich air reaches the alveoli, the oxygen must pass through several layers before it reaches the red blood cells. At the same time carbon dioxide passes into the alveoli to be breathed out.

The **Endocrine System**

Often overshadowed by the brain and nerves, the endocrine system also deals with control and coordination, in its case using chemical messengers called hormones. These travel in **the blood to control the rate at which various organs and tissues work. The study of hormones has had important medical implications over the centuries.**

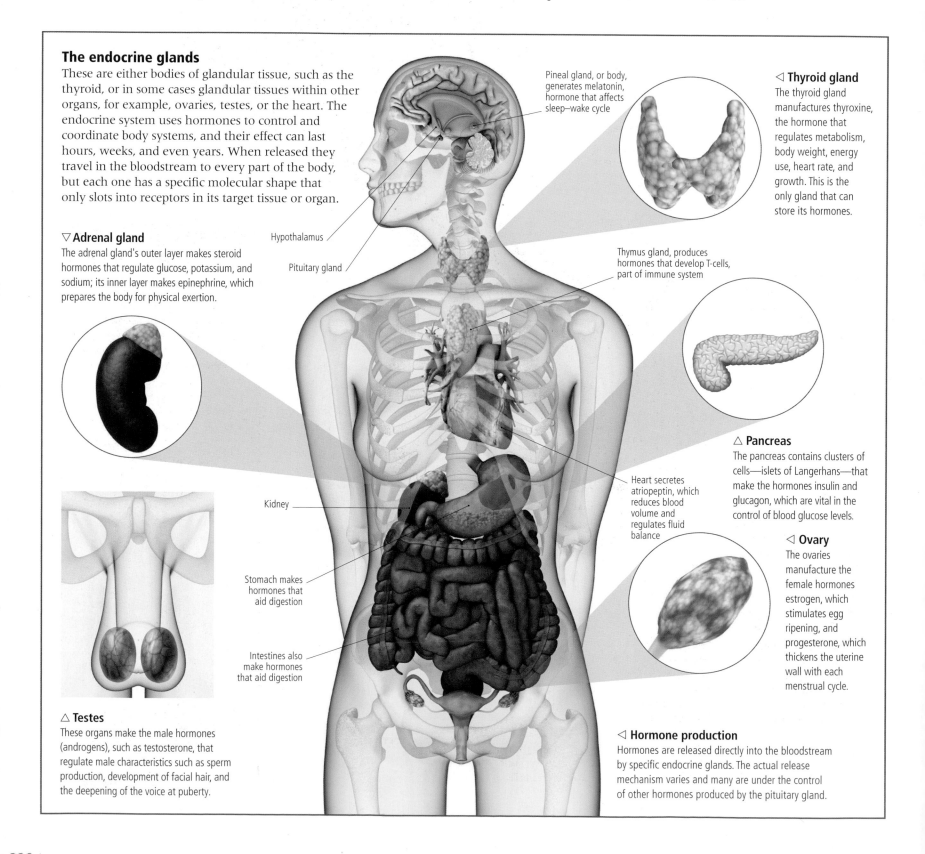

The endocrine glands
These are either bodies of glandular tissue, such as the thyroid, or in some cases glandular tissues within other organs, for example, ovaries, testes, or the heart. The endocrine system uses hormones to control and coordinate body systems, and their effect can last hours, weeks, and even years. When released they travel in the bloodstream to every part of the body, but each one has a specific molecular shape that only slots into receptors in its target tissue or organ.

Pineal gland, or body, generates melatonin, hormone that affects sleep–wake cycle

Hypothalamus

Pituitary gland

Thymus gland, produces hormones that develop T-cells, part of immune system

Heart secretes atriopeptin, which reduces blood volume and regulates fluid balance

Kidney

Stomach makes hormones that aid digestion

Intestines also make hormones that aid digestion

▽ Adrenal gland
The adrenal gland's outer layer makes steroid hormones that regulate glucose, potassium, and sodium; its inner layer makes epinephrine, which prepares the body for physical exertion.

◁ Thyroid gland
The thyroid gland manufactures thyroxine, the hormone that regulates metabolism, body weight, energy use, heart rate, and growth. This is the only gland that can store its hormones.

△ Pancreas
The pancreas contains clusters of cells—islets of Langerhans—that make the hormones insulin and glucagon, which are vital in the control of blood glucose levels.

◁ Ovary
The ovaries manufacture the female hormones estrogen, which stimulates egg ripening, and progesterone, which thickens the uterine wall with each menstrual cycle.

△ Testes
These organs make the male hormones (androgens), such as testosterone, that regulate male characteristics such as sperm production, development of facial hair, and the deepening of the voice at puberty.

◁ Hormone production
Hormones are released directly into the bloodstream by specific endocrine glands. The actual release mechanism varies and many are under the control of other hormones produced by the pituitary gland.

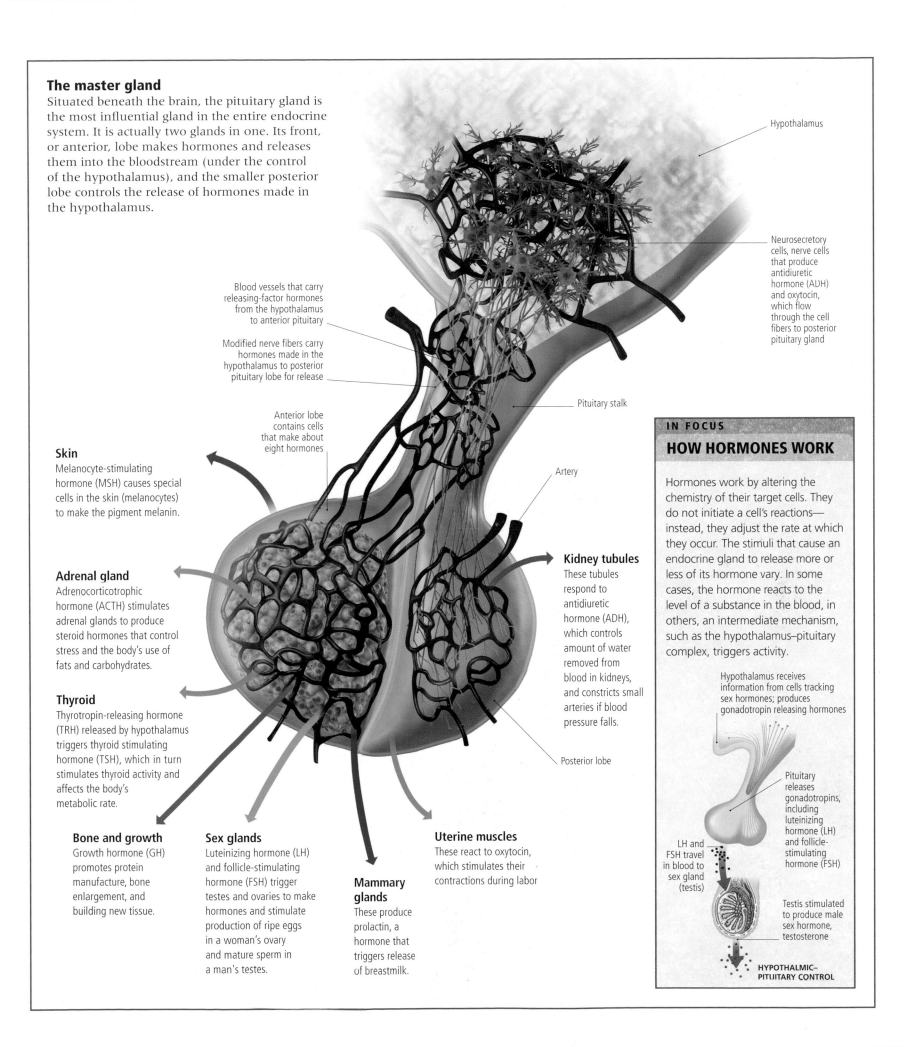

The master gland

Situated beneath the brain, the pituitary gland is the most influential gland in the entire endocrine system. It is actually two glands in one. Its front, or anterior, lobe makes hormones and releases them into the bloodstream (under the control of the hypothalamus), and the smaller posterior lobe controls the release of hormones made in the hypothalamus.

Hypothalamus

Neurosecretory cells, nerve cells that produce antidiuretic hormone (ADH) and oxytocin, which flow through the cell fibers to posterior pituitary gland

Blood vessels that carry releasing-factor hormones from the hypothalamus to anterior pituitary

Modified nerve fibers carry hormones made in the hypothalamus to posterior pituitary lobe for release

Pituitary stalk

Anterior lobe contains cells that make about eight hormones

Artery

Skin

Melanocyte-stimulating hormone (MSH) causes special cells in the skin (melanocytes) to make the pigment melanin.

Adrenal gland

Adrenocorticotrophic hormone (ACTH) stimulates adrenal glands to produce steroid hormones that control stress and the body's use of fats and carbohydrates.

Thyroid

Thyrotropin-releasing hormone (TRH) released by hypothalamus triggers thyroid stimulating hormone (TSH), which in turn stimulates thyroid activity and affects the body's metabolic rate.

Kidney tubules

These tubules respond to antidiuretic hormone (ADH), which controls amount of water removed from blood in kidneys, and constricts small arteries if blood pressure falls.

Posterior lobe

Bone and growth

Growth hormone (GH) promotes protein manufacture, bone enlargement, and building new tissue.

Sex glands

Luteinizing hormone (LH) and follicle-stimulating hormone (FSH) trigger testes and ovaries to make hormones and stimulate production of ripe eggs in a woman's ovary and mature sperm in a man's testes.

Mammary glands

These produce prolactin, a hormone that triggers release of breastmilk.

Uterine muscles

These react to oxytocin, which stimulates their contractions during labor

IN FOCUS

HOW HORMONES WORK

Hormones work by altering the chemistry of their target cells. They do not initiate a cell's reactions— instead, they adjust the rate at which they occur. The stimuli that cause an endocrine gland to release more or less of its hormone vary. In some cases, the hormone reacts to the level of a substance in the blood, in others, an intermediate mechanism, such as the hypothalamus–pituitary complex, triggers activity.

Hypothalamus receives information from cells tracking sex hormones; produces gonadotropin releasing hormones

Pituitary releases gonadotropins, including luteinizing hormone (LH) and follicle-stimulating hormone (FSH)

LH and FSH travel in blood to sex gland (testis)

Testis stimulated to produce male sex hormone, testosterone

HYPOTHALMIC– PITUITARY CONTROL

The **Digestive System**

This system consists of a long passageway that extends right through the body, as well as associated organs, including the liver and pancreas. The study of the digestive system first **began with the ancient Greeks, but it was only with 20th-century knowledge of molecular biology that it was revealed to be such a precisely controlled chemical system.**

Digestive anatomy

The digestive tract starts at the mouth and continues through the gullet, or esophagus, into the stomach. From there the "tube" progresses through the small and large intestines to the anus. Along the way food is first crushed by the teeth, then broken down into its constituent parts during chemical digestion. As it continues through the system, nutrients are extracted and the waste material disposed of. This system depends especially on healthy functioning of the nervous, hormonal, and immune systems.

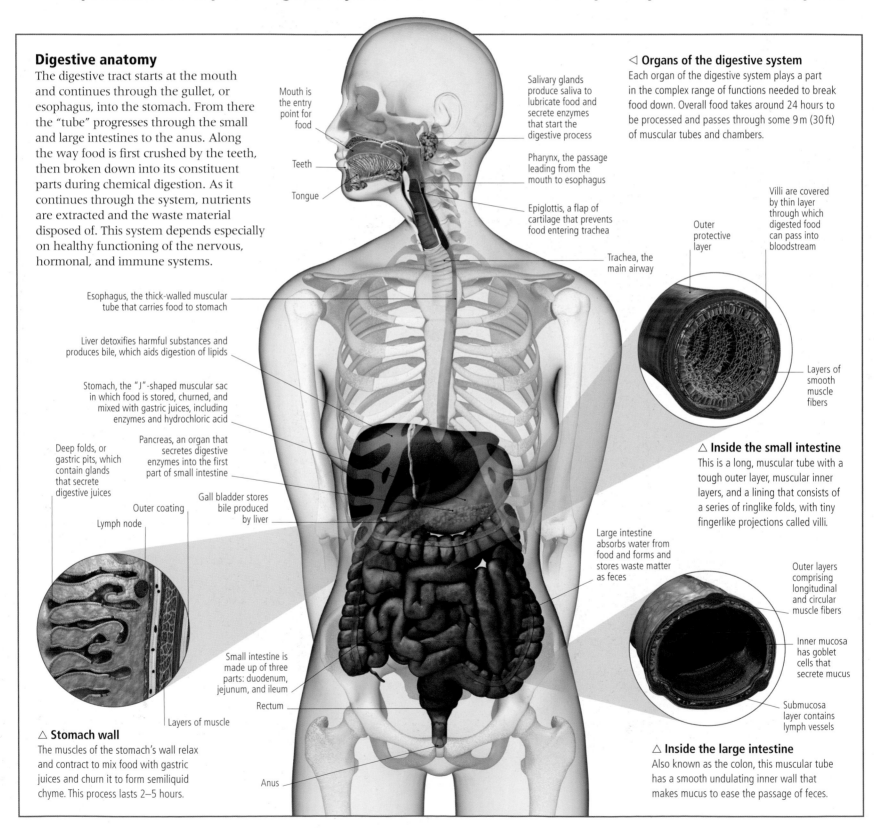

Mouth is the entry point for food

Teeth

Tongue

Salivary glands produce saliva to lubricate food and secrete enzymes that start the digestive process

Pharynx, the passage leading from the mouth to esophagus

Epiglottis, a flap of cartilage that prevents food entering trachea

Trachea, the main airway

Esophagus, the thick-walled muscular tube that carries food to stomach

Liver detoxifies harmful substances and produces bile, which aids digestion of lipids

Stomach, the "J"-shaped muscular sac in which food is stored, churned, and mixed with gastric juices, including enzymes and hydrochloric acid

Deep folds, or gastric pits, which contain glands that secrete digestive juices

Pancreas, an organ that secretes digestive enzymes into the first part of small intestine

Outer coating

Gall bladder stores bile produced by liver

Lymph node

Small intestine is made up of three parts: duodenum, jejunum, and ileum

Rectum

Layers of muscle

Anus

Large intestine absorbs water from food and forms and stores waste matter as feces

◁ Organs of the digestive system

Each organ of the digestive system plays a part in the complex range of functions needed to break food down. Overall food takes around 24 hours to be processed and passes through some 9 m (30 ft) of muscular tubes and chambers.

Villi are covered by thin layer through which digested food can pass into bloodstream

Outer protective layer

Layers of smooth muscle fibers

△ Inside the small intestine

This is a long, muscular tube with a tough outer layer, muscular inner layers, and a lining that consists of a series of ringlike folds, with tiny fingerlike projections called villi.

Outer layers comprising longitudinal and circular muscle fibers

Inner mucosa has goblet cells that secrete mucus

Submucosa layer contains lymph vessels

△ Stomach wall

The muscles of the stomach's wall relax and contract to mix food with gastric juices and churn it to form semiliquid chyme. This process lasts 2–5 hours.

△ Inside the large intestine

Also known as the colon, this muscular tube has a smooth undulating inner wall that makes mucus to ease the passage of feces.

Digestive organs of the upper abdomen

The liver, gallbladder, and pancreas play key roles in the production of the body chemicals that aid the processing of food. Weighing around 3⅓ lb (1.5 kg), the liver is the largest organ inside the body. It produces about 2 pints (1 liter) of bile a day, which passes via hepatic (liver) ducts into the gallbladder, where it is concentrated and stored for later release. The pancreas sits behind the stomach, under the liver, and also produces digestive juices that feed directly into the duodenum.

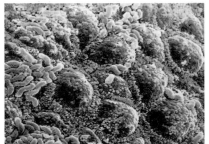

◁ Gut flora

Billions of microorganisms, mostly bacteria, live in the digestive tract—mainly the large intestine. Known as gut flora, they break down certain food components, especially cellulose, or plant fiber, that humans cannot digest. Gut flora feed on the undigested fiber in fecal material and provide additional nutrients to the body.

◁ Gallbladder

This organ holds about 1.7 fl oz (50 ml) of bile. It receives the bile via hepatic ducts from the liver and feeds it directly into the duodenum when needed.

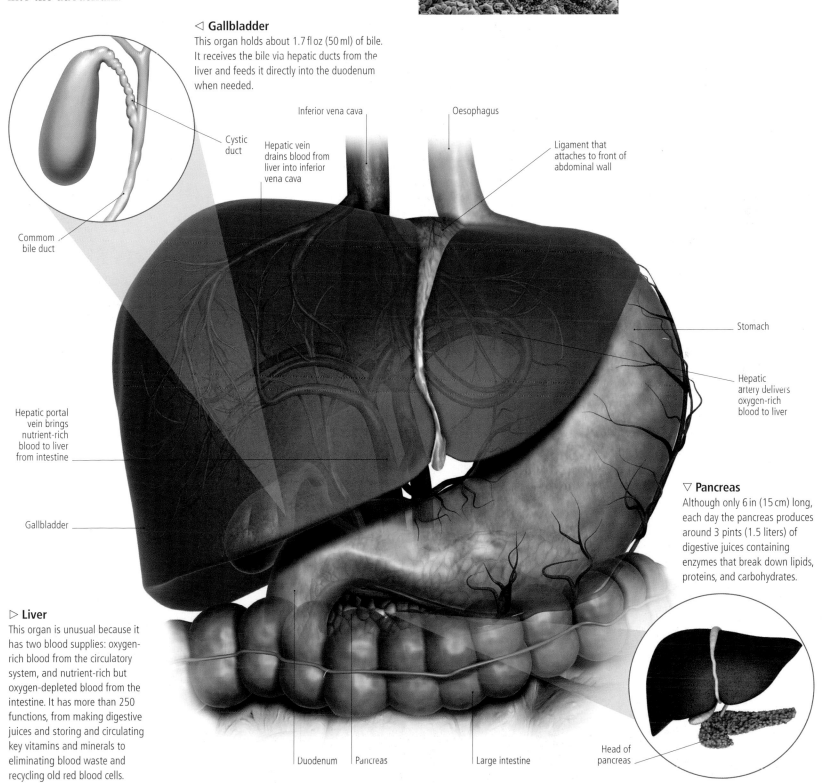

Cystic duct

Commom bile duct

Inferior vena cava

Hepatic vein drains blood from liver into inferior vena cava

Oesophagus

Ligament that attaches to front of abdominal wall

Stomach

Hepatic artery delivers oxygen-rich blood to liver

Hepatic portal vein brings nutrient-rich blood to liver from intestine

Gallbladder

▽ Pancreas

Although only 6 in (15 cm) long, each day the pancreas produces around 3 pints (1.5 liters) of digestive juices containing enzymes that break down lipids, proteins, and carbohydrates.

▷ Liver

This organ is unusual because it has two blood supplies: oxygen-rich blood from the circulatory system, and nutrient-rich but oxygen-depleted blood from the intestine. It has more than 250 functions, from making digestive juices and storing and circulating key vitamins and minerals to eliminating blood waste and recycling old red blood cells.

Duodenum

Pancreas

Large intestine

Head of pancreas

The **Immune System**

Several systems in the body are involved in helping protect it from everyday hazards, such as excessive heat or cold; injury; and the threat of microorganisms, **such as bacteria or viruses. However, it is the complex immune system that provides the main means of defense against invasion.**

Anatomy of the immune system

The lymphatic system is a key part of the immune system, but some organs, considered to make up the auxiliary immune system, also play an important role (see below). The active element of the lymphatic system is the lymph fluid, which starts as interstitial fluid that collects between cells all over the body and drains into special lymph vessels, or lymphatics. Lymph glands, or nodes, filter and store lymph along the routes. Lymph organs (thymus and spleen) and lymph tissue (tonsils and Peyer's patches) contain large numbers of infection-fighting white blood cells called lymphocytes (see opposite).

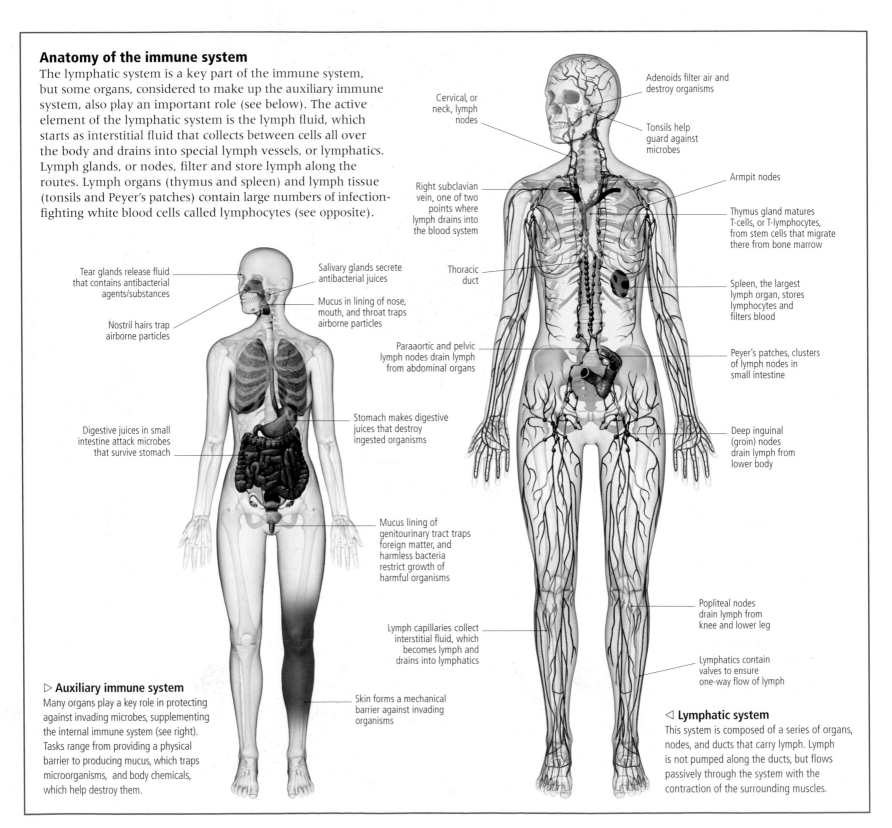

Tear glands release fluid that contains antibacterial agents/substances

Nostril hairs trap airborne particles

Salivary glands secrete antibacterial juices

Mucus in lining of nose, mouth, and throat traps airborne particles

Digestive juices in small intestine attack microbes that survive stomach

Stomach makes digestive juices that destroy ingested organisms

Mucus lining of genitourinary tract traps foreign matter, and harmless bacteria restrict growth of harmful organisms

Lymph capillaries collect interstitial fluid, which becomes lymph and drains into lymphatics

Skin forms a mechanical barrier against invading organisms

Cervical, or neck, lymph nodes

Right subclavian vein, one of two points where lymph drains into the blood system

Thoracic duct

Paraaortic and pelvic lymph nodes drain lymph from abdominal organs

Adenoids filter air and destroy organisms

Tonsils help guard against microbes

Armpit nodes

Thymus gland matures T-cells, or T-lymphocytes, from stem cells that migrate there from bone marrow

Spleen, the largest lymph organ, stores lymphocytes and filters blood

Peyer's patches, clusters of lymph nodes in small intestine

Deep inguinal (groin) nodes drain lymph from lower body

Popliteal nodes drain lymph from knee and lower leg

Lymphatics contain valves to ensure one-way flow of lymph

▷ Auxiliary immune system

Many organs play a key role in protecting against invading microbes, supplementing the internal immune system (see right). Tasks range from providing a physical barrier to producing mucus, which traps microorganisms, and body chemicals, which help destroy them.

◁ Lymphatic system

This system is composed of a series of organs, nodes, and ducts that carry lymph. Lymph is not pumped along the ducts, but flows passively through the system with the contraction of the surrounding muscles.

Lymph nodes

Vital to the body's defense system, these glands produce and store immune cells that protect the body from disease. They are scattered all over the body (see opposite), some singly and some concentrated into groups. Each node is formed from a mass of lymphatic tissue, which is divided into compartments (trabeculae). Lymph fluid is filtered in the nodes (and lymph organs) and most flows through more than one node before it is returned to the circulatory system.

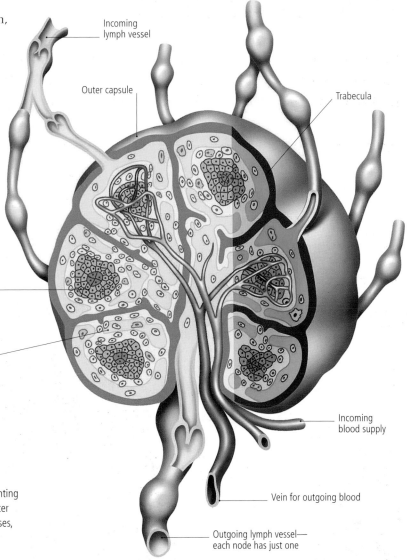

Incoming lymph vessel

Outer capsule

Trabecula

Germinal center where lymphocytes multiply and start to mature

Channels (sinuses) that narrow allowing white blood cells to attack intruders

Incoming blood supply

Vein for outgoing blood

Outgoing lymph vessel— each node has just one

▷ **Inside a lymph node**
Lymph nodes vary in size and swell when fighting infection or illness. Protected by a fibrous outer capsule, they contain many channels, or sinuses, where white blood cells ingest bacteria or foreign matter and debris.

WHITE CELL TYPES

There are many types of white cells. Monocytes and neutrophils engulf pathogens; lymphocytes are the main immune cells; basophils and eosonophiles are involved in allergic reactions. They all are derived from the bone marrow.

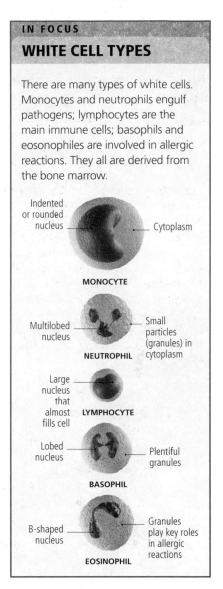

Indented or rounded nucleus

Cytoplasm

MONOCYTE

Multilobed nucleus

Small particles (granules) in cytoplasm

NEUTROPHIL

Large nucleus that almost fills cell

LYMPHOCYTE

Lobed nucleus

Plentiful granules

BASOPHIL

B-shaped nucleus

Granules play key roles in allergic reactions

EOSINOPHIL

VIRUSES AND BACTERIA

The two main categories of harmful microorganisms attacking the body are viruses and bacteria. Viruses cannot exist independently and the body relies on the immune system's response to fight them. However, bacteria have the cellular capacity to obtain energy, process nutrients, and reproduce, which also makes them vulnerable to chemical interference; a feature exploited by the introduction of antibiotics (see pp. 200–01).

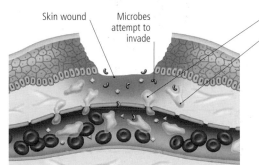

Bacteriophage virus

Bacterium

ELECTRON MICROGRAPH OF BACTERIOPHAGE VIRUSES ATTACKING A BACTERIUM

Immune response

White blood cells respond to invasion by different microorganisms. The system aims to create a condition of immunity in which the body is protected or is resistant to future attacks by that type of microorganism. The immune response may involve attacks on specific toxins or bacteria reaction to any kind of damage, such as injury. Damaged tissue releases chemicals that attract white blood cells, the blood capillaries become more porous to allow white cells to the area and the "battle" proceeds.

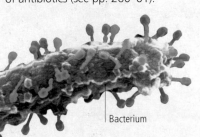

Skin wound

Microbes attempt to invade

White cells pass through capillary

White cells engulf microbes

◁ **Inflamed tissue**
The common signs of inflammation are redness, swelling, increased warmth, and discomfort—all signs that the fight against invading microbes is under way.

The **Urogenital System**

The urogenital, or genitourinary, system includes the organs involved in reproduction, and the processing of urine. They are grouped together because of their close proximity to each other and the fact that in men there is a common pathway—the urethra. These are also the systems that differ most between men and women.

The urinary system

This system regulates the volume and composition of body fluids, removes waste products from the blood, and expels them and surplus water from the body in the form of urine. The urinary system consists of a pair of kidneys, each containing a mass of microscopic filtering units that remove unwanted material and excess water from the blood. Long tubes, or ureters, carry urine from each kidney into a hollow, muscular organ—the bladder—for storage then excretion.

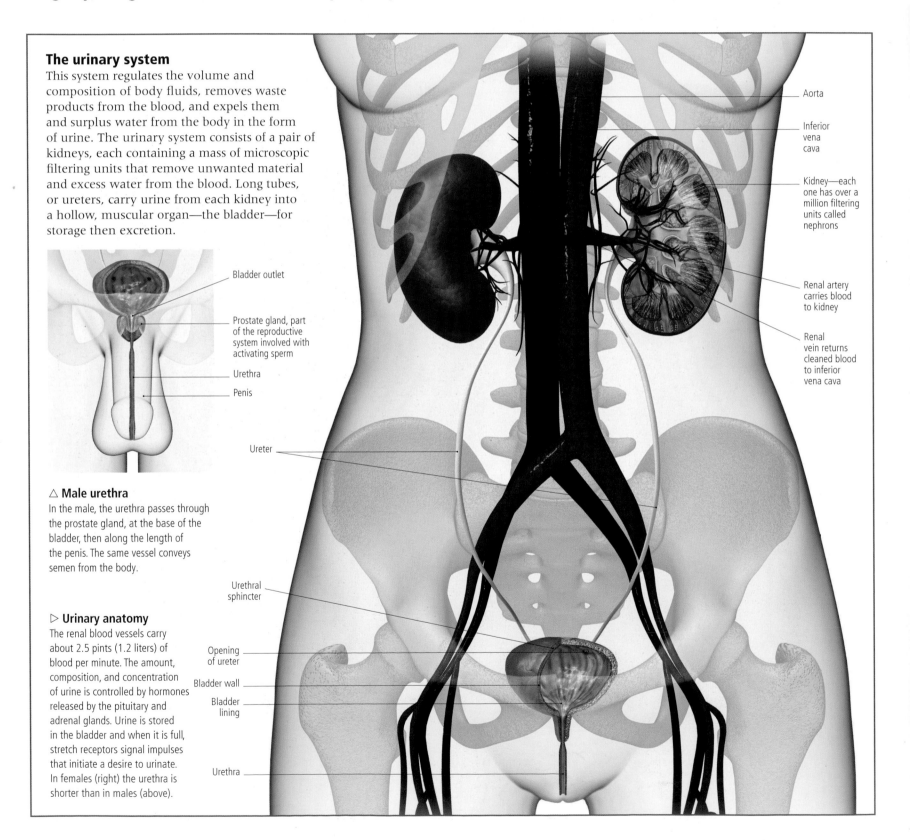

Bladder outlet

Prostate gland, part of the reproductive system involved with activating sperm

Urethra

Penis

Aorta

Inferior vena cava

Kidney—each one has over a million filtering units called nephrons

Renal artery carries blood to kidney

Renal vein returns cleaned blood to inferior vena cava

Ureter

Urethral sphincter

Opening of ureter

Bladder wall

Bladder lining

Urethra

△ Male urethra

In the male, the urethra passes through the prostate gland, at the base of the bladder, then along the length of the penis. The same vessel conveys semen from the body.

▷ Urinary anatomy

The renal blood vessels carry about 2.5 pints (1.2 liters) of blood per minute. The amount, composition, and concentration of urine is controlled by hormones released by the pituitary and adrenal glands. Urine is stored in the bladder and when it is full, stretch receptors signal impulses that initiate a desire to urinate. In females (right) the urethra is shorter than in males (above).

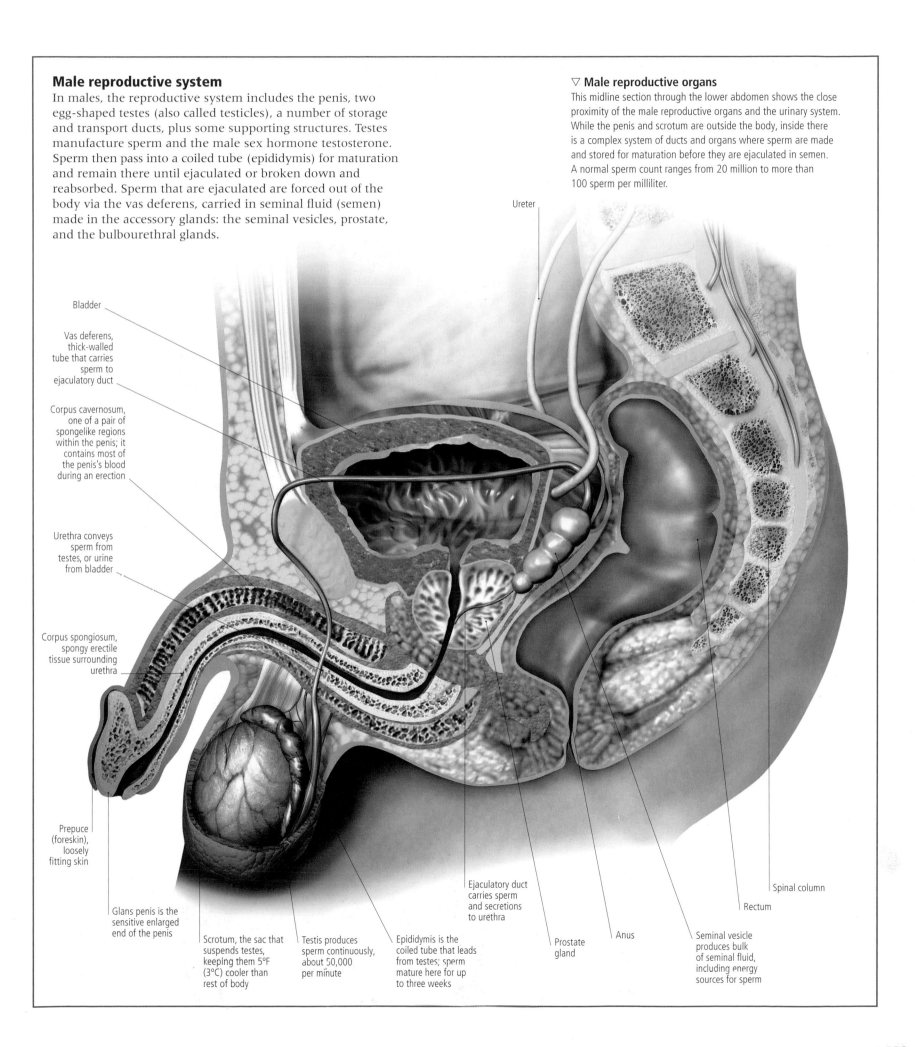

Male reproductive system

In males, the reproductive system includes the penis, two egg-shaped testes (also called testicles), a number of storage and transport ducts, plus some supporting structures. Testes manufacture sperm and the male sex hormone testosterone. Sperm then pass into a coiled tube (epididymis) for maturation and remain there until ejaculated or broken down and reabsorbed. Sperm that are ejaculated are forced out of the body via the vas deferens, carried in seminal fluid (semen) made in the accessory glands: the seminal vesicles, prostate, and the bulbourethral glands.

▽ Male reproductive organs

This midline section through the lower abdomen shows the close proximity of the male reproductive organs and the urinary system. While the penis and scrotum are outside the body, inside there is a complex system of ducts and organs where sperm are made and stored for maturation before they are ejaculated in semen. A normal sperm count ranges from 20 million to more than 100 sperm per milliliter.

Ureter

Bladder

Vas deferens, thick-walled tube that carries sperm to ejaculatory duct

Corpus cavernosum, one of a pair of spongelike regions within the penis; it contains most of the penis's blood during an erection

Urethra conveys sperm from testes, or urine from bladder

Corpus spongiosum, spongy erectile tissue surrounding urethra

Prepuce (foreskin), loosely fitting skin

Glans penis is the sensitive enlarged end of the penis

Scrotum, the sac that suspends testes, keeping them 5°F (3°C) cooler than rest of body

Testis produces sperm continuously, about 50,000 per minute

Epididymis is the coiled tube that leads from testes; sperm mature here for up to three weeks

Ejaculatory duct carries sperm and secretions to urethra

Prostate gland

Anus

Seminal vesicle produces bulk of seminal fluid, including energy sources for sperm

Rectum

Spinal column

Female reproductive system

The organs of the female reproductive system are sited entirely inside the body. They include two ovaries containing the female sex cells (eggs, or ova), a pair of fallopian tubes, and the uterus (womb). From puberty onward the ovaries release a mature egg once a month and the lining of the uterus thickens in preparation for pregnancy—all under the control of female hormones (see pp.286–87). The egg travels along the fallopian tube into the uterus and if fertilized develops into an embryo, then a fetus; if unfertilized, the egg and the uterine lining are lost via the vagina.

▽ Female reproductive organs

The main female reproductive organs are well protected by the bowl formed by the pelvic bones (see pp.276–77). Hormones from the pituitary gland and the reproductive organs play a major part in how this system functions in preparation for pregnancy, as well as during and after the birth. Developments in understanding this process have led to many new fertility treatments.

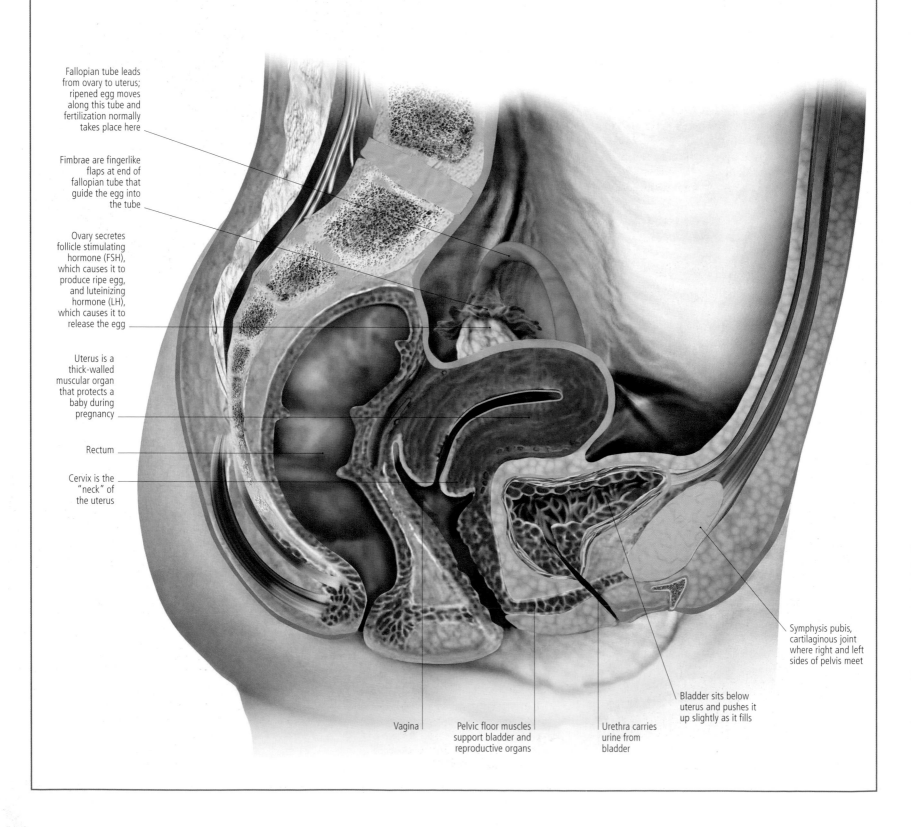

Fallopian tube leads from ovary to uterus; ripened egg moves along this tube and fertilization normally takes place here

Fimbrae are fingerlike flaps at end of fallopian tube that guide the egg into the tube

Ovary secretes follicle stimulating hormone (FSH), which causes it to produce ripe egg, and luteinizing hormone (LH), which causes it to release the egg

Uterus is a thick-walled muscular organ that protects a baby during pregnancy

Rectum

Cervix is the "neck" of the uterus

Vagina

Pelvic floor muscles support bladder and reproductive organs

Urethra carries urine from bladder

Bladder sits below uterus and pushes it up slightly as it fills

Symphysis pubis, cartilaginous joint where right and left sides of pelvis meet

Inheritance

The genitals include parts of the sexual or reproductive system. This produces egg and sperm cells that carry genetic information from parent to child—termed biological inheritance. Every human cell normally has 23 pairs of chromosomes (46 altogether), numbered 1 to 23, approximately largest to smallest. The fertilized egg has one of each pair from the mother and the other from the father. The 23rd pair is the dissimilar sex chromosomes, XX signifying female and XY male. Information, in the form of chemical codes, is carried by deoxyribonucleic acid (DNA) in the chromosomes. All chromosomes together carry an estimated 20,000 genes, forming the human genome. A gene consists of sequences of DNA needed to construct one particular protein (proteins are the body's main building blocks). Many body features are governed by more than one gene, termed polygenic inheritance. For example, eye color involves at least 15 genes, most on chromosome 15.

IN FOCUS

SEQUENCING DNA

DNA can be analyzed using electrophoresis. It is extracted from cells, purified, chemically broken into smaller fragments, and placed in a gel. When an electric current is passed through this gel the DNA fragments separate out according to size. When stained these show up as dark stripes like a bar code. Computers can read the bar codes to reveal the base pair sequences holding the strands of DNA together (see right).

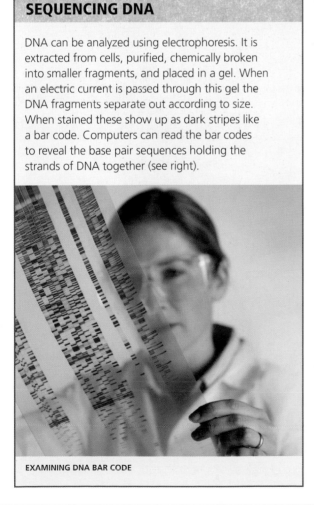

EXAMINING DNA BAR CODE

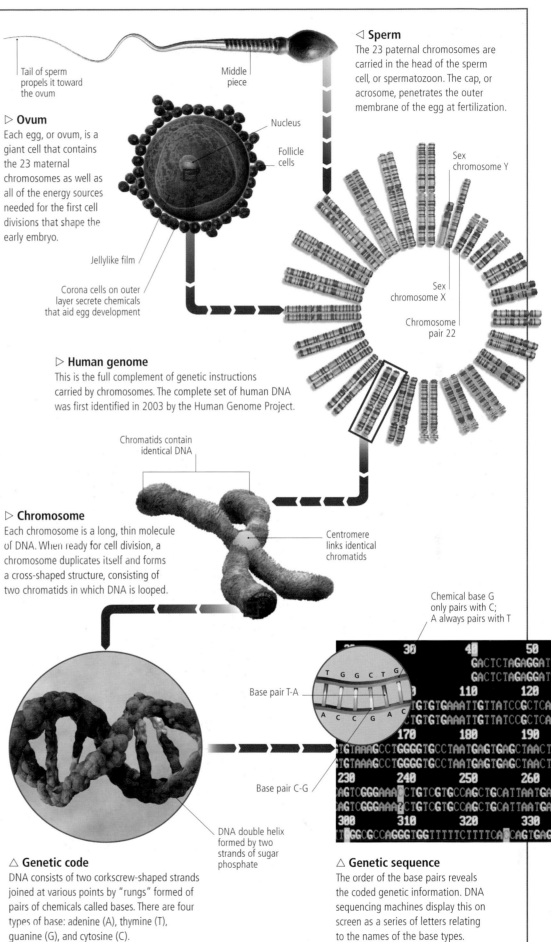

◁ **Sperm**
The 23 paternal chromosomes are carried in the head of the sperm cell, or spermatozoon. The cap, or acrosome, penetrates the outer membrane of the egg at fertilization.

Tail of sperm propels it toward the ovum

Middle piece

▷ **Ovum**
Each egg, or ovum, is a giant cell that contains the 23 maternal chromosomes as well as all of the energy sources needed for the first cell divisions that shape the early embryo.

Nucleus

Follicle cells

Jellylike film

Corona cells on outer layer secrete chemicals that aid egg development

Sex chromosome Y

Sex chromosome X

Chromosome pair 22

▷ **Human genome**
This is the full complement of genetic instructions carried by chromosomes. The complete set of human DNA was first identified in 2003 by the Human Genome Project.

Chromatids contain identical DNA

▷ **Chromosome**
Each chromosome is a long, thin molecule of DNA. When ready for cell division, a chromosome duplicates itself and forms a cross-shaped structure, consisting of two chromatids in which DNA is looped.

Centromere links identical chromatids

Chemical base G only pairs with C; A always pairs with T

Base pair T-A

Base pair C-G

DNA double helix formed by two strands of sugar phosphate

△ **Genetic code**
DNA consists of two corkscrew-shaped strands joined at various points by "rungs" formed of pairs of chemicals called bases. There are four types of base: adenine (A), thymine (T), guanine (G), and cytosine (C).

△ **Genetic sequence**
The order of the base pairs reveals the coded genetic information. DNA sequencing machines display this on screen as a series of letters relating to the names of the base types.

Sensory Organs

The sense organs inform and help to protect the body. The five main senses are sight, smell, touch, taste, and hearing. The structures that provide them are the eyes, nose, skin, tongue, and ears, respectively. Each of these structures contains specialized nerve receptors that send information to the brain via sensory neurons.

Skin, hair, and nails

The skin is the largest organ in the body and weighs about 6–9 lb (3–4 kg). It is very complex and comprises two main layers—the outer epidermis and the underlying dermis. It contains many different cell types, some of which are capable of producing hair and nail tissue. It can repair itself if injured and every month the outer layer is completely replaced, shedding cells at a rate of around 30,000 a minute; likewise, body hair and nails are also self-replacing.

IN FOCUS

HOW HAIR GROWS

A hair is actually a rod of dead cells filled with a protein called keratin. It grows for about three years, then enters a rest phase. Three to six months later the follicle activates and starts new growth, pushing the old hair out.

Dead hair

Epidermis

Dermis

Hair bulb

REST PHASE

Old dead hair forced out by new growth

New hair growth

GROWTH PHASE

◁ **Nail structure**
Composed of hard plates of keratin, nails grow continuously. The matrix below the visible part of the nail adds keratinized cells to the root and pushes the whole structure along the nail bed.

Matrix Nail root Cuticle Nail bed

Nail

Fat

Bone

▽ **Skin structure**
The skin's outer layer is chiefly protective, but the dermis contains many different tissues, with varied functions. Millions of microsensor nerve endings enable the sense of touch, and sweat glands and blood vessels help with temperature regulation.

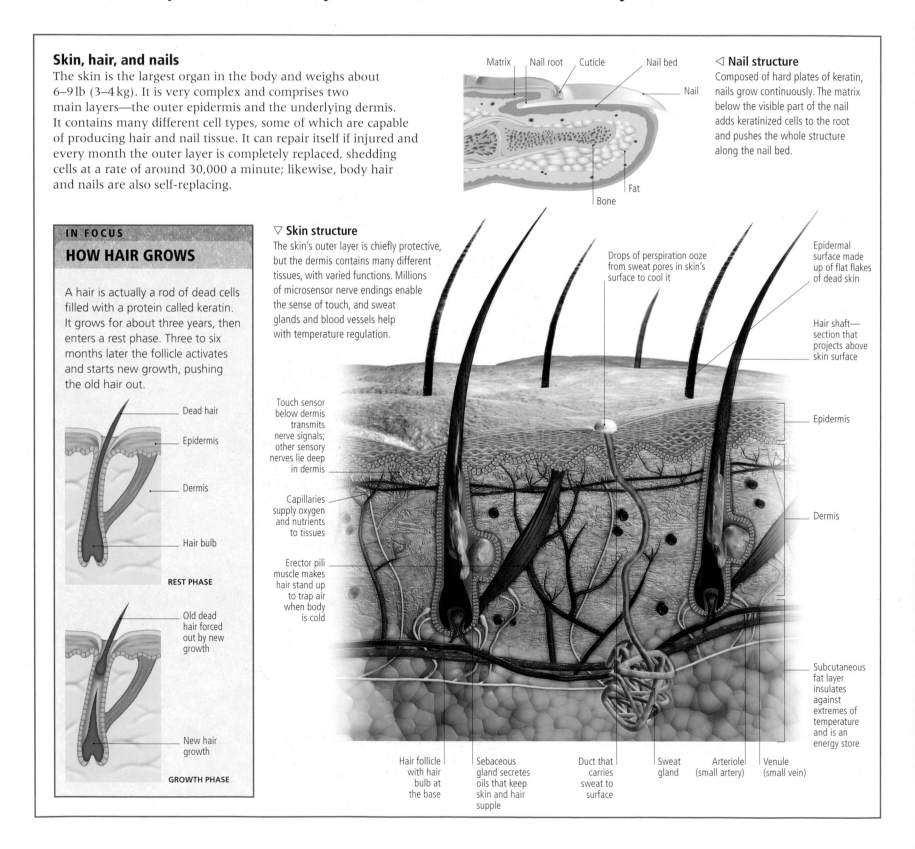

Drops of perspiration ooze from sweat pores in skin's surface to cool it

Epidermal surface made up of flat flakes of dead skin

Hair shaft—section that projects above skin surface

Epidermis

Dermis

Touch sensor below dermis transmits nerve signals; other sensory nerves lie deep in dermis

Capillaries supply oxygen and nutrients to tissues

Erector pili muscle makes hair stand up to trap air when body is cold

Subcutaneous fat layer insulates against extremes of temperature and is an energy store

Hair follicle with hair bulb at the base

Sebaceous gland secretes oils that keep skin and hair supple

Duct that carries sweat to surface

Sweat gland

Arteriole (small artery)

Venule (small vein)

Eyes and vision

Eyesight provides the brain with more information than all the other senses combined. It is estimated that more than half the information in the conscious mind is received through the eyes. Rays of light enter the eye through the domed cornea where they are partially bent, or refracted. A lens fine-focuses the image, then the light continues through the eye and shines an upside-down image on the retina where it is "translated" into nerve signals. These are transmitted via the optic nerve to the brain's cortex, which turns the image upright.

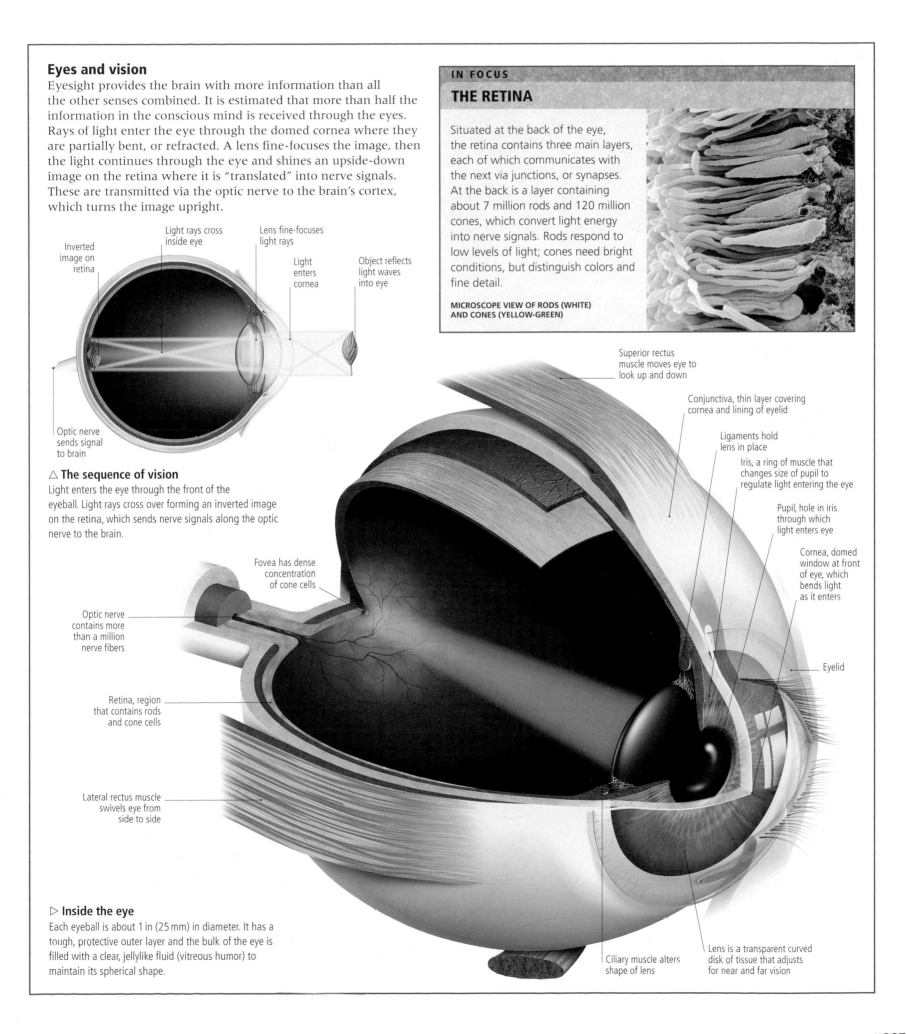

IN FOCUS

THE RETINA

Situated at the back of the eye, the retina contains three main layers, each of which communicates with the next via junctions, or synapses. At the back is a layer containing about 7 million rods and 120 million cones, which convert light energy into nerve signals. Rods respond to low levels of light; cones need bright conditions, but distinguish colors and fine detail.

MICROSCOPE VIEW OF RODS (WHITE) AND CONES (YELLOW-GREEN)

Inverted image on retina

Light rays cross inside eye

Lens fine-focuses light rays

Light enters cornea

Object reflects light waves into eye

Optic nerve sends signal to brain

△ The sequence of vision

Light enters the eye through the front of the eyeball. Light rays cross over forming an inverted image on the retina, which sends nerve signals along the optic nerve to the brain.

Superior rectus muscle moves eye to look up and down

Conjunctiva, thin layer covering cornea and lining of eyelid

Ligaments hold lens in place

Iris, a ring of muscle that changes size of pupil to regulate light entering the eye

Pupil, hole in iris through which light enters eye

Cornea, domed window at front of eye, which bends light as it enters

Eyelid

Fovea has dense concentration of cone cells

Optic nerve contains more than a million nerve fibers

Retina, region that contains rods and cone cells

Lateral rectus muscle swivels eye from side to side

▷ Inside the eye

Each eyeball is about 1 in (25 mm) in diameter. It has a tough, protective outer layer and the bulk of the eye is filled with a clear, jellylike fluid (vitreous humor) to maintain its spherical shape.

Ciliary muscle alters shape of lens

Lens is a transparent curved disk of tissue that adjusts for near and far vision

Ears, hearing, and balance

The ears are vital organs—not only do they enable us to hear, but they are also organs of balance. The parts that concern hearing and balance are in different sections of the ear, however, the function of both is based on sensitive hairlike nerve receptors. The ear is divided into three parts: the outer ear, which guides sound into the ear, the middle ear, where its structures amplify sound waves and then transfer them to the fluid-filled inner ear to become nerve signals.

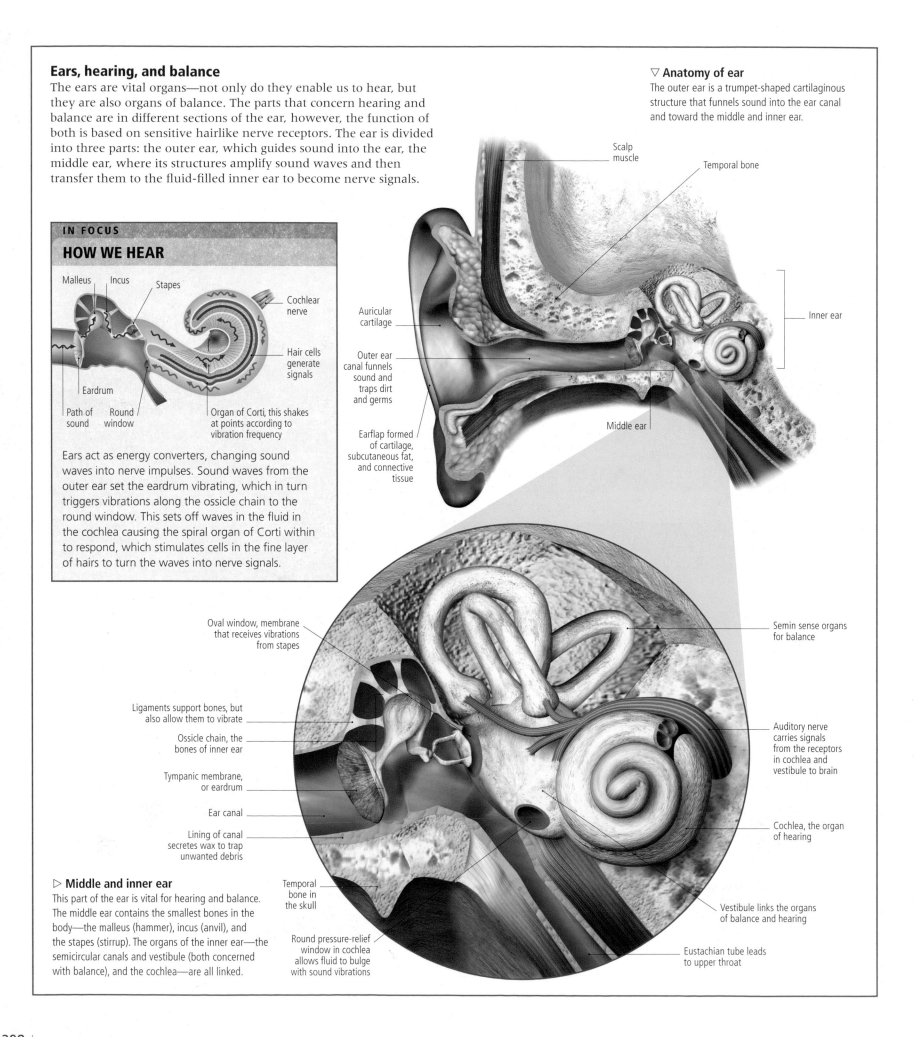

IN FOCUS
HOW WE HEAR

Malleus   Incus   Stapes

Cochlear nerve

Hair cells generate signals

Eardrum

Path of sound   Round window

Organ of Corti, this shakes at points according to vibration frequency

Ears act as energy converters, changing sound waves into nerve impulses. Sound waves from the outer ear set the eardrum vibrating, which in turn triggers vibrations along the ossicle chain to the round window. This sets off waves in the fluid in the cochlea causing the spiral organ of Corti within to respond, which stimulates cells in the fine layer of hairs to turn the waves into nerve signals.

▽ **Anatomy of ear**
The outer ear is a trumpet-shaped cartilaginous structure that funnels sound into the ear canal and toward the middle and inner ear.

Scalp muscle

Temporal bone

Inner ear

Auricular cartilage

Outer ear canal funnels sound and traps dirt and germs

Middle ear

Earflap formed of cartilage, subcutaneous fat, and connective tissue

Oval window, membrane that receives vibrations from stapes

Semin sense organs for balance

Ligaments support bones, but also allow them to vibrate

Auditory nerve carries signals from the receptors in cochlea and vestibule to brain

Ossicle chain, the bones of inner ear

Tympanic membrane, or eardrum

Ear canal

Lining of canal secretes wax to trap unwanted debris

Cochlea, the organ of hearing

▷ **Middle and inner ear**
This part of the ear is vital for hearing and balance. The middle ear contains the smallest bones in the body—the malleus (hammer), incus (anvil), and the stapes (stirrup). The organs of the inner ear—the semicircular canals and vestibule (both concerned with balance), and the cochlea—are all linked.

Temporal bone in the skull

Round pressure-relief window in cochlea allows fluid to bulge with sound vibrations

Vestibule links the organs of balance and hearing

Eustachian tube leads to upper throat

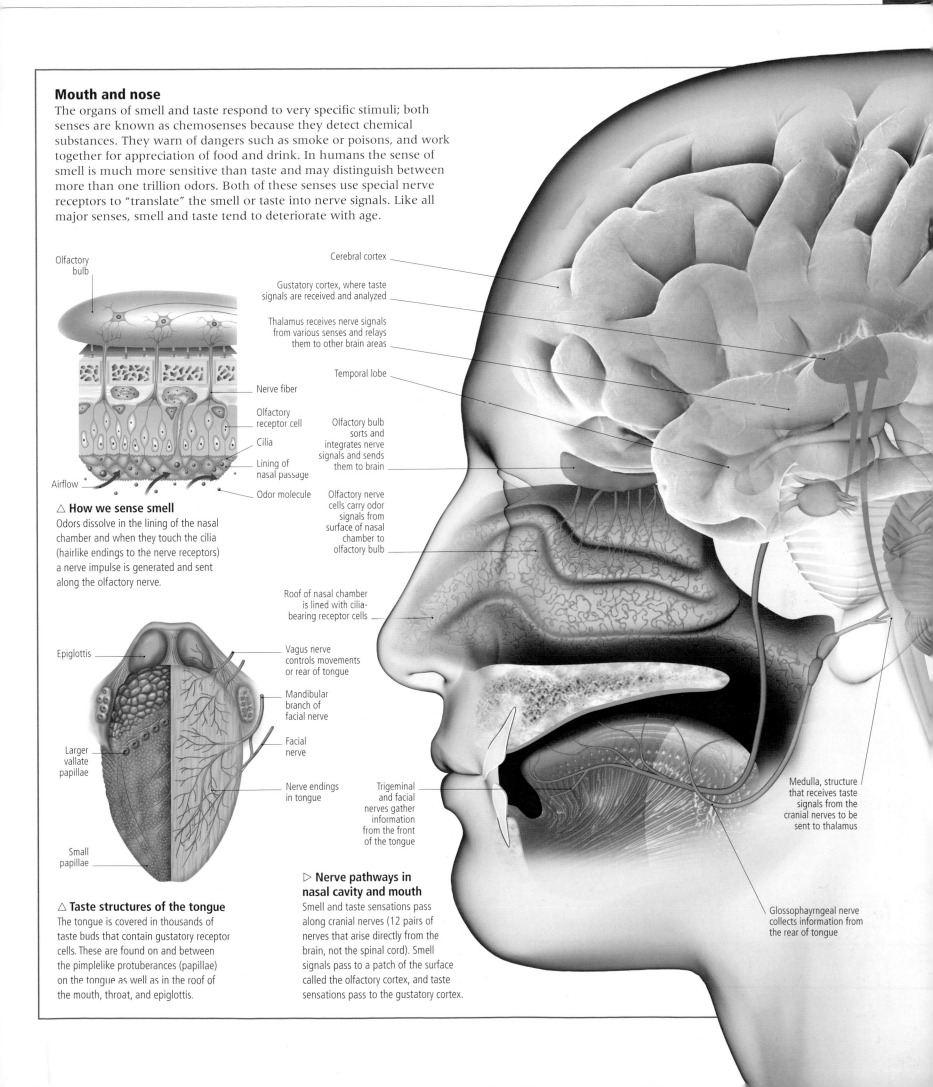

Mouth and nose

The organs of smell and taste respond to very specific stimuli; both senses are known as chemosenses because they detect chemical substances. They warn of dangers such as smoke or poisons, and work together for appreciation of food and drink. In humans the sense of smell is much more sensitive than taste and may distinguish between more than one trillion odors. Both of these senses use special nerve receptors to "translate" the smell or taste into nerve signals. Like all major senses, smell and taste tend to deteriorate with age.

Olfactory bulb

Nerve fiber

Olfactory receptor cell

Cilia

Lining of nasal passage

Airflow

Odor molecule

△ How we sense smell

Odors dissolve in the lining of the nasal chamber and when they touch the cilia (hairlike endings to the nerve receptors) a nerve impulse is generated and sent along the olfactory nerve.

Cerebral cortex

Gustatory cortex, where taste signals are received and analyzed

Thalamus receives nerve signals from various senses and relays them to other brain areas

Temporal lobe

Olfactory bulb sorts and integrates nerve signals and sends them to brain

Olfactory nerve cells carry odor signals from surface of nasal chamber to olfactory bulb

Roof of nasal chamber is lined with cilia-bearing receptor cells

Epiglottis

Vagus nerve controls movements or rear of tongue

Mandibular branch of facial nerve

Facial nerve

Larger vallate papillae

Nerve endings in tongue

Small papillae

Trigeminal and facial nerves gather information from the front of the tongue

Medulla, structure that receives taste signals from the cranial nerves to be sent to thalamus

Glossophayrngeal nerve collects information from the rear of tongue

△ Taste structures of the tongue

The tongue is covered in thousands of taste buds that contain gustatory receptor cells. These are found on and between the pimplelike protuberances (papillae) on the tongue as well as in the roof of the mouth, throat, and epiglottis.

▷ Nerve pathways in nasal cavity and mouth

Smell and taste sensations pass along cranial nerves (12 pairs of nerves that arise directly from the brain, not the spinal cord). Smell signals pass to a patch of the surface called the olfactory cortex, and taste sensations pass to the gustatory cortex.

Who's who

Alphanus I c.1015–1085

A notable physician and Latin poet, Alphanus I was born in Salerno, Italy. He taught at the university there and entered the Benedictine monastery at Monte Cassino. In 1058 he was appointed Archbishop of Salerno. Alphanus translated Greek medical texts into Latin and his most influential work, *De quattuor humoribus (On The Four Humors),* became the cornerstone of medical theory.

al-Nafis, Ibn 1213–1288

See p.49

al-Hariri, Ibn Ali 1054–1122

A wealthy Arab poet and silk trader, al-Hariri is best known for writing *Al Maqamat,* a book consisting of 50 short stories, each with a particular moral; the book was once memorized by scholars. The text of the book was based on the Qur'an, well-known proverbs, and classical poems. The book was later published in English as *The Assemblies of al-Hariri.*

al-Razi, Ibn 854–925 CE

Also known by his latinized name Rhazes, al-Razi was born in Rayy, Mesopotamia (now in Iran). He was a physician, philosopher, and alchemist who encouraged experimentation. He headed one hospital in Rayy, then two more in Baghdad. Among his many achievements was the first recorded clinical trial, which he undertook on patients with meningitis. His *Kitab al-Mansouri fi al-Tibb* (*Book of Medicine for Mansur*), written in 903 CE for the Persian Prince Mansur, was highly influential in the Islamic world and the West—it was translated into Latin in the 12th century by Italian scholar Gerard of Cremona.

Auenbrugger, Joseph Leopold 1722–1809

An Austrian physician, Auenbrugger discovered the diagnostic value of the percussion technique, which involved putting his ear to a patient's chest and lightly tapping on it to assess the texture of underlying organs. He published his findings, but he was largely ignored until René Laënnec adopted the technique with his new invention—the stethoscope.

Babinski, Joseph 1857–1932

Son of a Polish military officer who fled to France, Babinski was a clinical neurologist who initially studied under Jean-Martin Charcot at the Salpêtrière Hospital in Paris, France. He was the first person to differentiate hysteria from organic diseases. He wrote more than 200 papers on nervous disorders and became professor of neurology at the University of Paris. He described Babinski's sign, a reflex of the big toe that indicates a type of brain injury.

Banting, Frederick 1891–1941

In 1921 Canadian doctor Banting, together with Charles Best, discovered insulin—the life-saving treatment for diabetes. In 1922 the duo injected insulin into a diabetic. Banting worked in the laboratories of physiologist Professor John Macleod at the University of Toronto; Charles Best was Banting's research assistant. In 1923 only Macleod and Banting were awarded the Nobel Prize in Physiology or Medicine, so Banting shared his half of the award money with Best. Banting was appointed senior demonstrator in medicine at the University of Toronto, Canada. He continued research into silicosis, cancer, and the mechanism of drowning. His interest in aviation medicine led him to work with the Royal Canadian Air Force on a number of projects.

Barnard, Christiaan 1922–2001

A South African cardiac surgeon, Barnard became the first person to perform a human-to-human heart transplant in December 1967. He studied medicine at the University of Cape Town, then in Minneapolis, Minnesota, under Walt Lillehei. Returning to South Africa, Barnard introduced open-heart surgery at Cape Town's Groote Schuur Hospital. He retained an active interest in heart surgery until 1983, when he had to retire because of rheumatoid arthritis.

Becquerel, Antoine Henri 1852–1908

A French physicist who discovered radioactivity; the international system unit (SI) of radioactivity is named after Becquerel. He shared the Nobel Prize in Physics with Marie Sklodowska Curie and Pierre Curie in 1903.

Behring, Emil von 1854–1917

See p.158

Bernard, Claude 1813–1878

A French physiologist, Bernard is often considered to be the "father of physiology." He used rigorous experimental work to make important discoveries about the functioning of the liver, pancreas, and the nervous system; and also established the principle of biological regulation. Bernard's experiments on living animals were controversial but resulted in several breakthroughs in medicine. He was the first scientist to be awarded a public funeral in France.

Best, Charles Herbert 1899–1978

A Canadian medical scientist, Herbert helped discover insulin in 1921, after which it became a standard treatment for diabetes. Best worked closely with Frederick Banting, but missed out on winning the Nobel Prize in Physiology or Medicine in 1923. He went on to develop heparin, recognizing that it could be an important anticoagulant drug for preventing blood clotting.

Blackwell, Elizabeth 1821–1910

See p.141

Boveri, Thomas 1862–1915

German biologist Boveri, the second of four sons of a physician, studied at the University of Munich. By 1892 he was professor of zoology and comparative anatomy at the University of Wurzburg. In 1902 Boveri and American geneticist Walter Sutton both independently observed that chromosomes are the carriers of genetic material, which became known as the Boveri–Sutton chromosome theory. From his work on worms and sea urchins Boveri went on to demonstrate that each chromosome regulates embryo development in a different way so they must all be present for it to take place. He identified the centrosome, the specialized cell structure that regulates cell cycles. He also hypothesized that cellular processes cause cancer—that is, a tumor must begin with a single cell in which the chromosomal information is scrambled.

Broca, Paul 1824–1880

Physician, surgeon, and anatomist, Broca was born in southwest France, the son of a surgeon in Napoleon's service. He is best known for his discovery of the speech center in the frontal lobes of the brain—named Broca's area after him. Not only was this the first anatomical proof of localized brain function, but it also revolutionized the understanding of language processing.

Carpi, Jacopo Berengario da 1460–1530

The son of a barber-surgeon, Italian physician and anatomist Carpi studied medicine at the University of Bologna in Italy. Carpi wrote a number of treatises, but he is best known for *Anatomia Carpi* (*Carpi's Anatomy*)—the first anatomical text with illustrations—published in 1535. All the illustrations were based on Carpi's personal dissection experience and it was the first work since the time of Galen to display original anatomical information based upon personal investigation and observation; it undermined Galen's theories.

Carter, Henry Vandyke 1831–1897

A British anatomist, surgeon, and anatomical artist, Carter drew the illustrations for *Anatomy: Descriptive and Surgical* (later known as *Gray's Anatomy)* from 1856 to 1857. Born in Hull, England, he was the son of an artist. He studied in London, and was a student of human and comparative anatomy at the Royal College of Surgeons. He was also a demonstrator of anatomy at St. George's Hospital, London. In 1858 he joined the Indian Medical Service in Bombay, India, only returning to the UK after his retirement in 1888.

Charcot, Jean-Martin 1825–1893
See p.161

Chauliac, Guy de c.1300–1368
See p.69

Colombo, Realdo c.1516–1559
Born in Cremona, Italy, this anatomist and surgeon was the son of an apothecary. Colombo studied philosophy in Milan, then pursued his father's profession before becoming apprenticed to the Venetian surgeon Giovanni Antonio Lonigo. In 1538 he enrolled at the University of Padua, Italy, to study anatomy and medicine, where he became close friends with Andreas Vesalius. In 1544 he moved to the University of Pisa, as the "master of anatomy and surgery." He then moved to Rome in 1548, where he taught anatomy until his death. His only treatise, *De Re Anatomica (On Things Anatomical)* was published shortly after his death.

Constantine the African c.1020–died before 1099
A widely traveled Muslim scholar from Tunisia whose translations of Arab texts influenced European scholars, Constantine is thought to have studied medicine in Africa and Baghdad, and had knowledge of Mesopotamian, Greek, and Indian medicine. He became a Christian and entered the Monte Cassino monastery at Salerno, where he remained translating medical books from Arabic into Latin until his death.

Crick, Francis 1916–2004
A British biophysicist and neuroscientist, Crick determined the structure of deoxyribonucleic acid (DNA) in 1953, with his colleague James Watson. Crick, Watson, and Maurice Wilkins were jointly awarded the 1962 Nobel Prize in Physiology or Medicine for this work. In England, Crick studied in London, and then Cambridge. He later moved to the Salk Institute in California, where he spent the remainder of his career.

Davy, Humphry 1778–1829
Born in Penzance, southwest England, Davy was a chemist and inventor. He was apprenticed to a surgeon in Penzance, then became a chemist in the apothecary's dispensary. In 1798 he became medical superintendent of the Pneumatic Institution, Bristol, which was established to investigate the medical powers of gases, including nitrous oxide (laughing gas). In 1801 Davy joined the Royal Institution, first as lecturer, then professor of chemistry, and was later elected a fellow. He discovered a number of elements including potassium (the first metal isolated by electrolysis), sodium, and calcium. He also invented the Davy gas safety lamp for miners.

Democritus c.460–c.370 BCE
An ancient Greek natural philosopher, Democritus was born in Abdera to a wealthy noble family. He is known for his ideas on the formulation of the atomic theory of the Universe, but much of his work is inspired by that of his mentor and teacher— Leucippus, a 5th-century BCE Greek philosopher. Known to Aristotle as well, many consider Democritus to be the father of modern science.

Dioscorides, Pedanius 40–90 CE
See p.39

Duchenne, Guillaume Benjamin Armand 1806–1875
Also known as Duchenne du Boulogne, he descended from a family of fishermen.

Duchenne studied medicine in Paris before returning to Boulogne, France. He specialized in disorders of nerves and muscles, and experimented with the therapeutic qualities of electricity— showing how electrical stimulation of the brain could cause the facial muscles to contract. He also described the genetic disorder known as Duchenne muscular dystrophy.

Dunant, Jean Henri 1828–1910
See p. 267

Ehrlich, Paul 1854–1915
Born in Lower Silesia, now part of Poland, Ehrlich was the son of an innkeeper and distiller. He started his medical studies in Breslau and after obtaining his doctorate he began working in Berlin, Germany. During his career in hematology (study of blood), he developed stains that made it possible to distinguish between different cells— including bacteria. From this he pursued the concept of the "magic bullet"—a drug that could target specific infectious organisms or cancerous mutations, now known as chemotherapy. He was awarded the 1908 Nobel Prize in Physiology or Medicine for his contributions to the science of immunology.

Einthoven, William 1860–1927
See p.185

Empedocles c.490–c.430 BCE
A philosopher born in the ancient Greek city of Agrigentum (Acragas), Sicily (now in Italy), Empedocles is best known for formulating the four elements, or humors—earth, air, fire, and water—later formalized by Galen. He was a famous orator and had a reputation for his knowledge of curing disease and averting epidemics. Much of his work was written in verse and some fragments survive. According to Aristotle, Empedocles died at the age of 60, although some suggest he lived to more than 100.

Erasistratus c.304 BCE–c.250 BCE
An ancient Greek anatomist, Erasistratus was royal physician under Seleuceus I Nicator of Syria. He founded the school of anatomy in Alexandria, along with fellow physician Herophilus. He is credited with describing the valves of the heart, noting that the heart functioned as a pump, and distinguishing between arteries and veins, although he thought that arteries were full of air. He also described the cerebrum and cerebellum in the brain, observed a difference between motor and sensory nerves, and linked them to the brain.

Fabricius, Hieronymus 1537–1619
A pioneering Italian surgeon and anatomist, Fabricius was educated at the University of Padua, Italy, where he studied under Gabrielle Fallopio—and in 1565 succeeded him as the professor of anatomy. Fabricius's pioneering work on veins had a great influence on his student William Harvey, who went on to describe blood circulation. He identified the larynx as a vocal organ, and elevated embryology to an independent science. In his *Tabulae Pictae*, published in 1600, he described the cerebral fissure that separates the frontal and temporal lobes of the brain.

Falloppio, Gabrielle 1523–1562
Born in Modena, Italy, Falloppio studied medicine in Ferarra and went on to teach anatomy there and in Padua. He succeeded Realdo Colombo as professor of anatomy and surgery at the University of Padua and is credited with many discoveries in the anatomy of the eyes, ears, and nose, as well as the human reproduction system. He wrote a number of treatises, but only one, *Observationes Anatomincae (Anatomical Observations)*, was published in his lifetime.

Fleming, Alexander 1861–1955
See p.198

Florey, Howard Walter 1898–1968

An Australian pharmacologist and pathologist, Florey shared the 1945 Nobel Prize in Physiology or Medicine with Ernst Boris Chain and Alexander Fleming for his role in the development of penicillin (while Fleming discovered penicillin, it was Florey who carried out the first clinical trials). Born in Adelaide, Australia, Florey studied medicine in his home city, then moved to the University of Oxford, England. After periods in the US and at the University of Cambridge, England, he was appointed professor of pathology at the University of Sheffield, England, but returned to Oxford in 1935, where he carried out his work on penicillin.

Franklin, Rosalind 1920–1958

A British chemist and crystallographer, Franklin produced the X-ray diffraction image of deoxyribonucleic acid (DNA), the so-called "Photograph 51," which revealed a cross pattern that suggested DNA was helical in shape. This was the key evidence for James Watson and Francis Crick's double helix model. However, she did not share in the Nobel Prize for this achievement because she died before it was awarded.

Freud, Sigmund 1856–1939

See pp.182–83

Galen, Claudius c.129–c.216 CE

See pp.40–41

Gall, Franz Joseph 1758–1826

A German neuroanatomist and physiologist, Gall was the first to ascribe various mental functions to different parts of the brain. Born in Baden, he studied first in Strasbourg, France, and then in Vienna, Austria. Through observations about skull sizes and facial features Gall developed the theory of organology, which led to cranioscopy (later known as phrenology), a theory that asserts that the shape of a person's skull reveals intellectual and emotional characteristics. His work was condemned by religious leaders in 1802 and three years later he left Austria for France. He was the first person to identify gray matter (active

tissue) and white matter (conducting tissue) in the brain.

Galvani, Luigi 1737–1798

Born in Bologna, Italy, Galvani was the son of a goldsmith. Philosopher, physician, biologist, and physicist, he is best known for his theories on animal electricity and in particular his accidental discovery that electricity could make the muscles of a dead animal twitch. He trained as a physician at the University of Bologna, was made public lecturer in anatomy in 1762, and became its president in 1772.

Gaskin, Ina May 1940–

See p.135

Gerhardt, Charles Frédéric 1816–1856

See p.170

Golgi, Camillo 1843–1926

An Italian scientist, pathologist, and physician, Golgi spent much of his career studying the central nervous system. In 1873 he published *On the Structure of the Brain Gray Matter*, in which he described a revolutionary silver-nitrate, tissue-staining technique initially named the "black reaction"; Golgi staining or Golgi impregnation is still in use today. Golgi established the Institute of General Pathology in 1881 and was jointly awarded the Nobel Prize in Physiology or Medicine in 1906 with Santiago Ramón y Cajal.

Gräfe, Karle Ferdinand von 1787–1840

A German surgeon who pioneered plastic and reconstructive surgery, especially of the nose and eyelids. Gräfe studied in Halle and Leipzig, in Germany, and in 1811 was appointed professor of surgery and director of the Ophthalmological Institute in the University of Berlin. He developed his own techniques using modifications of methods used by Italian surgeon Gasparo Tagliacozzzi as well as ancient Indian practices. He performed one of the first operations for the treatment of cleft palate.

Gray, Henry 1827–1861

A British anatomist and surgeon, Gray is most famous for publishing the book *Anatomy: Descriptive and Surgical* (later known as *Gray's Anatomy*) which has became a standard medical reference work and is still in print today. Gray was born in London, England, and studied at St. George's Hospital. He was elected a Fellow of the Royal College of Surgeons at the young age of 27. The first edition of Gray's book *Anatomy* was published in 1858 and featured 350 illustrations by his friend and colleague Henry Vandyke Carter. A second edition followed in 1860.

Hahnemann, Samuel 1755–1843

See p.109

Harvey, William 1578–1657

Born in England, Harvey studied at the University of Padua, Italy. He had a deep conviction in the importance of experimental science and a particular interest in the circulation of blood. In 1628 he published the results of 30 years' work in which he established that blood is circulated around the body via the heart's pumping mechanism. He was appointed physician to Kings James I and Charles I of England. He also tended to the victims of the English Civil War (1642–1651).

Hayyan, Jabir ibn c.721–815

An Islamic alchemist, astronomer, physicist, and natural philosopher, Hayyan was also known as Jabir Geber or by the full name of Abu Musa Jabir ibn Hayyan, and described as the father of Arab chemistry. He turned alchemy into a science, invented many pieces of laboratory equipment, and identified chemical substances such as hydrochloric acid and nitric acid.

Herophilus c.335–c.280 BCE

A Greek physician who, along with Erasistratus, founded the school of anatomy in Alexandria, Egypt, Herophilus described anatomical organs and structures including the pancreas, liver, genitals, and eye. He wrote at least nine books and, according to notes from

Hippocrates, these included one for midwives, as well as treatises on the causes of sudden death and anatomy. All of his writing was lost in the destruction of the great library at Alexandria in 272 CE.

Hildegard of Bingen 1098–1179

See p.56

Hippocrates 460–370 BCE

See pp.36–37

Hopps, John Alexander 1919–1998

A Canadian biomedical engineer, Hopps was one of the pioneers of the heart pacemaker. Born in Winnipeg, Hopps studied at the University of Manitoba and in 1942 joined the National Research Council of Canada. In 1949 he moved to Toronto to work at the university's Banting Institute and, with Wilfred Bigelow and John Callaghan, designed and developed the first external heart pacemaker—used to pace a dog's heart in 1950. Hopps founded the Canadian Medical and Biomedical Engineering Society and was its first president.

Hounsfield, Godfrey 1919–2004,

See p.217

Ivanovsky, Dmitry 1864–1920

See p.167

Jenner, Edward 1749–1823

A British scientist and medical practitioner, Jenner demonstrated how inoculation with the cowpox virus could safely provide immunity from the deadly viral disease of smallpox—the world's first vaccine. Born in Gloucestershire, England, Jenner was the son of a vicar and the eighth of nine children. At the age of 14 he was first apprenticed to a local surgeon, then in 1770 to one of London's most distinguished surgeons, John Hunter, at St. George's Hospital. In 1773 he returned to Gloucestershire to become a family doctor and surgeon, and remained there until his death. He was elected Fellow of the Royal Society in 1788.

Kanada, probably lived in 2nd century BCE

A Hindu sage and philosopher, Kanada founded the philosophical school of Vaisheshika. His primary area of study was *Rasavadam*—a type of alchemy. He believed that all living things were composed of five elements: water, fire, earth, air, and ether.

Kelling, Georg 1866–1945

See p.189

Kelman, Charles 1930–2004

Born in New York, Kelman was an ophthalmologist who pioneered new, safer methods of cataract surgery as well as repairing retinal detachments. He studied at Tufts University, Massachusetts, and then in Geneva, Switzerland. He later returned to the US—first to Brooklyn, then Philadelphia's Wills Eye Hospital, and finally back to New York. Kelman was also an accomplished saxophonist and even performed at Carnegie Hall.

Koch, Robert 1843–1910

A German physician, Koch is considered to be a founder of modern bacteriology. His discoveries facilitated the development of the first chemicals designed to attack specific bacteria (Ehrlich's "magic bullet"). Building on Louis Pasteur's germ theory, Koch isolated the bacilli for anthrax (1877), tuberculosis (1882), and cholera (1883). He also developed Koch's postulates of germ theory that sets out the criteria to be met to establish whether specific bacteria cause specific diseases. Koch received the 1905 Nobel Prize in Physiology or Medicine for his tuberculosis-related discoveries. He was appointed professor of hygiene at the University of Berlin in 1885, and was director of the Institute for Infectious Diseases, Berlin, from 1891 to 1904.

Laënnec, René 1781–1826

The inventor of the stethoscope, Laënnec was born in Brittany, France. The son of a lawyer, his mother died when he was 5 years old and he was sent to live with a great uncle who was a priest. At the age of 12, Laënnec went to study with an uncle who worked on the faculty of medicine in the University of Nantes, France. In 1799 he went to Paris to continue his medical studies under several physicians including Jean-Nicolas Corvisart-Desmarets, who reintroduced the use of sound as a diagnostic aid; at the time this meant placing an ear against the patient's chest. Laënnec also developed an understanding of many other conditions, coined the term melanoma, and studied tuberculosis, the condition that eventually took his life.

Landsteiner, Karl 1868–1943

See p.176

Lane-Claypon, Janet 1877–1967

See p.126

Larrey, Dominique-Jean 1766–1842

See p.256

Leeuwenhoek, Antoni van 1632–1723

See p.92

Lister, Joseph 1827–1912

See p.154

Luzzi, Mondino de c.1270–1326

An Italian physician, anatomist, and surgeon, Luzzi was born to a Florentine family, but worked in Bologna; he is sometimes known as Mundinus. Luzzi famously reintroduced the practice of dissecting bodies for the study of anatomy. One of his most famous pupils was Nicólò Bertuccio, who succeeded him and in turn taught Guy de Chauliac.

Macewen, William 1848–1924

A Scottish surgeon, Macewen was not only a pioneer of modern brain surgery and the development of bone grafts, but also (influenced by the work of Joseph Lister) an innovator in the field of surgical practice—introducing antisepsis in operating rooms. He adopted the deep cleansing and disinfection of arms and metal instruments, also sterilization with steam; the use of surgical gowns; and the recently discovered anesthesia. Macewen studied at the University of Glasgow, Scotland,

became a full surgeon at the Glasgow Royal Infirmary in 1877, and was Regius Professor of Surgery at the university by 1892, a post previously held by Lister.

Malpighi, Marcello 1628–1694

See p.96

Marshall, Barry 1951–

See p.244

Meduna, Ladislas Joseph 1896–1964

Born in Budapest to a well-to-do Hungarian family, Meduna began his medical studies in 1914, but did not complete them until after World War I. He pursued a career in neurology at the Hungarian Institute for Brain Research and began looking into the structure of the pineal gland in the brain. He moved to the Psychiatric Institute and began research into psychotic conditions such as schizophrenia. There he developed a method of treating psychotic patients with chemically induced seizures. Meduna emigrated to Chicago, Illinois, in 1938 and remained there for the rest of his life.

Mesmer, Franz Anton 1734–1815

A German physician, Mesmer also had an interest in astronomy. He developed a theory that there was a natural transfer of energy between all animate and inanimate objects that he called animal magnetism—later called mesmerism—a form of hypnosis that could induce a trance. His idea had a substantial following, but never achieved scientific recognition.

Morton, William Thomas Green 1819–1868

Morton was the dentist who first publicly demonstrated the use of ether as a surgical anesthetic for a tooth extraction in 1846. Born in Massachusetts, Morton had various jobs before entering Baltimore Dental College in 1840. However, he left two years later without completing his degree and went to work with another dentist, Horace Wells. In 1844 Morton entered Harvard Medical School, where he was introduced to the anesthetic properties of ether, but again he left without

graduating. He did receive an honorary degree from Washington University of Medicine in 1852.

Nightingale, Florence 1820–1910,

See p.142

Nobel, Alfred 1833–1896

A Swedish chemist, Nobel invented dynamite and its less sensitive form nitroglycerine, and also founded the Nobel Prizes. He amassed a fortune and willed most of it to the Nobel Prize annual awards in physics, chemistry, economics, physiology or medicine, literature, and peace.

Papanicolaou, George N. 1883–1962

See p.204

Pappenheim, Bertha (Anna O) 1859–1936

See p.251

Paracelsus 1493–1541

Born in Switzerland as Philippus von Hohenheim, Paracelsus's adopted name meant equal to or greater than Celsus—a Roman scholar of medicine. Paracelsus traveled widely in Europe, Africa, and the Middle East, and had many interests including alchemy, astrology, medicine, psychology, and toxicology. A prolific writer, he established the use of chemistry in medicine, introducing laudanum, sulfur, lead, and mercury as medical remedies, and gave a clinical description of syphilis. Paracelsus was an outspoken opponent of the merely academic knowledge of medicine that was acquired at universities.

Paré, Ambroise 1510–1590

See pp.78–79

Pasteur, Louis 1822–1895

See pp.148–49

Pott, Percivall 1714–1788

See p.230

Ramón y Cajal, Santiago 1852–1934

A Spanish pathologist, histologist, and neuroscientist, Ramón y Cajal is considered by many to be the founder of modern neuroscience. He was responsible for identifying the type of nerve cell that controls the slow waves of contractions that move food along the intestine, and an expert in hypnotism, which he used to help his wife during labor. Ramón y Cajal and Camillo Golgi were jointly awarded the Nobel Prize in Physiology or Medicine in 1906 in recognition of their work on the nervous system.

Röntgen, Wilhelm Conrad 1845–1923

See p.172

Ross, Ronald 1857–1932

See p.175

Ruggiero, Trotula de c.1090–unknown

An Italian physician born to a wealthy family, also known as Trotula de Salerno, she is considered one of the first expert gynecologists. Trotula studied at the new medical school at Salerno—the first of the Western world—and went on to become a professor. She is known for having written two particularly important works, the most famous of which is *Passionibus Mulerium Curandorum* (*The Diseases of Women*), sometimes called *Trotula Major*. It is a 63-chapter compendium covering everything from anatomy and sex to menstruation, and it was written to educate her male colleagues. The second work—*Trotula minor*—covered more general information.

Salk, Jonas Edward 1914–1995

An American medical researcher, Salk discovered a vaccine for the debilitating, and often fatal, disease polio. Benefiting from the work of Harvard's John Franklin Enders, who had developed a way to grow polio in test tubes, Salk began a human trial of the vaccine in 1955. When it was made public, he became famous overnight and polio was virtually eradicated. In 1963 he founded the Salk Institute for Biological Studies and in 1977 was awarded the US Presidential Medal of Freedom.

Sanger, Margaret 1879–1966

See pp.226–27

Saunders, Cicely 1918–2005

See p.263

Schleiden, Matthias 1804–1881

See p.150

Seishu, Hanaoka 1760–1835

A Japanese surgeon, Seishu gained an extensive knowledge of Chinese herbal medicine and Western surgical techniques that he learned from the Dutch. Seishu studied medicine in Kyoto, but due to the self-imposed isolation of Japan at the time, there were few foreign medical texts available. Seishu is noted for combining Dutch and Japanese surgery and bringing modern techniques to Japan.

Semmelweis, Ignaz 1818–1865

See p.139

Servetus, Michael 1511–1553

Spanish theologian, physician, cartographer, and Renaissance humanist, Servetus is also known as Miguel Servet. He wrote several treatises on medicine and human anatomy and was the first European to describe the function of pulmonary (heart–lung) circulation correctly in his book *Christianismi Restitituto* (*The Restoration of Christianity*). However, his theological works were considered heretical and he was burnt at the stake for his views.

Shibasaburo, Kitasato 1853–1931

A Japanese physician and bacteriologist, Shibasaburo was educated at Kumaoto Medical School and Tokyo Imperial University. He studied under Robert Koch in Berlin, Germany, from 1885 to 1891, where he met and worked with Emil von Behring. Shibasaburo was the first person to grow the tetanus bacillus, and together he and von Behring discovered that injecting dead or weakened disease-causing bacteria into an animal caused its blood to make antibodies. In 1894 he identified the bacterium that caused

the outbreak of bubonic plague in Hong Kong and later described the organism that causes dysentery.

Simpson, James Young 1811–1870

A Scottish obstetrician, Simpson discovered the anesthetic properties of the gas chloroform in 1847, which became a popular method of pain relief in childbirth. Born in West Lothian, Simpson completed his studies at the age of 18, but was not allowed a license to practice medicine for another two years. He designed the air tractor, an early version of the ventouse (vacuum extraction), and improved the design of forceps to this day known as Simpson's forceps.

Sina, Ibn 980–1037

Also known by his latinized name Avicenna, Ibn Sina was born in Uzbekistan. A prolific scholar, he made important contributions to medicine as well as mathematics, chemistry, astronomy, psychology, and geology. Ibn Sina claimed to have treated patients by the time he was 16. He is known to have written 450 works, some of which have survived. His major work *Al Qanun fi al-Tibb (The Canon of Medicine)* (see pp.52–53) written when he was in the court of Shams al Dawla of Hamadan, became a standard text in European universities.

Snow, John 1813–1858

See pp.124–25

Sydenham, Thomas 1624–1689

See pp.90–91

Tourette, Georges Gilles de la 1857–1904

A French physician, Tourette described a syndrome he called "maladie de tics" (illness of tics), which was later named Gilles de la Tourette's illness in his honor by his mentor Jean-Martin Charcot. Born in southeastern France, he began his medical studies in Poitiers, then moved to Paris where he studied under influential neurologist Charcot at the Salpêtrière Hospital. Tourette studied and lectured in psychotherapy, hysteria, and the medical

and legal ramifications of mesmerism. After a series of tragic events he developed mood swings and depression, and eventually died in a mental hospital in Switzerland.

Tuo, Hua c.140–c.208

An ancient Chinese physician and surgeon, born in Bozhou province, Tuo lived in the late Eastern Han Dynasty. He is best known for the use of acupuncture and a concoction made from wine and máfèisàn (a herbal formulation made from hemp), as a general anesthetic. He also pioneered hydrotherapy and did innovative work with physical therapy using a series of exercises known as the frolics of the five animals.

Vesalius, Andreas 1514–1564

See p.75

Virchow, Rudolf Karl Ludwig 1821–1902

See p.152

Warren, Robin 1937–

See p.244

Watson, James Dewey 1928

An American molecular biologist, geneticist, and zoologist, Watson discovered the double helix structure of deoxyribonucleic acid (DNA) in 1953 with Francis Crick, and shared the Nobel Prize with both Crick and Maurice Wilkins in 1962. Watson wrote many science text books and was widely honored for his work. Crick was also associated with the US National Institutes of Health. He played a leading role in establishing the Human Genome Project and was appointed head of the project from 1990 to 1992.

Zhongjing, Zhang 150–219 CE

See p.26

Glossary

A

Acute condition One that begins abruptly and may last for a short time, in contrast with a chronic condition.

Agonist drug Simulates a natural substance and replicates its effects in the body.

AIDS Abbreviation for acquired immune deficiency syndrome, a deficiency of the immune system that can occur as a result of infection with HIV.

Alchemy A medieval practice that tried, among other things, to find a way to change common metals, such as lead, into gold and create an elixir for eternal life.

Allele Form or version of a gene, for example there are several versions of the genes for eye color.

Alveolus (plural: **alveoli**) A tiny air sac in the lungs.

Amino acid A simple organic compound that contains one or more amino groups and one or more carboxyl groups. Amino acids are the chemicals that make up proteins.

Anesthesia Medical inducement for pain relief or complete loss of sensation. May be in part of the body (local anesthesia) or in all of the body (general anesthesia).

Analgesia Form of pain relief.

Anatomist Person who studies the structure of living things. Human anatomy is the study of the human body.

Angina (or **angina pectoris**) Chest pain that occurs when the blood supply to the muscles of the heart is restricted, usually because the arteries supplying the heart become hardened and narrowed. Importantly the pain eases with rest.

Antagonist drug Blocks the action of a natural substance in the body.

Anthrax A serious, potentially fatal, bacterial infection that affects livestock but can spread to humans through contact with affected animals, or inhaling spores from contaminated animal fibers.

Antitoxin An antibody that counteracts a toxin, or poison.

Antibiotic Drug that is used to kill or inhibit the growth of bacteria, usually those causing infections.

Antibodies Proteins produced in the body by white blood cells to mark foreign particles or antigens and stimulate the immune response.

Antigen A substance that stimulates the body to produce antibodies and an immune response.

Antimicrobial Substance capable of killing microbes that cause infection or stopping them from multiplying.

Antiretroviral drugs Drugs used for the treatment of retrovirus infections, principally HIV.

Antisepsis Prevention of infection by inhibiting or arresting the growth and multiplication of microbes.

Antiseptic An antimicrobial substance applied to living tissue skin to reduce the risk of infection, which works by killing microbes that may be present.

Apothecary A term used in medieval times, to refer not only to the place where remedies were dispensed, but also to the person who dispensed them.

Arteriole Smaller blood vessel that leads from the arteries and links to the capillaries.

Artery Blood vessel that carries blood away from the heart.

Aseptic technique The performance of a medical or laboratory procedure under completely sterile conditions (that are free of all living microorganisms).

Atherosclerosis Disease of the arteries characterized by deposits of fatty material on the linings of artery walls.

Atria The two chambers of the upper heart that receive blood from veins.

Autoclave Steam-heated container used for sterilizing medical instruments at high temperature and pressure.

B

Bacteriophage Parasitic virus that infects bacteria; also called a phage.

Bacterium (plural: **bacteria**) A single-celled microscopic organism that does not have a membrane-enclosed nucleus or other organelles and is too small to see.

Base pairs Complementary pairs of nucleotide bases that link the two sides of the double spiral, or helix, of a DNA molecule. The order of base pairs spells out the DNA code.

Bile Dark green/yellowish substance that aids the digestion of fats in the small intestine. Produced by the liver, it is stored and released by the gallbladder. Also the yellow and black "bile" in humorism, a system of medicine that thrived in Europe from around 500 BCE to the 19th century.

Biopsy The taking of a tissue or fluid sample for analysis.

Bipolar disorder A long-term mental condition characterized by alternating periods of depression and elation; previously called manic depression.

Black Death An outbreak of the highly contagious bubonic plague that began in Central Asia in the 1330s, arriving in Europe in 1347. After five years the outbreak had killed approximately 60 percent of Europe's population.

Outbreaks in Europe continued into the 18th century.

Blastocyst Hollow ball of cells that is an early stage in the development of an embryo.

Blood circulation The continuous flow of blood around the body via the heart and blood vessels.

Blood pressure The pushing force of blood as it is pumped around the body by the heart; it can be detected at points where arteries are close to the skin's surface.

Blood type/group Any of several types—A, B, AB, or O—into which an individual's blood can be classified based on the antigens on the surface of red blood cells.

Brain stem A stalk of nerve tissue that forms the lowest part of the brain and links to the spinal column.

Bubonic plague Highly contagious disease that causes fever and painful swelling of the lymph glands— sometimes called buboes, hence its name. Other symptoms include spots on the skin, that turn black, so it was also known as the Black Death. Today, the disease mainly affects rodents, but can be transmitted from person to person by fleas.

C

Cancer A group of diseases characterized by the abnormal and unrestrained growth of cells in body organs or tissues.

Capillary Minute blood vessel with thin walls through which nutrients and waste products pass to and from body tissues.

Cardiac dilatation A condition in which the heart cavities become enlarged, causing the outer muscular wall (myocardium) to become thinner.

Cardiology The study of the heart and circulatory system; a cardiologist is a doctor who specializes in cardiology.

Cardiovascular system The system comprising heart, blood vessels—arteries, capillaries, and veins—and blood.

Cartilage Firm, flexible tissue found in various forms in the body, for example, in the larynx and respiratory tract, the external ear, and on the articulating surfaces of bones at joints.

Cautery The practice of cauterization, or burning, of part of a body to remove or close off the area, for example in an attempt to stop bleeding or remove unwanted growth.

Cell The smallest unit of an organism that can exist on its own—the building blocks of the body. Humans have more than 250 different types of cells.

Central nervous system The collection of nerves in the brain and spinal cord that acts to control the body.

Centrosome An organelle near the nucleus of an animal or plant cell that contains the centrioles from which the spindle fibers develop in cell division.

Cerebellum Area at the back of the brain, below the cerebrum, whose primary role is to control movement and maintain balance.

Cerebrum The largest part of the brain in humans, responsible for most conscious thought and activity. In humans it is divided into two cerebral hemispheres (left and right) and surrounds most of the rest of the brain.

Chakra In Indian Ayurvedic medicine a *chakra* is a spinning center of energy aligned along the middle of the body; there are seven *chakras* altogether.

Chemotherapy Treatment that uses drugs to target/kill cancer cells (also called cytotoxic drugs).

Chickenpox Also called varicella, this is a common infectious disease caused by the varicella-zoster virus, and is characterized by a rash and a fever.

Cholera An infection of the small intestine that causes severe watery diarrhea. It results from ingesting water or food that contains the bacterium *Vibrio cholera*.

Chromatid One of two threadlike strands into which a chromosome divides during cell division—each one contains a double-helix of DNA.

Chromosome A structure made of DNA and protein that is found in cells. A chromosome contains the genetic information (in the form of genes) for an organism; humans have 23 pairs.

Chronic condition A persistent medical condition that usually lasts six months or longer and may result in long-term change in the body.

Cilium (plural: **cilia**) A tiny "hair" projecting from a cell, usually on the surface of a tissue or small organism. Cilia line surfaces inside the body, for example the respiratory tract.

Congenital A physical abnormality or condition that is present from birth and can be the result of environmental or genetic factors.

Congestive heart failure A condition in which the heart is not pumping effectively; this can be the result of coronary artery disease or persistent high blood pressure.

Conjunctiva The mucous membrane that covers the front of the eye and lines the inside of the eyelids.

Contagion A living thing, usually a microbe, that can be passed between people to cause disease.

Contrast medium A substance through which X-rays cannot pass.

Coronary artery disease A condition that results when a waxy substance (plaque) builds up inside the coronary arteries that supply the heart muscle, causing them to narrow and restrict blood flow.

Cranial nerves Twelve pairs of nerves that arise directly from the brain, not via the spinal cord, and pass through openings in the skull, for example, the optic nerves and the auditory nerves.

Crookes tube An experimental electrical discharge tube with a partial vacuum invented by 19th-century British physicist William Crookes.

Crystal A solid whose constituent atoms, ions, or molecules are arranged in a regularly repeating pattern. A crystal lattice is the repeating pattern of atoms or ions that forms a crystal.

Crystallogram The pattern formed on a photographic plate by passing X-ray beams through a crystal.

Crystallography The study of atomic and molecular structure.

CT scan/CAT scanning Computerized (axial) tomography is an imaging technique that uses weak X-rays to record thin 2-D slicelike views through the body, then combines them to make 3-D images.

Culture medium A nutritious substance, sometimes called a growth medium, to support the growth of microorganisms, for example in a laboratory.

Cystic fibrosis Hereditary disorder of the exocrine glands that causes excess mucus production, which leads to clogging of the airways.

Cytoplasm The liquid contents of a cell, apart from organelles.

D

Defibrillator A device for restoring the rhythmic beating of the heart by the administration of a controlled electric shock.

Dermis Inner layer of skin composed of connective tissue interspersed with hair follicles, sweat glands, sebaceous glands, blood and lymph vessels, and sensory receptors that detect pressure, temperature, and pain.

Diabetes A disease that results from lack of or insufficient insulin production by the pancreas. Type 1 diabetes is caused by lack of insulin; in Type 2 diabetes insulin is produced, but the body is not able to use it properly.

Diagnosis Identification of an illness from its symptoms (what the person describes) and signs (what is observed).

Diastole/diastolic The period when all the chambers of the heart are relaxed and the heart is filling with blood. Diastolic pressure is the second number of a blood pressure reading, for example 120/80 or 120 over 80.

Diffraction The phenomenon that occurs when a light wave hits an object or passes through a gap and splits, or bends.

Digestion The breaking down of food into simpler molecules that can be utilized by the body.

Digestive system The digestive tract (mouth, esophagus or gullet, stomach, small and large intestines), plus associated organs—liver, pancreas, and gallbladder.

Diphtheria Highly infectious disease characterized by a fever, severe cough, and a grey coating over the infected areas, particularly the throat and tonsils.

Disinfect Make something clean and free from infection, especially with the use of germ-killing chemicals.

Distillation A process by which a pure liquid is separated from a mixture of liquids.

Deoxyribonucleic acid (DNA) A long thin double-helix-shaped molecule that makes up the chromosomes found in almost all cells. It contains the encoded genetic information of living organisms.

Dosha In Ayurvedic medicine a *dosha* is an energy believed to circulate round the body. There are three *doshas*—*vata* (wind), *pitta* (bile), and *kapha* (phlegm)—and good health and well-being occur when all three doshas are well-balanced.

Double helix A pair of parallel helices that intertwine around a common axis, for example in the structure of a DNA molecule.

Dysentery An intestinal infection that causes diarrhea and severe abdominal pain. It results from infection by either *Shigella* bacteria (shigellosis) or the parasite *Entamoeba histolica* (amoebic dysentery).

E

Electrocardiogram (ECG) A noninvasive test that measures and records the electrical activity in the heart.

Electroencephalogram (EEG) A noninvasive test that measures and records the electrical activity of the brain.

Electron A subatomic particle with a negative electric charge.

Electron microscope A microscope that uses a beam of electrons to produce a magnified image of an object. In a transmission electron microscope (TEM) electrons pass through a thin section of a specimen; with a scanning electron microscope (SEM) electrons bounce off the surface to give a 3-D image.

Elephantiasis A disease that occurs in the tropics characterized by massive swelling and thickening of the legs, arms, and scrotum, with thickening and darkening of the skin. It is mostly caused by chronic lymphatic obstruction due to an infection by a parasitic roundworm.

Embryo The first stage of development of newly conceived offspring. In human beings, it covers the first eight weeks of pregnancy.

Endocrine system The glands and cells of this system make and control the production of the body's chemical messengers—hormones. The main structures are: the hypothalamus, pituitary, thyroid, thymus, adrenal glands, pancreas, ovaries (female) and testes (male). The heart, stomach, and intestines also produce hormones.

Endocrinology The branch of medicine concerned with hormones and the endocrine glands. An endocrinologist is a doctor who specializes in conditions affecting this system.

Endometrium Mucous membrane that lines the uterus/womb.

Endorphins Protein molecules produced by the body that relieve pain by activating opiate receptors in the nervous system.

Endoscope A viewing instrument inserted into the body though a natural orifice or a surgical incision. Endoscopes may be flexible or rigid, and comprise a light source, and a series of lenses or a miniature camera. Surgical instruments may be passed inside endoscopes to carry out operations or take samples.

Enlightenment A philosophical movement of the 18th century characterized by belief in the human power of reason and innovations in political, religious, and educational doctrine.

Enzymes Substances secreted by organs in the body that speed up or slow down the rate of chemical changes, for example when food is digested.

Epidemic An outbreak of a contagious disease in which the incidence rate is much higher than expected, but it is confined to a particular region.

Epidemiology The study of diseases, how common they are, their causes and effects, and how they can be controlled.

Epidermis The outer layer of skin made up of keratin and dead cells. As dead cells are worn away they are replaced by new ones from the base of the epidermis.

Epilepsy A tendency to have recurrent seizures. In many cases the cause is unclear—it may be genetic or the result of injury, illness, or metabolic disorder.

Evaporation The process by which a liquid, for example water, becomes a gas.

Excretion The elimination of waste by organisms.

F

Fetus An unborn baby after the first eight weeks of pregnancy.

Fluorescent light Tube containing low-pressure, mercury-vapor gas that has a coating of phosphor on the inner side. An electric current "excites" the gas, producing short-wave ultraviolet light that causes the phosphor coating to glow.

Fluorescent screen A glass screen, one face of which is coated with a salt that emits light under the action of X-rays, or cathode rays.

Fungus Member of a group of unicellular, multicellular, or syncytial spore-producing organisms that feed on organic matter. They include molds, yeasts, mushrooms, and toadstools.

G

Gallstones Lumps of solid matter composed mainly of cholesterol or bile pigments that can form in the gallbladder or its ducts.

Galvanometer Instrument for detecting the existence of small electrical currents and determining their strength.

Gamete A sex cell—the sperm in males and the ovum in females.

Gastroenterology The study of the conditions that affect the digestive system; a gastroenterologist is a doctor who specializes in gastroenterology.

Gene Basic unit of heredity in living things, typically a segment of DNA or RNA that provides the coded instructions for a particular protein.

Gene map A plot of the sequence of genes along a strand of DNA.

General practitioner A primary care physician, who treats patients' minor disorders and refers them to a specialist for treatment of more serious conditions.

Genetic code Sequence of nucleotide bases on DNA that codes for a particular gene.

Genetic engineering The process of artificially modifying the characteristics of an organism by manipulating its genetic material.

Genetic fingerprinting Analysis of a DNA sample to identify who it belongs to.

Germ A harmful microbe, such as a virus, bacterium, fungal spore, or protist.

Gland Specialized cells or groups of cells that produce and secrete a specific substance, such as a hormone or digestive enzyme.

Glucagon Hormone that stimulates the liver to turn stored glycogen into glucose when blood sugar levels are low; the opposite effect of insulin.

Glucose A simple sugar that is the main carbohydrate source of energy in most living cells.

Glycogen A form of glucose stored in animal cells, made mainly in the liver and muscles.

Golgi staining A method of staining nerve tissue with silver nitrate so that it can be viewed under a microscope.

Gynecology The branch of medicine that deals with the function and disease of the reproductive system of girls and women.

H

Hematology The branch of medicine that deals with the diagnosis and treatment of disorders of blood; a hematologist is a doctor who specializes in this area.

Hemoglobin Protein in red blood cells that combines with oxygen from the lungs so it can be carried around the body.

Hemorrhage The escape of blood from a blood vessel, usually after injury; a hemotoma (bruise) is an accumulation of blood from a torn blood vessel that remains within the tissues.

Heart valves Structures within the heart that ensure blood only flows through it in one direction. There are four altogether: two ventricular valves (mitral and tricuspid) between the upper and lower chambers and two semilunar valves (aortic and pulmonary), which ensure blood flow out to the rest of the body and the lungs respectively.

Hepatologist A doctor specializing in conditions that affect the liver.

Histology The study of the microscopic structure of tissues and cells.

HIV Abbreviation for human immunodeficiency virus, a retrovirus that is the cause of AIDS. HIV gains access to the body through blood contact, infected needles, or sexual intercourse.

Hormone A substance (chemical messenger) produced in an endocrine gland that controls a specific biological process or activity in the body.

Human genome The complete set of genes for a human—there are approximately 20,000 genes.

Humors Body fluids or temperaments (blood/sanguine, yellow bile/choleric, black bile/melancholic, and phlegm/phlegmatic). Early physicians believed in the concept of the four humors, also known as humorism, which stated that well-being depended on balancing these four body fluids, or humors.

I

Immune system The body's natural defense network that protects against infection and other diseases. It includes the thymus gland, spleen, white blood cells, lymph ducts and vessels, and lymph—the fluid that passes along them.

Immunity The ability of an organism, or the body, to resist or fight a particular infection or toxin by the action of antibodies or white blood cells.

Immunization Rendering a person resistant to attack from microbes, which would otherwise cause infectious disease, usually by inoculation.

Immunosuppressant Substance that reduces the workings of the immune system, for example to prevent rejection of transplanted organs.

Implant An item surgically inserted into the body. It may be living (for example, bone marrow cells), mechanical (hip replacement), electronic (heart pacemaker), or a combination of all three.

In vitro fertilization (IVF) An artificial method of conception in which egg cells are fertilized by sperm outside the womb, in vitro (in glass).

Infection A disease caused by invading microbes such as bacteria, viruses, protists, or similar life forms.

Inheritance The range of natural characteristics and potential passed on by parents or ancestors to offspring.

Inoculation In immunization, the introduction of disease-causing organisms into the body in a mild or harmless form to stimulate the production of antibodies that will provide future protection against the disease.

Insulin A hormone, produced by the Islets of Langerhans in the pancreas, that regulates the use of glucose in the blood. Lack of it causes Type 1 diabetes; the body's inability to use it can result in Type 2 diabetes.

Intestine The longest part of the digestive tract—from the stomach to the anus. It consists of the small intestine (duodenum, jejunum, and ileum) where most food is broken down and absorbed, and the shorter large intestine.

Iris The colored part of the eye that surrounds and controls the size of the pupil.

JK

Joint An area of the body where bones meet. They are normally held together by bands of fiber called ligaments.

Keratin One of the main proteins in skin, hair, and nails.

Keyhole surgery Surgery performed through a very small incision, using special instruments and an endoscope.

Kidney One of a pair of internal organs that filter waste products and excess water from the blood.

L

Laporascope A type of endoscope inserted directly into the abdomen through an incision.

Larynx Structure in the neck at the top of the trachea (windpipe) that contains the vocal cords.

Laser surgery Surgery performed with a laser beam, for example, reshaping the cornea to improve eyesight.

Leeching The application of a living leech to the skin in order to initiate blood flow or deplete blood from a localized area of the body.

Lens The structure near the front of the eye that fine-focuses vision.

Lesion An abnormality, such as an ulcer, in body tissue or an organ.

Leucocyte A general term for any white blood cell.

Ligament A short, elastic band of fibers that connects two bones or cartilage at a joint.

Ligature A cordlike item used to tighten or constrict, for example a filament or fine thread used to tie up a bleeding artery during surgery.

Limbic system A collection of structures in the centre of the brain that play a vital role in the control of the automatic (autonomic) body functions, emotions, and sense of smell.

Lymph The excess fluid that collects in the tissues as blood circulates through the body; it contains mainly white blood cells.

Lymphatic system An extensive network of tubes, small organs, and glands that drains lymph from the body's tissues into the bloodstream.

Lymphocytes White blood cells that protect against infection, for example by producing antibodies.

M

Magnetic resonance imaging (MRI) A form of computerized scanning that uses a powerful magnetic field and radio pulses to visualize 2-D slices through the body, then combines them to create a 3-D image.

Malaria A disease caused by the parasitic protozoa *Plasmodia* and spread by the bite of female *Anopheles* mosquitos. It causes flulike symptoms: a high temperature (fever), shaking chills, headaches, muscle aches, and fatigue. Vomiting, nausea, and diarrhea may also occur. In severe cases, it causes kidney failure, confusion, seizures, comas, and sometimes death.

Matter Anything that has mass and occupies space—it can be liquid, solid, or gas.

Median A "line" down the middle of the body, in the plane that divides the body into right and left halves.

Meiosis The type of cell division that results in daughter cells that have half the number of chromosomes of the parent cell. Meiotic division results in the production of egg and sperm cells.

Melanin A brown pigment found especially in the skin, hair, and eyes.

Meridian According to Chinese medicine, meridians are a network of pathways through which life energy, or *qi*, flows.

Metabolism The sum of the physical and chemical processes that takes place within the body, from digesting food to using energy for muscle action.

Metastasis The spread of cancerous cells from one part of the body to the other.

Microbe Any living organism that is too small to be seen by the naked eye.

Microscope An instrument that produces magnified images of very small objects.

Microscopy The process of examination by microscope—often to make a diagnosis.

Mineral A naturally occurring and usually inorganic solid.

Mitochondria Sausage-shaped organelles, found in cells, that contain genetic material and make energy available for a cell to live and function.

Mitral valve The valve between the two chambers of the left side of the heart.

Molecular structure Arrangement, type, position, and direction of the bonds that link atoms within a molecule.

Molecule The smallest unit of an element or compound that contains at least two atoms bonded together; water (H_2O), for example, has three atoms: two hydrogen and one oxygen.

Motor nerves Nerve fibers that carry impulses (electrical signals) to the muscles or glands.

Mucous membrane Soft, pink, skinlike layer that lines many cavities, tubes, and ducts in the body. The mucous membranes contain millions of goblet cells that secrete a fluid called mucus.

Mucus Thick slimy fluid secreted by mucous membranes that moistens, lubricates, and protects cavities, tubes, and ducts in the body.

Musculoskeletal system The body's bones, joints, and muscles form this system.

Myelin A fatty material found especially around nerve fibers.

Myocardial infarction Often called a heart attack, this is a condition that occurs when one or more of the arteries that supply blood to the heart muscle (coronary arteries) become blocked, so depriving the area beyond the blockage of blood.

Myocardium Special muscle type only found in the heart; its fibers form a network that contracts spontaneously.

Myofibril Stretchy threads found in muscle cells.

N

Natural philosophy Ancient term used right up to the 19th century to describe the practice of studying natural science, which included medicine.

Nephrology Study of conditions affecting the kidneys; a nephrologist is a doctor who specializes in this branch of medicine.

Nephron One of the million or so minute purification and filtration units in the kidney.

Nerve A sheathed bundle of threadlike projections, or fibers, of nerve cells (neurons) that carry electrical impulses between the brain, spinal cord, and body tissues.

Nervous system The body system made up of the brain, spinal cord, and nerves.

Neurology The study of the nervous system; a neurologist is a doctor who specializes in this branch of medicine.

Nucleotides Chemical subunits or base of deoxyribonucleic acid (DNA) and ribonucleic acid (RNA) that function as code letters for genetic information.

Nucleus The part of a cell in which genetic information is stored.

Nutrients Substances in food that are used by living organisms for growth, maintenance, and reproduction.

Nutrition The processes in which an organism takes in food and uses it for growth and maintenance.

O

Obstetrics The branch of medicine concerned with pregnancy and childbirth; a doctor who specializes in this subject is an obstetrician.

Oncology The branch of medicine concerned with cancers and similar diseases; an oncologist is a doctor who specializes in oncology.

Ophthalmology The branch of medicine concerned with the study and treatment of disorders and diseases of the eye; a doctor who specializes in this area is an ophthalmologist.

Organ Main body part or structure with a specific function, for example the heart, brain, liver, or spleen.

Organelles Specialized membrane-bound structures within a cell.

Orthopedics The study of bones and joints.

Ovary One of two structures, each at the end of a fallopian tube, that makes egg cells or ova.

Ovulation The release of an ovum or egg cell from the ovary about midway through a woman's menstrual cycle.

Ovum (plural: **ova**) The egg cell.

P

Pandemic A very large-scale outbreak/epidemic of a contagious disease that affects the human population in a huge geographical area, such as a continent.

Pathogen A microbe that causes disease or other harm.

Pathology The study of disease—its causes, mechanism, and effects on the body. Pathologists conduct autopsies to determine cause of death and ascertain the effects that a disease or treatment has had.

Parasite An organism that lives in or on another living creature.

Pediatrician A doctor who specializes in the diagnosis and treatment of disorders affecting children.

Penicillin An antibiotic, or group of antibiotics, produced naturally by certain blue molds; they are now usually prepared synthetically. First discovered in 1928, penicillin was one of the first antibiotic agents and is still widely used today.

Pertussis Also known as whooping cough, this is a highly contagious, potentially fatal infectious disease that causes severe bouts of coughing that often end with a "whooping" sound, hence its name.

Pessary 1. A small soluble block that is inserted into the vagina to treat infection or as a contraceptive.

2. An elastic or rigid device that is inserted into the vagina to support the uterus.

Pharmacist A person who prepares and dispenses drugs; today, a pharmacist must be professionally qualified.

Pharmacology The study of drugs and how they act on the body.

Philosophy The study of the fundamental nature of knowledge, reality, and existence through logical reasoning; early physicians and scientists were referred to as natural philosophers.

Photographic plate A flat sheet coated with a light-sensitive chemical.

Phrenology An 18th-century practice in which detailed study was made of the shape and size of a skull, which was thought to be an indication of a person's character and mental abilities.

Physician A person who practices medicine—especially one who specializes in diagnosis and medical treatment rather than surgery.

Physiology The study of the normal function of living organisms and their parts.

Pituitary gland Known as the master gland, the pituitary gland is the most important gland in the endocrine system. It regulates and controls the activities of most other endocrine glands and many body processes.

Placebo A chemically inert substance given instead of a drug. Many new drugs are tested against a placebo preparation.

Plasma The liquid part of the blood.

Platelets Cells in the blood that are vital to the blood-clotting process.

Pneumonia Inflammation of the air sacs and smaller air passages in the lungs due to infection or inhalation of irritants.

Poliomyelitis Commonly called polio, this is an infectious viral disease that in severe cases can attack the brain and spinal cord.

Polymath A person with expertise in many wide-ranging subjects.

Positron emission tomography (PET) A form of computerized scanning that uses rays given off by substances put into the body to identify very busy (metabolically active) cells and tissues.

Pox Skin eruptions or sacs that leave pitted pockmarks on the skin; the term has been applied to a wide range of diseases—from acne to syphilis.

Prosthesis An artificial item used as a substitute or replacement body part.

Proteins Huge molecules made up of chains of amino acids—proteins are the building blocks of the body.

Protists Single-celled eukaryotic microorganisms (the cell has a nucleus), some of which are parasites and cause disease.

Psychology Scientific study of the human mind and its functions, especially those affecting behavior in a given context.

Psychotherapy The treatment of mental disorders by psychological rather than medical means.

Pulse Rhythmic expansion and contraction of an artery as blood is forced through it.

Pupil The opening in the iris that allows light into the eye; it opens (dilates) and closes (contracts) under the control of the iris.

Q

qi In Chinese culture *qi*, or ch'i, "breath, air," is an active or energy-based part of a living thing.

Quinine A bitter crystalline compound found in cinchona bark used as a tonic and once prescribed as a treatment for malaria.

R

Rabies An acute viral infection of the nervous system, also known as hydrophobia (fear of water), which primarily affects animals, but can be passed to humans by a bite or lick over a wound.

Radiation Emission or transmission of energy in the form of waves or particles through space or through a material medium.

Radiotherapy The treatment of disease, especially cancer, using localized X-rays or similar forms of radiation.

Red blood cells Biconcave, disk-shaped cells that contain hemoglobin. There are 4–5 million in 1 cubic mm (0.06 cubic in) of blood.

Renaissance Meaning "rebirth," this term describes the revival of arts, literature, science, and learning in Europe from the 14th and 15th centuries.

Renal Relating to the kidneys.

Reproductive system The organs involved in reproduction. It is the part of the body that most differs between males and females.

Respiration 1. Bodily movements of breathing. 2. Gas exchange of oxygen for carbon dioxide in the lungs. 3. Similar gas exchange in the body tissues. 4. Breakdown of molecules to release their energy for cellular action.

Retina Light-sensitive layer that lines the inner side of the back of the eye. The retina converts optical images into nerve signals that travel to the brain via the optic nerve.

Rhinoplasty Plastic or cosmetic surgery performed on the nose.

Rickets A condition that arises from vitamin D deficiency and which affects bone development. Bones become soft and weak and if the condition is untreated it may lead to deformities such as bowed legs.

Rictus (of limbs) Severe "twisting" pain in the limbs that can arise from scurvy, which is a disease resulting from lack of vitamin C.

Ribonucleic acid (RNA) In most organisms, a molecule that decodes DNA's instructions to make proteins and control this process.

S

Scarlet fever A contagious disease caused by the streptococcal bacteria that results in high fever, severe sore throat, vomiting, and a rash of tiny red spots.

Schizophrenia Long-term mental condition that causes a range of symptoms including hallucinations, delusions, muddled thoughts, and extreme behavior changes.

Sense organs The structures that provide the five main senses of sight, smell, touch, taste, and hearing. They are the eyes, nose, skin, tongue, and ears. They detect information from outside the body and transmit it to the brain.

Sensory nerve Nerve that carries sensory information from the tissues toward the spinal cord and brain.

SI (Système International) unit A unit in the international system of measures based on the meter, kilogram, second, ampere, kelvin, candela, and mole.

Skeleton The frame of bone and cartilage that supports the body and protects its organs.

Smallpox A highly contagious viral infection causing fever, red rash, blisters, and bleeding in its severest form, hence its name the "red plague." Smallpox has now been eradicated following worldwide vaccination programs.

Sperm One of the male reproductive cells released in semen during ejaculation, which must enter an ovum for fertilization to take place.

Sphygmograph A mechanical device used to measure blood pressure commonly used in the 19th century—it gave a reading on a piece of paper. It has been replaced by the sphygmomanometer cuff and meter used today.

Spinal cord A bundle of nerves running from the brain down through the spinal column.

Spinal nerves Thirty-one pairs of nerves that carry motor and sensory signals between the spinal cord and body tissues.

Sterilization 1. The removal of life forms from an object. 2. Medical procedure to prevent reproduction.

Stethoscope An instrument used for listening to body sounds, particularly from the heart, lungs, and digestion.

Subcutaneous Situated or applied just under the skin.

Sublimation Chemical process in which a solid turns into a gas without going through a liquid stage.

Superficial veins Veins that run very close to the surface of the skin.

Syphilis A chronic bacterial infection that is contracted mainly through sexual intercourse, but can also be passed to a developing fetus.

Systole/systolic The phase when the heart muscle contracts to pump blood out of its chambers to the lungs or the rest of the body; systolic pressure is the first number of a blood pressure reading, for example 120/80 or 120 over 80.

T

Talking cure A method of treating psychological disorders or emotional difficulties by talking to a therapist or counselor, either one-to-one or in groups.

Tendons Bands of fiber that attach muscles to bones.

Tetanus A disease of the central nervous system marked by rigidity and spasm of the voluntary muscles. It is caused by infection of a wound with spores of *Clostridium tetani* bacteria.

Tissue Groups of similar cells that carry out the same function, such as muscle tissue, which can contract.

Tissue typing The identification of antigens in the tissue of a donor and recipient before procedures such as organ transplantation, in order to minimize the possibility of rejection caused by antigenic differences.

Tourniquet A device used to prevent blood flowing through a vein or artery, typically by a tight bandage or cord.

Toxicology The study of toxic or harmful substances.

Toxin A harmful substance, especially one produced by certain bacteria, and some animals and plants.

Trachea Air passage between the throat (pharynx) and the lungs.

Transcription The copying of sequences of genes from DNA to RNA.

Transfusion The transfer of blood from a donor to a recipient.

Translocation Movement of a segment of a chromosome from one location to another, either on the same chromosome or to another chromosome.

Transplant The taking and implanting of tissue or organs from one part of the body to another, or from a donor to a recipient.

Tricuspid valve The valve between the two chambers of the right side of the heart.

Tuberculosis (TB) An infectious bacterial disease characterized by the growth of nodules (tubercles) in the tissues, especially the lungs.

Tumor A growth or lump of abnormal cells that may be malignant (cancerous) and spread throughout the body or benign (noncancerous) with no tendency to spread.

Typhoid fever An infectious disease contracted by ingesting food or

water contaminated by the bacteria *Salmonella typhi*.

Typhus Any of a group of illnesses caused by rickettsiae and spread by insects or similar animals. Potentially fatal, symptoms of typhus include headache, back and limb pain, followed by high fever, a rash, and confusion.

U

Ultrasound Sound with a frequency above that which the human ear can detect.

Ultrasound scan Diagnostic technique in which high-frequency sounds are passed into the body; reflected echoes are analyzed by computer to build up an image of the organ or structure.

UNESCO Abbreviation for United Nations Educational, Scientific, and Cultural Organization. Set up in 1945, UNESCO encourages international peace and respect for human rights and has its headquarters in Paris, France. Its motto is "Building peace in the minds of men and women."

Urinary system The system comprising organs of the body that form and eliminate urine from the body: kidneys, ureters, bladder, and urethra.

Urine A yellowish waste fluid made in the kidneys, stored in the bladder, and discharged through the urethra.

V

Vaccination Deliberate introduction of a weakened disease-causing substance to provide immunity against the disease.

Vaccine A preparation of weakened or neutralized germs, or harmful products, that makes the body become immune to the germs.

Vacuum tube Sealed glass tube containing virtually no gases, to allow freer passage of electrons (electric current).

Variolation An early method of immunizing patients by infecting them with smallpox

pustules taken from someone with a mild form of the disease.

Vector Organism that transmits disease.

Veins Blood vessels that carry blood from all parts of the body back to the heart.

Ventricle A chamber or compartment, usually fluid-filled. For example, two of the large cavities of the heart (cardiac ventricles) and four cerebral ventricles in the brain.

Venule Smaller blood vessels that link the capillaries to the veins to carry blood back to the heart.

Vessel A duct or tube carrying blood or other fluid through the body.

Virus Smallest type of harmful microbe, consisting of genetic material wrapped in a protective coating; it can only multiply by invading other living cells.

Vitamin Organic compound found in foods that is essential for good health. There are 13 vitamins: A, C, D, E, K, B12, and seven grouped under vitamin B complex.

W

Wavelength Distance between successive crests of a wave, especially points in a sound wave or electromagnetic wave.

White blood cells Any of the colorless blood cells that play a part in the body's defensive immune system.

World Health Organization (WHO) Specialized agency within the United Nations concerned with public health around the world. The WHO was established in 1948 and its headquarters are in Geneva, Switzerland.

X

X-ray 1. A type of electromagnetic radiation with a wavelength shorter than ultraviolet radiation. 2. A photographic

or digital image of the internal composition of something, especially a body part, taken with X-rays.

Y

Yellow fever A serious viral disease spread by mosquitoes that affects the liver and kidneys causing fever and jaundice.

Z

Zygote Cell produced when an ovum is fertilized by a sperm.

Index

Page numbers in *italics* indicate a caption to an illustration, and those in **bold** a main entry.

A

AbioCor *235*
abortion 226
accoucher see man-midwife
acetylsalicylic acid 170
Acremonium 201
acupressure 29
acupuncture 14, **28–29**
  points on body *29*
Acute and Chronic Diseases, On 39
Adams, George 96
Adler, Alfred 251
adrenal glands *286*, *287*
adult stem cells 270
Aedes aegypti 267
Aeneas *38*
Aerohaler 215
AESOP 254
Ätiologie, der Begriff und die Prophylaxis des Kindbettfiebers, Die (The Etiology, Concept, and Prophylaxis of Childbed Fever) 139
agglutination 176–77
agni 30, 31
Agnodice 140, *141*
Agrippa, Cornelius 65
ah'men 15
AIDS *see* HIV/AIDS
air, foul-smelling 121
akasha 30
al-Baytar, Ibn 51
Albucasis *see* al-Zahrawi
albumenometer 116
alchemy *49*, **64–65**, **70–71**
Alfanus I, Archbishop of Salerno 55
al-Hariri, Ibn Ali 300
Al-Judari Wal Hasabah (Concerning Smallpox and Measles) 51
Al-Kitab 'l-jami' fi 'l-aghdiya wa-'l-adwiyah al-mufraddah (The Comprehensive Book of Foods and Simple Remedies) 51
Allen, Edgar 205
allergies to drugs **208–9**
Allgemeines Krankenhaus 106
Allium sativum see garlic
al-Nafis, Ibn 49, 51, 83
aloe vera *62*
Alphanus I 300
Al-Qanum fi al-Tibb (The Canon of Medicine) **52–53**
Al-Qanum fi al-Tibb (The Canon of Medicine) 51
al-Razi, Ibn *48*, 49–50, 55, 61, 86, 101, 228, 300
Altmann, Richard 150
al-Zahrawi 51, *202*
Alzheimer, Alois 260, *261*
Alzheimer's disease **260–61**
ambulances 256
American hospitals 107
AMI *see* myocardial infarction, acute
aminoglycosides 201
amoxicillin 201
amphenicols 201
ampicillin 201
amputation *78*, 78, 128, 154

amulets 14, *18*
anesthesia 124, **128–29**, **130–31**
  for cataract surgery 86
  Chinese 26
  electric *132*
  for trepanning 17
anaphylaxis 208
Anathomia Corporis Humani (Anatomy of the Human Body) 61
Anatomy 145
anatomy 160, **72–75**
  from autopsy *152*
  brain 160, *281*
  cardiovascular *282*
  digestive *288*
  Galen's ideas on 40–41, 160
  hands *145*
  heart *282*
  medieval to Renaissance *55*, **60–61**
  microanatomy 96, 160
  respiratory system **284–85**
  theater *73*
  urinary *292*
Anatomy Lesson by Velpeau 119
Anatomy of the Bee, as Revealed by the Microscope 92
ancient history medicine 12
  see also Chinese medicine; Egyptian medicine; Greek medicine, ancient; Islamic medicine; Roman medicine, ancient
Anderson, William French 248
angina 185
angina pectoralis 206
angiogenesis 229
angiogenesis inhibition 231
angioplasty, coronary *206*
angiotensin converting enzyme (ACE) inhibitors 219
Anglesey leg 237
anna vaha 31
Anopheles mosquito 174–75
  A. albimanus 174
anthrax 146, 149, 158
antibiotics **200–201**
  resistance **258–59**
  for TB 156
antigens 176–77
antihistamines **208–9**
antiretroviral drugs 243
antiseptics 139, **154–55**
Antonine plague 38, 41
anxiety disorders 251
ap 30
apothecary **62–63**
apothecary jar *59*, *62*
Aretaeus of Cappadocia 190, 191
Aristotle 160
arms
  artificial *194*, 237, *238*
  bionic 237, *238*
  prosthetic *238*
arsenic 187
artava vaha 31
artemisinin 175
arterial narrowing *206*, 206
Arthrobot 252
arthroplasty, knee *238*
arthroscopy *188*
Asclepiades of Bithnyi 33
Asclepios *32*, 32, 29, 39, 40

ascorbic acid *see* vitamin C
Ashtanga Hridayam 30
Ashtanga Sangraha 30
Ashurbanipal, King 24, *25*
aspirin **170–71**, 218
asthi 31
asthma 115, 209, 214–15
asû 24
asylums 162–63, **164–65**
atherosclerosis 22, 244
atomizers 215
atrial fibrillation 185
auscultation 115
Auenbrugger, Joseph Leopold 300
Aurelio, Marco 152
Aurelius, Marcus, Emperor 41
auscultation 115
Australian Royal Flying Doctor Service 252–53
autoclave *154*
auto-injector *202*
autopsy 61, **152–53**
Avicenna 50, 51, **52–53**
Ayurveda **30–31**
azidothymidine (AZT) 243
Aztecs
  dying from epidemics 88, *88*
  herbal medicine 15
  trepanning 16

B

Babinski, Joseph 300
Bacillus 122
  B. anthracis 146
back pain, acupuncture for 29
Bacon, Roger 65
bacteriology 93, 153, 167, **200–201**, *291*
  bacterial culture 147
  bacterial resistance 258
  cancers linked to bacteria 244
  cholera 122
bacteriophage 167, *291*
balance 298
Baliff, Peter 237
Balkan War, First *123*
Bally, William *104*
Bandages, On 39
Banting, Frederick 190, 191, 300
barber-surgeons 59, **76–77**, 118
barium X-ray imaging 172
Barnard, Christiaan *234*, 235, 300
Bartisch, Georg *87*
barû 24
basil, holy 31
basket Boyle anesthesia machine *130*
Bassi, Agostino 146
Bassi, Laura 140
Bayer 218
Bayliss, William 185
Beatson, Thomas 231
Beck, Aaron 251
Becquerel, Antoine Henri 300
behavioral therapy 251
Behring, Emil von 158
belladonna 108
Bellevue Hospital, New York *256*
Benedict VIII, Pope 118
Benedictine Abbey, Monte Cassino 54
Benedictines 56

Berger, Jeffrey 170
Berlichingen, Goetz von 237
Bernard, Claude 300
Bertuccio, Nicola 61
Best, Charles Herbert *190*, 191, 300
beta-amyloid 260, *261*
beta-lactams 200
Beydeman, Alexander *109*
bezoar stone 79
bhumi 30
Bian Que 29
Bicêtre Hospital asylum *162*
Bichat, Marie-François 96, 150
bile
  black 33, 34, *58*, *250*
  yellow 33, 34
bionic arm *238*
bionic eyes 237
biopsy *253*
bipolar disorder 251
birch bracket fungus 14
birth 136
birth control 226–27
Bishop, J. Michael 229
Black Death **66–67**, 68
Black Bile, On the 41
black reaction *96*
Blackwell, Elizabeth 141
bladder cancer 244
Blaiburg, Philip 235
bleeding, herbs for *14*
blood
  agglutination 176–77
  cancers 152
  capillaries 96
  circulation 49, **82–83**, **84–85**, *283*
  clotting 152, 170, 176
  donation 194
  flow, assessment 217
  glucose test 191
  groups **176–77**
  as humor 33, 34
  letting *33*, 34, 35, *58*, 59
  making cells 277
  plasma 194, 194–95
  serum 176–77, *177*
  serum albumin 194
  transfusion **176–77**, 194, *195*
  vessels *282*
  what is it? *283*
  see also white blood cells
bloody flux 90
Blundell, James 176, *177*
Boccaccio, Giovanni 67, 69
body-snatchers 118–19
body systems **274–75**
Boethus, Flavius 40–41
Bona Dea 39
bone forceps *42*
bone growth *287*
bone lever *42*
bone marrow
  cells 270, *271*, 277
  transplant 270
bone structure *277*
Borel, J. F. 235
Borel, Peter 92
Bostock, John 208
Bouestard, Jacques 17
Bourgery, Jean-Baptiste *86*
Boveri, Thomas 300

Boveri stain *160*
Bovert, Daniel *208*, 208
bowel cancer 230
Bower Manuscript 30
Boyle, Robert 71
Boyle's apparatus *130*
Bozzini, Philipp *188*, 189
Brahe, Tycho 80
brain **160–61**
　Alzheimer's disease *260*
　anatomy 75, *281*
　biopsy 252
　depression and *251*
　function 58
　nervous system *280*
　primitive *281*
　scan *232*
Branca, Gustavo 81
Brand, Henning 71
breast cancer
　drugs for 230
　epidemiology 126
　genes 230
　screening 205, *205*
breathing *285*
Brès, Madeleine 141
Breuer, Josef 183, 251
British hospitals 107
Broca, Paul 300
bronchial asthma 215
bronchiole *284*
Brotzu, Giuseppe 201
Brown, Louise 240
Brunner, Johann 190
buboes 66
bubonic plague *see* Black Death
Buch der Bündth Ertznei (Book of Directions for Bandaging) 81
Buchner, Joseph 170
buchu *15*
Budd, William 125
Burch, George J. 185
Burke, William 118, 119
burns 81
Butenandt, Adolf 205
butterfly cannula *202*
bypass surgery 207

C

C-A-B protocol 257
Calcar, Jan van 74
Calne, Roy 235
Campani, Giuseppe *92*
Campani's microscope *92*
cancers **228–31**
　bacteria and **244–45**
　cell theory 151
　drugs 219
　lung *229*
cannula *202*
Canon of Medicine 50
Cantigas de Santa Maria (Canticles of Holy Mary) 56
capillaries 96
carbolic acid 154, *155*
carcinogens 228
cardiac *see* heart
cardiovascular system *275*, **282–83**
Carna 39
Carpi, Jacopo Berengario da 300
Carter, Henry Vandyke 145, 300–301
Cary-Gould microscope *95*
cataract surgery 24, **86–87**
catheter, male *42*

Catholic Church role in medicine 56
Causae et Curae (Causes and Cures) 58
Çavuşoğlu, M. Cenk 189
CD4-helper cells 243
celery, wild *62*
cells 93
　cell theory **150–51**
　cycle regulators 231
　HeLa *244*
　turning cancerous *228*, *229*
　see also stem cell therapy
Celsus, Aulus 80, 86, 260
Centers for Disease Control and Prevention 127
centesimal scale 108–9
cephalosporins 201
Cephalosporium acremonium 201
cerebral ventricles 160
Cerebri Anatome (Anatomy of the Brain) 160
Cerrahiyyetu 'l-Haniyye (Imperial Surgery) 140
cervical cancer 204, 230, 231, *244*, 244
Cesalpino, Andrea 83
cesarean section *134*, 134
Chadwick, Edwin 126
Chain, Ernest 198
chakras 30, 31
Chamberlain filter *166*, 166
chamomile 14, *15*
chancres 186
Channel Theory 20
Charaka 30
Charaka Samhita 30
Charcot, Jean-Martin 160–61, 183, 251
charités 107
Charles V, Emperor 75
Chauliac, Guy de 61, *61*, 69, 72
chemistry **70–71**
　see also alchemy
chemotherapy 187, *230*, 230–31
chestnut 58
childbed fever 135, **138–39**
childbirth **134–35**
　bleeding after 176, *177*
　forceps *139*
　Medieval times 56
China Root Epistle 72, 75
Chinese medicine
　acupuncture 28–29
　anesthesia 128
　childbirth 134
　early medicine **26–27**
　malaria 174
　smallpox 101
　trepanning 16
Chirurgia Magna (Great Surgery) 61, *61*, 69
Chlamydia pneumoniae 244
chloramphenicol 201
chlorina liquida 139
chlormethine 230
chlorofluoroocarbon inhalers 215
chloroform 124, 128–29, *129*, *130*
Chloroform and Other Anesthetics, On 125
chlortetracycline 201
cholera 121, **122–23**, **124–25**, 126, 147, 149, 158
cholera beds *122*, 123
Cholera Tramples the Victor and the Vanquished Both 121
Christianismi Restitutio (The Restoration of Christianity) 83
chromosomes 150, *295*
cimetidine 219
cinchona 89, *89*, 108, *108*, 174
cinchona bag *88*
Clarence carriages 107
Clayfield's mercurial holder *130*

Clement VI, Pope 69
clinical medicine 107
Clinton, Hillary 224
cloning *270*
Clostridium difficile 259
clot busters 257
Clover, Thomas *129*
clyster *42*
coca 15
cognitive therapy 251
Cohn, Edwin 194
Cohn, Ferdinand 93
Cohnheim, Julius 153
Colombo, Readlo 83, 301
Colton, Frank 205, 224
Columbian Exchange 89
Columbus, Christopher 101
Coming of a Physician to His Patient, The 55
Commodus, Emperor 41
Complete Human Anatomy Treatise including Surgical Treatments 86
computerized tomography (CT) 217, *232*
Comstock Laws 226
condoms 224, 227
congenital disorders 80
congenital heart abnormalities 207
conjunctivitis 96
Constantine the African 55, 301
consumption 156
contraception 226–27
contraceptive pill 218, **224–25**
convulsive therapies 163
Cook, James 98
Cormack, Allan McLeod 217
Corner, George 205
coronary angioplasty *206*
coronary artery disease, transplant 235
Corvisart, Jean-Nicolas 207
couching 86
Courtois, Bernard 154
Cowley, R Adams 256, 257
cowpox 102, 103, *103*
CPR 257
Crick, Francis 213, 301
Crimean War *127*, 142
Crocco, John 145
crystallography 172
C-scale 108–9
CT scans 22, 217
Cuitláhuac 88
Culpeper microscope *95*
Cumming, Alexander 96
cupping 34, 35
Curtorum Chirurgia per Insitionem, De (On the Surgery of Mutilation by Grafting) 81
cyclosporine 235
cystoscope 189
cytology, exfoliative 204

D

da Vinci, Leonardo 72, 83
da Vinci Surgical System 252, 254
Dale, Henry 208
Dally, Clarence 172
Damadian, Raymond *216*, 232
Danchell, Frederick 122
datura 214, 215
Daunton, Martin 122
Daviel, Jacques 86
Davy, Humphry 128, 301
Day, George 260
DDT (dichlorodiphenyltrichloroethane) 175
de Osorio, Ana 89

Deaconess Institutions 107
DeBakey, Michael *234*
defibrillator 185, *206*, 257
deformities, repair **80–81**
dehydration 123
DEKA arm 237
dementias **260–61**
Democritus 70, 301
demons 24
dentistry **132–33**
　ancient Roman 39
　instruments *132*
dentures 238
Denys, Jean-Baptiste 176
deoxyribonucleic acid (DNA) *see* DNA
depression 163, *250*, 251
Desormeaux, Antoine 189
Desoutter brothers 237
Deter, Auguste 260
Dhanvantari, Lord *31*
dhatus 31
d'Herelle, Félix 167
diabetes **190–91**
Diagnosis of Uterine Cancer by the Vaginal Smear 204
diagnostic instruments **116–17**, *135*, *204*
diagnostic procedures 90–91
　early medicine 26
　genetic testing 249
　Islamic medicine 50–51
　see also imaging
diaphragms 227
diazepam (Valium) 218
Dickens, Charles 107
digestive problems 14
digestive system *274*, **288–89**
dilator, obstetric *42*
Dioscorides, Pedanius 39, *39*, 108
diphtheria 153, *158*, 158
Diseases and Cures of Women, On the 140
Diseases of Women 140
dispensaries 107
dissection of corpses 33, 61, *61*, **118–19**
Dissertatio epistolaris (Dissertation on Letters) 91
distillation 70–71
divining *18*, 24
Dix, Dorothea 163
Dix Livres de la Chirurgie (Treatise on Surgery) 79
Djerassi, Carl 205, 224
DNA **246–47**
　bacterial *200*
　electrophoresis 249
　HIV 243
　sequencing *246*, *295*
　structure **212–13**
　testing **248–49**
Dobson, Matthew 190
Doisy, Edward 205
dolls as diagnostic tools *135*
Dolly the sheep *270*
Domagk, Gerhard 200
Doppler, pulsed *217*
doshas 30, 31
Down syndrome 249
Downs, Fred *237*
dreams 182–83
Drexler, Eric 265
drills 14, *132*
　drugs *see* medicines *and specific drugs, e.g.* aspirin
du Bois-Reymond, Emil 185
Dubois, Jacques 72

Duchenne, Guillaume-
  Benjamin-Armand 160–61, 301
Duchesne, Ernest 198
Duggar, Benjamin 201
Dunant, Jean-Henri 267
duodenal ulcers 244
dyes 96–97
dysentery *90*

E

ears *298*
  administration of medicine via *31*
Ebers papyrus 20–21, 134, 228
Ebola virus disease 159, **268–69**
  ECG *see* electrocaradiogram
ECT 163, *163*
eczema 209
education for medicine 49, **54–55**, 107
Edwards, Robert 240
Edwin Smith papyrus 20, 21, 228
ego *182*
Ego and the Id, The 183
Egyptian medicine **20–21**, 54
  cancers 228
  childbirth 134
  diabetes 190
  inhalers 214
  polio *210*
  smallpox 101
  willow use for pain relief 170
  women working in 140
Ehrlich, Paul *187*, 187, 218, 230, 301
Einthoven, Wilhelm *184*, 185
Eisenhower, Dwight D. 194
Elam, James 257
elbows: how they work *279*
electricity and ECG 184
electric shocks 185
electro-anesthesia *132*
electrocardiogram (ECG) **184–85**
electroconvulsive therapy (ECT)
  163, 163
electron microscope *95*
electrophoresis of DNA 249
elements, five 30–31
*Elements According to Hippocrates,
  On the* 41
elephantiasis, in mummies 22
elixir of youth *65*
El-Sidrón 14, *15*
embryonic stem cells 270
emergency medicine **256–57**
Empedocles 33, 301
Empiric School 39
Encode 246
endocrine system *274*, **286–87**
end-of-life care **262–63**
endoscopes *116*, *188*, 189
Endovelicus 39
Enovid 227
*Enquiry into the Causes and Effects
  of Vaccinae, An* 102
Enterovirus 210
ephedra 208
Ephesus 134
Epidaurus *33*
epidemics **88–89**
  cholera **122–23**, 126
  polio 210
  smallpox **100–101**
  studying 90
  yellow fever 69
  see also pandemics; plagues
epidemiology 124, **126–27**

epilepsy 24
  trepanning for 17
epinephrine 215
Epipens *202*
Epistolae responsoriae (Letters and Replies) 91
Epitome 75
Erasistratus *33*, *82*, 301
Erxleben, Dorothea 140
erythromycin 201
ether 124, 128–29, *130*
ethics: early medicine 55
Euget-Les-Bain, Auphon 215
*Exercitatio Anatomica de Motu Cordis et
  Sanguinis in Animalibus (An Anatomical
  Exercise on the Motion of the Heart and
  Blood in Animals)* 83, *83*, 85
*Exercitationes Duae Anatomicae de Circulatione
  Sanguinis* 83
exorcism *18*, 24
Experiments on the Generation of Insects 146
extended-spectrum beta-lactamase
  resistant bacteria 259
eyes *297*
  artificial *236*, 237, *238*
  retinal implant *236*, 237
  surgery **86–87**

F

Facia, Bartolomeo 81
Fallopio, Gabrielle 224, 301
Farmer, John 151
Febris 39
Fehling, Hermann von 191
feminism *225*
Fetti, Dominico *250*
Fewster, John 102
Feynman, Richard 265
Filetto, Raimondo 174
first aid 195, 256
fish, brain cell 160
Five Arabic Treatises on Alchemy 71
fixatives 96
flagellation against plague 69
fleas 66, *93*
Fleming, Alexander 198, 258
Flemming, Walther 150
Fliedner, Theodore 107, 142
Fliess, Wilhelm 183
Flood, Robert *70*, 71
Florentine Codex 88
Florey, Howard Walter 198, 302
flu *see* influenza
fluid replacement therapy 123
Folli, Francesco 176
Fool's Tower *162*
foot, electric prosthetic *238*
foot and mouth disease 166
forceps *42*, *139*
formalin 96
Francis, Thomas 197
Franklin, Rosamund 213, 302
free association 183
freezing cancers 230
freezing tissue 153
Frerichs, Friedrich von 190
Freud, Sigmund 161, 163, **182–83**, 204,
  251
frontotemporal dementia 261
Frosch, Paul 166
*Functions of the Brain and of Each of
  its Parts, On the* 104

G

galangal *62*
Galen, Claudius 39, **40–41**
  anatomy 72, 73, 160
  blood circulation 82
  cancers 228
  humors 34
Gall, Franz Joseph 104, 302
gallbladder removal 253
Gallo, Robert 242
Galvani, Luigi 184, 302
galvanometer *184*, 184, 185
Garcia, Manueal *116*
garlic 30, 31, *62*
Garrett Anderson, Elizabeth *140*, 141
gas attack 192
gas exchange in lungs *285*
gases, foul-smelling 121
Gaskin, Ina May 135
gastrointestinal imaging *172*
Geber 65
Geist, Emil 192
gene therapy 249, 265, 270
genetics
  cancerous cells *228*
  coding *295*
  DNA sequencing 246–47
  DNA structure 213
  horizontal gene transfer 258
  inheritance *295*
  smallpox 101
  testing *245*
  viruses *167*
urogenital system **292–95**
genome project **246–47**
gentamicin 201
Gerardus Cremonensis *48*
Gerhardt, Charles Frédéric 170
germ theory 125, **146–47**, 198
Gill, William *119*
ginger *62*
gladiators *40–41*, 41
gloves for surgery 155
glucose, excess 190
glycopeptides 201
Goetz, Robert 207
goiter 70
Golgi, Camillo *97*, 174, 302
Golgi stain 96
gonorrhea 186
gout 90, 91
Gräfe, Karle Ferdinand von 81, 302
Grassi, Giovanni 174–75
Graunt, John 126
graves, caged *119*
Gray, Henry 145, 302
Gray's Anatomy 145
Great Ormond Street hospital 107
Great Pestilence *see* Black Death
Great Plague of Athens 66
Great Plague of London 68, 69, 90
Great Plague of Marseilles 66
Greek medicine, ancient **32–33**, 56
  alchemy 70
  cancers 228
  childbirth 134
  diabetes 190
  homeopathy 108
  smallpox 101
  women working in 140
Grünenthal, Chemi *218*
Grünpeck, Joseph 186
Guide to Childbirth 135
Guild Book of the Barber Surgeons of York 35

Guillemeau, Jacques 134
Guinea Pig Club 81
Gula *24*
Gurdon, John 270
Gynecology 39
Gynaikeia (Gynecology) 134

H

H&E stain 97
Haberlandt, Ludwig 224
Hahnemann, Samuel 108, 109
hair *296*
hallucinogenics 14
Halstead, William 230
Hamman, Edouard 73
Hammurabi 24, *24*
Handbook of Psychiatry 260
hands
  anatomy *145*
  artificial *78*, 237, *238*
  imaging *172*
  tendons *279*
  washing 138
Hare, William 118, 119
Hare's stethoscope *116*
Harrington drill 132
Hartwell, Leland 231
Harvey, William 33, 41, 82–83, **84–85**, 302
Hasson, Harrith 189
Hausen, Harald zur 231, 244
Hawking, Stephen 270
hay fever 208
Hayyan, Jabir Ibn 302
Hayyan, Jamir Ibn 65
healing power items *18*
hearing *298*
heart
  anatomy 75, *282*, *283*
  artificial *235*, 238
  attack 185
  bypass surgery 207, 252
  cardiopulmonary resuscitation 257
  damage 207
  defibrillator 185, *206*, 257
  dilatation 206
  disease **206–7**
  drugs 219
  electric currents *184*, 185
  enlarged *207*
  restarting 206
  transplants 207, **234–35**
  valve replacements *238*
  valves 85
heart attacks, aspirin and 170
heart-lung machines 235
Heine, Jakob 210
HeLa cells *244*
Helicobacter pylori 244, *245*
hematopoietic stem cell transplantation 270
hematoxylin and eosin 97
hemp *128*
Henle, Jakob 146
Henry VIII 118
herbal medicine **62–3**
  alchemy *70*
  ancient 14, 15
  asthma 214
  for Black Death 67
  Indian medicine *31*
  Islamic 51
  Medieval 56, 58
  Mesopotamian 24
  rabies 168
  Thomas Sydenham 91

Herberden, William 206
Herophilus 33, *33*, 302
Herrick, James B 206
Hewitt drop bottle *130*
Hideyo Noguchi 158
highly active antiviral therapy (HAART) 243
Hildegard of Bingen 56, 58, *140*, 140
Himsworth, Harold 191
hip replacement surgery 252
Hippocrates **36–37**
   antiseptics 154
   blood circulation 82
   cancers 228
   diseases 126
   early surgery 17
   endoscopy 189
   homeopathy 108
   humanity 164
   humors 34
   inhalers 214
   malaria 174
   TB 156
Hippocratic Corpus 33
Hippocratic Oath *36*
Hippocratic Treatises On fractures and On Joints 37, *37*
hip prostheses *238*
His, Wilhelm 96
histamines 208–9
histology 96–97, 160
histopathology 97
HIV/AIDS *9*, 159, **242–43**
Hockett, Tobey 201
Hodierna, Giovanni 92
Hoffman, Felix *170*
Hoffmann, Friedrich *116*, 206
homeopathy **108–9**
Homo neanderthalensis 14
homosexuals, HIV/AIDS and 242–43
hook *42*
Hooke, Robert 92–93, *93*, 150
Hooke's microscope *95*
Hope, James 206
Hôpital de la Charité *107*
Hopps, John Alexander 302
hops *62*
hormonal therapy for cancer 231
hormone production *286*, 287
hormone replacement therapy (HRT) 205
Horsley, Victor 161
hospices **262–63**
hospitals 56, **106–7**
   ancient Rome 39
   for influenza *196*
   Islamic 49
   plague 69
Hôtel-Dieu 78
Hounsfield, Godfrey 217
Houses of Life 54
HPV *244*, 244
HRT 205
Huang-di 26
Huangdi Neijing (Yellow Emperor Classic of Internal Medicine) 26, 29, 82, 174
Hubbard, Louisa 135
Huggins, Charles 231
Hughe's stethoscope *116*
Human Genome Project 248
human papilloma virus (HPV) *244*, 244
Humani Corporis, De (On the Fabric of the Human Body) 61, 72–73, 74, 74–75, *75*, 83, 118
humors 30, 33, **34–35**, 37, 39, 41, 58–59, *162*
Hunt, Tim 231

Hunter, John 206, 230
Hustin, Albert 177
hydrofluoroalkene inhalers 215
hygiene 38
hypnosis 128, 160, 251
hypodermic syringe *130*
hypophosphatemia 71
hypothermia, topical 234
hysteria 163, 183, 204

I

iboga 15
ibuprofen 219
id *182*
IgE 209
imaging **172–73**, **216–17**, **232–33**
Imhotep 20, 32
immune response *291*
immune system 102–3, *274*, **290–91**
   antibodies 176
   against epidemics 88
   against smallpox 101
   transplants and 234
immunization *see* vaccination
immunoglobulin E (IgE) 209
immunosuppressant drugs 235
immunotherapy, anticancer 230
implants **236–37**
in vitro fertilization **240–41**
Incas: trepanning 16
incubators *136*
Indian medicine
   alchemy 70
   Ayurveda 30–31, 214
   cancers 228
   diabetes 190
   inhalers 214
   malaria 174
   nose reconstruction *80*, 81
   prostheses 236
   smallpox 101
   trepanning 16
infection 153
   ancient Egyptian medicine 21
   childbed fever 138–39
   Mesopotamia medicine 24
   penicillin against 198
   in wartime 192
inflammatory diseases 14, *244*, *245*
influenza 192, **196–97**
Inhalation of the Vapour of Ether, On the 125
inhalers **214–15**
Injuries of the Head, On the 17
Institute of Tropical Diseases 175
instruments *see* dentistry: instruments; diagnostic instruments; laboratory instruments; surgery: instruments
insulin *190*, **190–91**
insulin pen and cartridge *202*
International Committee of the Red Cross 267
International Office of Public Hygiene 267
International Red Cross 142
Interpretation of Dreams, The 183
InTouch robots 252
intracytoplasmic sperm injection (ICSI) *240*
intraocular lens (IOL) 86
iodine 154, *155*
iodoform 154
ipecacuanha 15
iron lung 210
Ishizaka, Kimishige and Teruko 209

Islamic medicine **48–51**
   alchemy 65, 70
   anesthesia 128
   cancers 228
   cataract surgery 86
   childbirth 136
   hospitals *106*, 106
   schools of medicine 55
   smallpox 101
   women working in 140
islets of Langerhans 191
Ivanovsky, Dmitry 166, *167*

J

Jacobaeus, Hans Christian 189
jala 30
Jamir ibn Hayyan 70
Janssen, Hans and Zacharias 92
Jarvik-7 artificial heart *238*
Jefferson, Thomas 102
Jenner, Edward 101, 102, 158, 166, 302
Jesty, Benjamin 102
Jex-Blake, Sophia 141
Jing Xiao Chan Bao (Treasured Knowledge of Obstetrics) 134
joints *277*
Jung, Carl 183, 251

K

Kahun papyrus 20
Kamen, Dean 237
Kanada 70, 303
Kantrowitz, Adrian *234*
kapha 31
Kaposi's sarcoma 242
karkinos 228
Kaulbach, Wilhelm von *165*
Kehrer, Ferdinand 134
Kelling, Georg 189
Kelman, Charles 86, 303
keyhole surgery 188–89, 252
kidney
   anatomy 75
   role 41
   transplants 234
   tubules *287*
King, Edmund 176
Kitab al-Hawi fi al-Tibb (The Comprehensive Book on Medicine) 49
Kitab al-Jadari wa 'l-Hasba (Treatise on Smallpox and Measles) 101
Kitab al-Kimya (Book of Composition of Alchemy) 65
Kitab al-Mansouri fi al-Tibb (The Book on Medicine Dedicated to al-Mansur) 49
Kitab al-Shifa (The Book of Healing) 51
Kitab at-Tasrif (The Method of Medicine) 51
Klebs, Edwin 153
knee, arthroplasty *238*
Knights Hospitaller *262*, 262
knives, surgical *42*
Koch, Robert 121, 122, *146*, 146–47, 156, 198, 303
Kolletschka, Jakob 138
Korean War *195*
Kouwenhoven, William *206*
Kraepelin, Emil 260
Kussmaul, Adolph 210

L

laboratory instruments *149*
Laënnec, Mériadec 115
Laënnec, René **114–15**, 206, 303
Laënnec's pearls 115
Laennec's stethoscope *116*
Laguess, Gustave-Edouard 191
Laguna, Andres 83
Laidlaw, Patrick Playfair 208
Lancisi, Mara 206
Landois, Leonard 176
Landsteiner, Karl 176–77
Lane-Claypon, Janet 126
laparoscopy 189, 252
Lapyx *38*
Larrey, Dominique-Jean 256
laryngoscope *116*
lashun 31
lasuna 31
laudanum *91*, 91
Laudanum sydenhamii 91
Laue, Max von 172
laughing gas 128
Laveran, Charles 174
laws regarding medicine 40–41
League of Nations 267
leeches 58, 59
Leeuwenhoek, Antoni van 92, 93, 150
Leeuwenhoek's microscope *95*
legs, artificial 236–37, *238*
leprosy 58
leukemia 152
Levine, Philip 177
Lewishohn, Richard 177
Lewy body dementia 261
Liber Simplicis Medicine (Book of Simple Medicine) 140
Lichtleiter *188*, 189
lifestyle
   coronary disease and 206
   Hippocrates and 37
ligatures *78*
light conductor *188*, 189
limbs, artificial 236–37
Lind, James **98–99**
Lindbergh operation *253*
lipopeptides 201
Lippershey, Hans 92
Lister, Joseph 139, 154, 198
Liston, Robert 128
liver: anatomy 75
livestock feed with antibiotics 258, *259*
lobotomy 163
Loeffler, Friedrich 166
London hospitals 106–7
Long, Crawford 128
Lower, Richard 176, 234
Lucretius 208
Lumway, Norman 234
lung disease
   cancer *229*, 230
   smoking and *127*, 127
   TB 156
lungs
   capillaries 96
   CT scans *217*
   in respiratory system 284–85
Luzzi, Mondino de 72, 303
lymphatic system *290*
lymph nodes *291*
Lyonnet's microscope *95*
lysozymes 198
Lyssavirus 168

M

McCafferty, John 235
Macewen, William 155, 161, 303
MacFarlane, Frank 197
McIndoe, Archie 81
Mackenzie, James 206
Maclagan, Thomas 170
macrolides 201
  mad dog disease *see* rabies
  madness *see* mental illness
mafeisan 26
magnetic resonance imaging (MRI) 216,
  217, **232–33**
Mahon, Henry Walsh *98*
Maison de Charenton 163
majja 31
malaria 89, 121, 159, **174–75**
malformations, repair **80–81**
Mallon, Mary *126*
Malpighi, Marcello 83, **96**, 96
Malpighian corpuscles 96
Malpighian layer of skin 96
Malpighian tubules 96
mammary glands *287*
mammograms *205*, 205
mamsa 31
Mandela, Nelson 243
mandible: anatomy 75
mandragora 128
mandrake *128*
mania 250, 251
man-midwife 134, *134*, 136
Marker, Russell 224
Marshall, Barry 244
masks
  anesthetic *130*
  shamanism *18*
masmassû 24
Massachusetts Eye and Ear
  Infirmary 107
massage, Qigong *26*
mast cell *209*
Materia Medica, De 39, *39*, 108
material property investigation
  70
Maternité de Paris *136*
maternity hospitals 107
Matteucci, Carlo 184
Maudsley, Henry 165
Mayer, Adolf 166
Mayo, Charles 132
meadowsweet 170
measles 50, 159
medas 31
Médecins Sans Frontiers 267
medical publishing **144–45**
Medicina, De 80, 86
medicine chest *108*
medicines
  for cancer 230
  computer modeling *219*
  development 51, **218–19**
  nanomedicine 265
  reactions to 208
Medieval medicine **56–57**
Meduna, Ladislas Joseph von
  163, 303
meiosis 151
Meister, Joseph 168
melancholia *250*
memory loss 260
Menkin, Miriam 240
menopause 205
menstruation 59

mental illness **162–65**, **250–51**
  Alzheimer's disease and dementias
    260–61
  trepanning for 17
Merck 218
mercury to treat syphilis 186
meridians 26, 29
Mering, Joseph von 190
Merit-Ptah 140
Merz, Heinrich *165*
Mesmer, Franz Anton 128, *160*, 160, 303
mesmerism 128, 160
Mesopotamia medicine **24–25**, 134, 228
metastasis 228
methicillin-resistant *Staphylococcus aureus*
  258, *259*
*Méthode Curative de Playes et Fractures de la
  Tête Humaine (Treatment Method for
  Wounds and Fractures of the Human
  Head)* 78
*Méthode de Traiter les Playes Faites par les
  Arquebuses et Autres Bastons à Feu, La
  (Method of Curing Wounds Caused by
  Arquebus and Firearms)* 79
Methodic School 39
*Methodus curandi febres (The Method of
  Curing Fevers)* 90
methotrexate *230*
Metrodora 140
miasma theory 67, 69, **120–21**, 124, 126,
  146, 154
miasms 109
microanatomy 96
microbiology **146–47**
Micrographia 93, *93*
microgrippers *253*
microprocessor control 237
microscopes 152–53
microscopy **92–93**, **94–95**, *94–95*, **96–97**,
  230
microtome 96
midwives 56–57, **134–35**, **136–37**
Miescher, Friedrich 150–51
migraine, trepanning for 17
Milan, Duke of 69
milk pasteurization *149*
mind map *182*
minerals and alchemy *70*
Minkoff, Laurence *216*
Minkowski, Oskar 190
Minnitt gas-air analgesia apparatus *130*
mint *62*
mitochondria 150
mitosis 150
MNSs blood group system 177
Mobile Army Surgical Hospitals 195
Mode of Communication of Cholera, On the
  125
modern medicine timeline **222–23**
Mohl, Hugo von 150
molds used against infection 198
monasteries 56
Mondino de Luzzi 61
monoclonal antibodies 230
Montagnier, Luc 242
Montagu, Lady Mary *102*, 102
Moore, John 151
Moorfields Hospital 107
moral treatment 163
Morgagni, Giovanni Batista 152, 153, 228
morning-after pill 225
morning sickness 218
morphine *194*
Morrison, James 132
mortality records 126, *127*
Morton, William Thomas Green 128, 303

Morton ether inhaler *130*
mortsafes 119
mosaic disease 166
mosquito 174, *174*
Motolinía, Toribio 89
Motu Cordis, De 83, 83, 85
mouth *298*
moxibustion *27*, 29
MRI *216*, 217
MRSA 258, 259, *259*
Mudge, John 215
Müller, Johannes 97, 228
Müller, Paul 175
mummies 14, *20*, **22–23**, 207, 228
mumps 159
Murdoch, Colin 202
muscular system *55*, *75*, *274*, **278–79**
mustard gas 230
mutra vaha 31
Mycobacterium tuberculosis 147, *147*, 156
myocardial infarction, acute (AMI) 185

N

Nägeli, Karl von 150
nails *296*
nanobots 253, 265, *265*
nanomedicine **264–65**
Narrentum 162
National Committee on Federal Legislation
  for Birth Control 227
National Health Service 127
*Nature and Structural Characteristics of Cancer,
  On the* 97
Nature of Man, The 35
Neal, Robert *143*
Neanderthals 14, *15*
nebulizers **214–15**
needles 202
needling **28–29**
Nelmes, Sarah *103*
neomycin 201
Neosalvarsan 187
nervous system *275*, *280*, **280–81**
  brain **160–61**
  function 58
  nasal cavity and mouth *299*
Nesperennub *22–23*
neurofibrillary tangles 260, *261*
neurology **160–61**, 251
neurosis 183
neurosurgery 161
New York Call 227
Nicon, Aelius 40, 41
Nightingale, Florence 107, 121, *127*, *142–
  43*
Ninevah tablet *25*
nitrogen, liquid 230
nitroimidazoles 201
nitrous oxide 128, *130*
Nitze, Maximilian Carl-Friedrich 189
Nitze cystoscope *189*
Nobel, Alfred 303
Nobili, Leopoldo 184
Noetzli, Jean 260
nose *298*
  reconstruction *80*, 80–81
Notes on Nursing 121
Nurse, Paul 231
nursing **142–43**, *192*
  palliative care 263
  training 107

O

O, Anna 183
O'Brien, 'Irish Giant' 119
*Observationes Medicae (Observations of
  Medicine)* 90
*Observations diverses sur la stérilité, perte de
  fruits, fécondité, accouchements et maladies
  des femmes et enfants nouveaux-nés
  (Various Observations on the Sterility,
  Fruit loss, Fertility, Childbirth and Diseases
  of Women and Newborn Infants)* 134
obstetric dilator *42*
Ocimum sanctum see basil, holy
octli 15
Oedipus complex 183
estrogen 205, 224
Oeuvres, Les 236
'Omnis cellula e cellula' 152
oncogenes *228*, 229
oncology 228
*Ophthalmodouleia Das ist Augendienst (In the
  Service of the Eyes)* 87
ophthalmoscope *116*
opium 62, 91, 128, *130*
Oporini, Joannis 74
oral contraceptive 227
orchids 14
Organon of Healing, The 109
osteotome *42*
otoscope *116*
Ötzi the Iceman 14
ovaries *286*
ovum *295*
oxazolidinones 201, 259

P

pacemakers *238*
Pacini, Filippo 122
Padua 72
pediatric hospitals 107
Pagenstecher, Johann 170
pain relief 170
  see also anesthesia
Palese, Michael 254
palliative care **262–63**
Pan-American Sanitary Bureau 267
Pancoast, William 240
pancreas *286*
  artificial *190–91*
  diabetes and 190–91
pandemics
  cholera 123
  flu 192, **196–97**
  see also epidemics
Pap test *204*, 204–5, 230
Papanicolaou, George N 204, 230
papilloma virus 231
Pappenheim, Bertha 251
Paracelsus 65, 70, *91*, 108, 303
paralysis from polio 210
Paré, Ambroise 17, 76, **78–79**, 134, 236
Parkington, John 198
Parkinson's disease 244
Parr, Thomas *82*
Pasteur, Louis 139, 146, **148–49**, 154, 158,
  166, 168–69, 198
pasteurization 148–49
Patent Coffin 119
pathology **152–53**
Paul of Aegina 231
Paul VI, Pope 224

pavana 30
Pavy, Frederick 191
peanut allergy 209
Pelletier, Pierre-Joseph 89
penicillin 187, 194, **198–99**, *200*, 200–201, 258
Penicillium 198
Per-Ankh 54
percussor *116*
Perls, Max 97
personality 182, 183
pestle and mortar *50, 62*
PET 217
Pfolsprundt, Heinrich von 81
phacoemulsification 86
pharmaceutical industry **218–19**
pharmacology 51
phenol 139
Philip II, King of Spain 75
Phipps, James *102*
phlegm 33, 34
phrenology **104–5**, 160
phthiasis 156
Physica 58
pill silverer *62*
Pincus, Gregory 224, 240
pineal gland, role 41
Pinel, Philippe 162, 163, **164–65**
Pirodon, Louise-Eugene *119*
Pirquet, Clemens von 208
pituitary gland *287*
placebo effect 109
Plague of Justinian 66
plagues
    Antonine 38, 41
    in Athens 37
    Black Death **66–67**, 68
    Great Plague of Athens 66
    Great Plague of London *68, 69*, 90
    Great Plague of Marseilles *66*
    Plague of Justinian 66
    preventing **68–69**
    red 101
    vaccination against *159*
    white 156
    yellow fever 69, *69*
    see also epidemics; pandemics
plasma *283*
Plasmodium genus 174
plastic surgery 81
Platearius, Matthaeus *54*
platelets *283*
Pliny 39
Pneumocystis carinii 242
podalic version 78
polio 158, 167, **210–11**
Polybus 35
polyuria 190
pomander *68*
Portier, Paul 208
positron emission tomography (PET) 217
postmortem instruments *153*
pot marigold *62*
Pott, Percivall 229, 230
Pound, D J *143*
pox 100
Pox, Great **186–87**
Practical Treatise on the Domestic Management and Most Important Diseases of Advanced Life 260
prana vaha 31
pregnancy **134–35**
    see also childbirth; contraceptive pill
prehistory medicine 12
privthi 30
progesterone 205, 224

prostate
    cancer 231
    surgery 252
prostheses **236–37, 238–39**
    arms *194, 238*
    foot *238*
    hands *78, 238*
    hip *238*
    knee *238*
    noses *81*
protooncogenes *228*
psychiatry 160
psychoanalysis 161, 163, 182–83
psychodynamic theory 251
psychosexual development 183
public health 38, **126–27**
publishing, medical **144–45**
puerperal fever **138–39**
pulque 15
pulverisateur *214*
PUMA robot 252
purisha vaha 31
Pussin, Jean-Baptiste 163, **164–65**
Pythagoras 260

Q

qi 26, 27, 29, 82
Qianjun Yaofang (Prescriptions Worth a Thousand Gold) 26
Qigong massage 26
quarantine 69
quartet systems *34*
quinine 89, 174, 194
quinolones 201
Qur'an 51

R

rabies 149, 158, 167, **168–69**
radiotherapy 230–31, *231*
Raistrick, Harold 198
rakta 31
Ramesses II, Pharaoh *100*
Ramesses V, Pharaoh *100*
Ramón y Cajal, Santiago 97, *97*, 303–4
rasa 31
rat, black 66
Re Anatomica, De (On Things Anatomical) 83
Recklinghausen, Friedrich von 153
    reconstructive surgery *see* surgery, reconstructive
Recueil Des Traités de Médecine (Collection of Medical Treatises) 48
red blood cells *283*
Red Crescent 267
Red Cross *142*, **266–67**
Redi, Francesco *146*, 146
Reformation medicine: hospitals 106
Regaud, Claudius 230
rehydration therapy 123
religion, role in medicine 56
reliquaries *56*
Renaissance medicine timeline **46–47**
repression 183
reproductive system *275*
    female **294–95**
    male *293*
respiratory diseases 244
respiratory system *275*, **284–85**
resurrection men **118–19**
retina *297*
    implant *236*, 237
Revival medicine timeline **46–47**

Rhazes *see* al-Razi, Ibn
rhesus factor 177
rhinitis, seasonal allergic 208
Rhinoplastik 81
    rhinoplasty *see* nose reconstruction
ribonucleic acid (RNA) 243
ribosomes 201
ribs: anatomy 75
Ricard, Philippe 186
Richet, Charles 208
Ridley, Harold 86
rifampicin 201
Rig Veda 236
Robert of Chester 65
Robinson, Henry Peach *156*
Robodoc 252
robots use in medicine 189, **252–53**, **254–55**, *265*
Roche 219
Rock, John 224, 240
Rod of Asclepius 32
Rokitansky, Karl 152
Roman medicine, ancient **38–39**, 56
    hospitals 106
    prostheses 237
    surgical instruments **42–43**
Röntgen, Willhelm Conrad 172–73
Roosevelt, Franklin D *210*
rose, China *62*
rosemary *62*
Ross, Ronald 174, 175
Rösslin, Eucharius 134
Roux, Émile 168
Royal Hospital for Diseases of the Chest 107
Ruggiero, Trotula de 55, 140, 304

S

Sabin, Albert 211
Sabolich, John 237
Safar, Peter 257
saffroclovesn 62
Sahachiro Hato 187, 218
St. Anthony's Fire 69
St. Christopher's Hospice 262–63
St, John's wort *62*
St. Thomas' Hospital *107*, 142
St. Vincent de Paul 262
St. Vitus' Dance 69
sal ammoniac 49
Sales-Girons, Jean *214*
salicin 170
salicylic acid 170
Salk, Jonas Edward 197, *210*, 211, 304
Salmonella Typhi 201
Salpêtrière Hospital 161, 165
Salvarsan 187, 218
Salvarsan kit *187*
Sanderson, John 198
Sanger, Margaret 141, **226–27**
Sanitary Commission 142
sanitation *see* public health
sanitoria 156
SARS 267
Saunders, Cicely 262–63
scalpel *42*
Sceptical Cymist, The 71
Schatz, Albert 201
Schaudinn, Frtiz 187
Schiller, Walter 204
schistosomiasis 244
Schlieden, Matthias 150
schools of medicine *see* education for medicine
Schopenhauer, Arthur 71

Schwangeren Frauen und Hebammen Rosengarten, Der (The Rose Garden for Pregnant Women and Midwives) 134
Schwann, Theodor *150*, 150
screening for cancers 229–30
scrotal cancer 230
Scuola Medica Salernitana **54–55**
scurvy **98–99**
Sedibus et Causes Morborum per Nantomen Indagatis, De (On the Seats and Causes of Diseases as Investigated by Anatomy) 152
Seishu, Hanaoka 129, 304
Sekhmet 20
Semmelweiss, Ignaz **138–39**, 146, 154
sensory system *275*, **296–99**
sepsis 154
Serefeddin, Sabuncuoglu 140
Sergius, Marcus 237
serum albumin 194
Servetus, Michael 83, 304
Severe Acute Respiratory Syndrome (SARS) 267
Severus, Septimus, Emperor 41
sex glands *287*
Sexual Discrimnation Act *225*
Seymour, Robert *121*
shalya chikitsa 30
shamanism 14, **14–15**, **18–19**
Shanghan Han Za Bing Lun (Treatise on Febrile, Cold and Miscellaneous Diseases) 26
Shao, Emperor 16
shears *42*
shell shock 163
Shennong, Emperor 208
Shibasaburo, Kitasato 67, 158, 304
shukra 31
sickle cell anemia 249
Siegle, Emil 215
Signs of Fractures, On 39
silkworm disease 149
simian immunovirus 243
Simplici Medicina (The Book of Simple Medicine), De 54
Simpson, James Young 129, 304
Sina, Ibn *50*, 51, 55, **52–53**, 228, 304
single-photon emission computed tomography (SPECT) 217
Sisters of Charity 262
skeletal system *74, 274*, **276–77**
skin *287, 296*
skin prick test 209
skin testing kit *208*
skull 74
sleeping drugs 218
sleeping sickness *151*
'slim' disease 243
Sloane, Hans 102
smallpox 50, *50, 88*, **100–101**, **102–3**, *126*
Smallpox Eradication Campaign 101
smear test 204
Smellie, William 135, 136
Smith, Gayle 268
Smith papyrus 80
smoking and cancer 127, *127*, 229
Snow, John 121, 122, **124–25**, 126, 146
Soho *125*
somatic stem cells 270
Soranus of Ephesus 39, 134
Souttar, henry 207
Spallanzani, Lazzaro 146
Spanish flu 196
spatha *42*
SPECT 217
speculum 204
    vaginal *42*, 135
sperm *295*

sphygmomanometer *116*
spinal cord *280*
Spiritual Midwifery 135
spontaneous generation 146
srota mano vaha 31
srotas 31
staining techniques 96, *96–97*, 160
stanya vaha 31
Staphylococcus *198*, 198
    S. aureus 258, *259*
Starling, Edward 185
Starling, Ernest 205
steam-cleaning 155
Stelluti, Francesco 92
stem cell therapy **270–71**
stent surgery *206*
stentrode 237
Steptoe, Patrick 240
Sterneedle gun *202*
stethoscopes **114–15**, *116–17*
Stevenson, Robert Louis 119
Stock, Harald 218
stomach disorders 219, 244, *245*
Storck, Anton von 108
Straet, Jan van der *70*
streptogramins 201
Streptomyces
    S. aureofaciens 201
    S. griseus 201
streptomycin 67, 156, 201
string galvanometer *184*, 185
Studies on Hysteria 183, 251
Subitaneis Mortibus (On Sudden Death), De 206
succussion 108–9
sulfanilamide 200
sulfonamides 194, 200
Sun Simiao 26
superbugs **258–59**
superego *182*
surgery
    ancient Egyptian 21
    ancient Roman 39
    barber-surgeons **76–77**, 78–79
    cardiac bypass 207
    for cataract **86–87**
    dissection 33, 61, *61*, 118–19
    early **16–17**
    Indian 30–31
    Islamic medicine 51
    minimally invasive **188–89**, 252
    neurosurgery 161
    plastic 81
    reconstructive **80–81**
    robotic **254–55**
    sepsis and 154–55
    stent *206*
    trepanning 14, **16–17**, 21, 188
surgical instruments
    ancient Egyptian *21*
    ancient Rome 39, **42–43**
    childbirth *139*
    instruments 176, *177*, 265
    Islamic 51
    for keyhole surgery 188–89
    postmortem *153*
    sterilizing *154*
Susruta 30, 190
Susruta Samhita 30, 80, 86, 174, 228
swan-neck experiment *149*
swine flu 267
Sydenham, Thomas **90–91**
Sylvius, Jacobus 72
synovial joints *277*
syphilis 80–81, 88, **186–87**, 218, 224
syringes **202–3**

T

Tabulae Anatomicae Sex (Six Anatomical Plates) 74
Tagamet 219
Tagliacozzi, Gaspare 81
talking therapy 163, 183, **250–51**
tamoxifen 231
Tasma, David 263
taste *299*
teeth, fossilized 14
teixobactin 259
tejas 30
teleconsultation *252*
telemanipulation 252
telemedicine **252–53**
teleradiology 253
telerehabilitation 253
telescopes 92–93
temperaments, four *35*
Temple of Peace, fire at *40*, 41
tendons *279*
Teniers, David *76*
testes *286*
tetanus 158
tetracycline *200*, 201
thalidomide *218*, 218
Theodoric of Lucca 59
Theriaca Andromachi 67
thermometer *116*
thigh tourniquet *42*
Three Essays on the Theory of Sexuality 183
thrombolytic drugs 257
thrombosis 153
Thucydides 101
thulasi 31
thyroid gland *286*, 287
tile cautery *42*
TNM staging of cancer 229–30
tobacco causing cancer 229, *230*
tobacco mosaic virus 166
tomography 216
tongue *299*
Tourette, George Gilles de la 161, 304
tourniquet *42*
Tractatus de Padagra et Hydrope (The Treatise on Gout and Dropsy) 91
Traditional Birth Attendant 136
Traité de L'Auscultation Médiate (A Treatise on the Diseases of the Chest) 115
transfusion **176–77**
    transmission of disease *see* germ theory
transplants, heart 207
    see also prostheses
trastuzumab 230
Travers, Frederick 139
Treatise on the Diseases of the Heart and Great Vessels 206
Treatise on the Human Body 55
Treatise of Scurvy, A 98–99
Treatise on the Theory and Practice of Midwifery, A 135
Treatments for Women 140
trepanning 14, **16–17**, 21, 188
trepans 17
Treponema pallidum 186, 187
triage 143, 195, 256
trocar *202*
Trypanosoma brucei 151
Tu Youyou 175
tuberculin 147
    syringes *202*
tuberculosis 147, *147*, 148–49, **156–57**
    resistant 258

Tuke, William 163
Tulp, Nicolaes 152, *152*
tulsi 31
tumi 16
tuning fork *116*
Tuo, Hua 16, 26, 304
Turpin, Dick 119
twentieth century medicine
    timeline **112–13**
Twort, Frederik 167
typhoid *126*, 201
typhus 158, 194

U V

Über eine Neue Art von Strahlen (On a New Kind of Rays) 172
ultrasound scans *217*, 217
University College London 107
Urban II, Pope 262
urethra, male *292*
urinary system *275*, **292–93**
uterine muscles *287*
Utriusque Cosmi Historia (History of the Two Worlds) 70, 71
vaccination **158–59**, 166
    cervical cancer 244
    how vaccines work *159*
    influenza 197
    mass 126
    Pasteur and 148–49
    polio *210*, 211
    rabies 168–69
    smallpox **102–3**, 126
Vaccination Act 159
Vaccinia virus 158
Vagbhata 30
vaginal speculum *42*, 135
valetudinaria 106
Valium 218
vancomycin 201
vancomycin-resistant enterococci (VRE) 258–59
Vanity of the Arts and Sciences, The 65
Variola virus 100, *101*
    variolation *see* vaccination
Varmus, Harold 229
vascular dementia 261
vascular suturing 234
vayu 30
veins, blood letting 59
Vejovis 39
Velpeau, Alfred 119
veress, Janos 189
Vero Telescopii Inventore, De (The True Inventor of the Telescope) 92
vertebroplasty 189
vervain *62*, 168
Vesalius, Andreas 41, 61, **72–75**, 83, 118, 160
Vibrio cholerae 122, 147
Victoria, Queen 125
Vienna General Hospital *162*
Vierordt, Karl 206
Villermé, Louis 126
Virchow, Rudolf Karl Ludwig 151, 152, 228–29
virology **166–67**
viruses **166–67**, *291*
    cancer and 244
    HIV 243
    influenza 196, *196*
    mutation *196*
    vs bacteria *201*
vision *297*

vitamin C deficiency **98–99**
Volta, Alessandro 184
Voluntary Aid Detachments 143
Voyages Faits en Divers Lieux, Les (Journeys in Diverse Places) 78
Vuillemin, Jean Paul 198

W

Waksman, Selman 201
Waller, Augustus 185
Warren, Robin 244
wartime medicine 127, 142, **192–93**, **194–95**, 199
Washkansky, Louis 235
water contamination 125
water-testing kit 122
Watson, James Dewey 213, 304
Watten, Raymond 123
Watten cot 123
Weigert, Carl 153
Wells, Horace 128
Westmacott, John 145
Whiston, Surgeon Captain *115*
white blood cells 96, *283*, *291*
"White Lady" painting *14*
white plague 156
Whole Works Of That Excellent Practical Physician Dr Thomas Sydenham, The 90
Wiener, Alexander 177
Wilkins, Maurice 213
Willis, Thomas 160, 190
willow use for pain relief 14, *170*, 170
Winston, Robert 240
Wissowzky, A 97
Woman Rebel, The 226
women in medicine **140–41**
    ancient times 55
    Medieval times 56–57
    midwifery 136–37
    nursing 142–43
    wartime 192
Women's Cosmetics 140
women's health **204–5**
    see also childbirth *and specific issues, e.g.* menstruation
World Health Organization 127, 267
World War I 192
World War II **194–95**, *199*
Wright, Almroth 198
wu-xing 26

X Y Z

X-rays **172–73**, 216, 232
    for cancer 230
    of mummies 22
    in wartime 192
Yamanaka, Shinya 270
yarrow 14, *14*, 15
Yayoi, Yoshioka 141
yellow fever 69, *69*, 167
Yersin, Alexandre 67
Yersinia pestis 67
yin-yang 26, *26*
Yu Hoa Long *101*
Zan Yin 134
zang-fu 26
ZEUS 254
Zhongjing, Zhang 26
Zika virus 159, 267
Zimmerman, Michael 22
zinc chloride 154
ZMapp 268

Acknowledgments

Dorling Kindersley would like to thank the following people for their assistance in the preparation of this book:

Alexandra Beeden for proofreading; Michele Clarke-Moody for compiling the index; Simar Dhamija, Konica Juneja, Rashika Kachroo, Divya PR, and Anusri Saha for design assistance; Suefa Lee and Ira Pundeer for editorial assistance; and Myriam Megharbi for picture research assistance.

The author would like to thank the following for advice in various medical specialties: Michael McManus, cardiopulmonary; Professor Chris Thompson FRCPsych FRCP MRCGP, psychiatry; Andrew Parker DGDP BDS, dentistry; James Halliday, pharmacology; Gerald Prior and Michael Stevenson, otolaryngology.

PICTURE CREDITS
The publisher would like to thank the following for their kind permission to reproduce their photographs:

(**Key**: a-above; b-below/bottom; c-center; f-far; l-left; r-right; t-top)

2 Corbis: Christie's Images. **4 Alamy Stock Photo:** The Art Archive / Gianni Dagli Orti (br). **Science Photo Library:** Sheila Terry (tr). **Wellcome Images http://creativecommons.org/licenses/by/4.0/:** Wellcome Library, London (c). **5 akg-images:** (tr). **Corbis:** (br). **Dorling Kindersley:** Army Medical Services Museum (cr). **Getty Images:** DEA PICTURE LIBRARY (bl). **Wellcome Images http://creativecommons.org/licenses/by/4.0/:** Wellcome Library, London (tl); Science Museum, London (cl). **6 123RF.com:** photka (c). **Alamy Stock Photo:** akg-images (br); The Art Archive / Gianni Dagli Orti (bl). **Corbis:** (tl); Centers for Disease Control - digital version copyright Science Faction / Science Faction (tr). **7 Alamy Stock Photo:** World History Archive (tl). **PunchStock:** Image Source (cl). **Science Photo Library:** James King-Holmes (br); BSIP, RAGUET (tr); Spencer Sutton (bl). **8–9 Science Photo Library:** Maurizio De Angelis. **10–11 Alamy Stock Photo:** Ivy Close Images. **12 Corbis:** Gianni Dagli Orti (ca); Frederic Soltan (br). **Getty Images:** Rob Lewine (clb). **Wellcome Images http://creativecommons.org/licenses/by/4.0/:** Science Museum, London (cra). **13 Alamy Stock Photo:** The Art Archive / Gianni Dagli Orti (bl). **Getty Images:** Time Life Pictures (cla). **Wellcome Images http://creativecommons.org/licenses/by/4.0/:** Wellcome Library, London (c, bc). **14 akg-images:** Jürgen Sorges (ca). **Bridgeman Images:** South Tyrol Museum of Archaeology, Bolzano, Italy / Wolfgang Neeb (b). **14–45 Wellcome Images http://creativecommons.org/licenses/by/4.0/:** Wellcome Library, London (t/Tab). **15 Science Photo Library:** Mauricio Anton. **16 Getty Images:** Science & Society Picture Library (clb, r). **17 Alamy Stock Photo:** The Art Archive (t). **Science Photo Library:** NLM / Science Source (bl). **18 Getty Images:** Werner Forman / Universal Images Group (cl); Science & Society Picture Library (br). **Glasgow City Council (Museums):** (tc). **SuperStock:** Science and Society (cr). **Wellcome Images:** Mark de Fraeye (tl). **Wellcome Images http://creativecommons.org/licenses/by/4.0/:** Science Museum, London (fcr). **19 Corbis:** Luca Tettoni (cr). **Dorling Kindersley:** Cecil Williamson Collection (tr). **Getty Images:** Werner Forman (tc); Rob Lewine (br). **20 Alamy Stock Photo:** The Print Collector (bl). **A. Nerlich/Inst. Pathology Munich-Bogenhausen:** (cra). **Science Photo Library:** National Library Of Medicine (br). **21 Alamy Stock Photo:** The Art Archive / Gianni Dagli Orti. **22–23 Press Association Images:** John Stillwell. **24 akg-images:** Erich Lessing (tr). **Corbis:** Gianni Dagli Orti (bl). **25 Bridgeman Images:** Zev Radovan. **26 akg-images:** Pictures From History (c). **Wellcome Images http://creativecommons.org/licenses/by/4.0/:** Wellcome Library, London (bc). **27 Alamy Stock Photo:** The Art Archive. **28–29 Alamy Stock Photo:** The Art Archive / Gianni Dagli Orti. **29 Alamy Stock Photo:** The Art Archive / Gianni Dagli Orti (cl). **30 akg-images:** Roland and Sabrina Michaud. **31 Alamy Stock Photo:** Jochen Tack (tr). **Corbis:** Frederic Soltan (br). **32 Corbis:** Gianni Dagli Orti (b). **32–33 Science Photo Library:** Gianni Tortoli (t). **33 Getty Images:** DEA / G. DAGLI ORTI (tc). **Wellcome Images http://creativecommons.org/licenses/by/4.0/:** Wellcome Library, London (br). **34 akg-images:** Erich Lessing (bl). **Wellcome Images http://creativecommons.org/licenses/by/4.0/:** Science Museum, London (br). **35 Science Photo Library:** British Library. **36 Alamy Stock Photo:** Heritage Image Partnership Ltd (clb). **Getty Images:** Time Life Pictures (l). **37 Bridgeman Images:** Greek School, (11th century) / Biblioteca Medicea-Laurenziana, Florence, Italy / Archives Charmet (cr). **iStockphoto.com:** imagestock (tl). **38 Corbis:** Leemage. **39 Alamy Stock Photo:** Everett Collection Inc (tr). **Getty Images:** Science & Society Picture Library (bc). **Wellcome Images http://creativecommons.org/licenses/by/4.0/:** Wellcome Library, London (cla). **40 akg-images:** (l). **41 Alamy Stock Photo:** INTERFOTO (cr). **Science Photo Library:** Sheila Terry (bl). **42 Dorling Kindersley:** The Trustees of the British Museum (tl, tr); Thackeray Medical Museum (tr/Spatha). **Courtesy of Historical Collections & Services, Claude Moore Health Sciences Library, University of Virginia:** (ftl). **Wellcome Images http://creativecommons.org/licenses/by/4.0/:** Wellcome Library, London (tc, r). **43 Dorling Kindersley:** The Trustees of the British Museum (crb); Thackeray Medical Museum (cla, cla/EAR SPECILLUM). **Courtesy of Historical Collections & Services, Claude Moore Health Sciences Library, University of Virginia:** (tl, tc, r, b, cl). **Wellcome Images http://creativecommons.org/licenses/by/4.0/:** Wellcome Library, London (ca, clb); Science Museum, London (cb). **44–45 Science Photo Library. 46 Bridgeman Images:** Historisches Museum, Bingen, Germany / Bildarchiv Steffens (c). **Dorling Kindersley:** The Science Museum, London (br). **Getty Images:** DEA / G. DAGLI ORTI (bc); DEA PICTURE LIBRARY (cla). **47 Getty Images:** Science & Society Picture Library (br). **TopFoto.co.uk:** 2003 Charles Walker (bl). **Wellcome Images http://creativecommons.org/licenses/by/4.0/:** Wellcome Library, London (cl); Science Museum, London (cr). **48 Bridgeman Images:** Pictures from History. **48–109 Wellcome Images http://creativecommons.org/licenses/by/4.0/:** Science Museum, London (t/Tab). **49 Alamy Stock Photo:** Art Directors & TRIP (bc). **Dorling Kindersley:** Natural History Museum, London (tr). **50–51 Getty Images:** DEA / G. DAGLI ORTI. **50 Wellcome Images http://creativecommons.org/licenses/by/4.0/:** Science Museum, London (tr). **52–53 Alamy Stock Photo:** The Art Archive / Gianni Dagli Orti. **54 Bridgeman Images:** National Library, St. Petersburg, Russia (bl). **54–55 Getty Images:** DEA PICTURE LIBRARY. **55 Bridgeman Images:** University of Bologna Collection, Italy (br). **The Art Archive:** Bodleian Libraries, The University of Oxford (tr). **56 Alamy Stock Photo:** The Art Archive / Gianni Dagli Orti (bl). **Getty Images:** Heritage Images / Hulton Archive (cra). **57 Getty Images:** DEA / G. DAGLI ORTI. **58 Alamy Stock Photo:** PBL Collection (tl). **58–59 Bridgeman Images:** Bibliotheque Nationale, Paris, France / Archives Charmet (b). **59 Getty Images:** DEA PICTURE LIBRARY (crb). **60–61 Getty Images:** DEA / M. SEEMULLER / Contributor. **62 123RF.com:** lehui (cb). **Bridgeman Images:** Private Collection / Archives Charmet (br). **Getty Images:** Science & Society Picture Library (bc). **Science Photo Library:** (fbr). **63 Getty Images:** DEA / G. Nimatallah (bl). **64–65 Mary Evans Picture Library:** INTERFOTO / Bildarchiv Hansmann. **66 Corbis. 67 Alamy Stock Photo:** The Art Archive (br). **68 Getty Images:** Hulton Archive (b). **SuperStock:** Science and Society (tr). **69 Corbis:** (tc). **Wellcome Images http://creativecommons.org/licenses/by/4.0/:** Wellcome Library, London (bc). **70 Corbis:** Heritage Images (c). **SuperStock:** Buyenlarge (bl). **70–71 Wellcome Images http://creativecommons.org/licenses/by/4.0/:** Wellcome Library, London (t). **71 Corbis:** The Gallery Collection (tr). **Science Photo Library:** British Library (bl). **72 Bridgeman Images:** Royal Collection Trust © Her Majesty Queen Elizabeth II, 2016. **73 Bridgeman Images:** Musee des Beaux-Arts, Marseille, France (t); University of Padua, Italy (br). **74 Corbis:** (l). **Science Photo Library:** (tc). **75 Corbis:** Christie's Images (tl). **Getty Images:** UniversalImagesGroup (tr). **Science Photo Library:** CCI Archives (br). **76–77 Corbis:** Burstein Collection. **78 akg-images:** (l). **79 Science Photo Library:** Sheila Terry (tl). **Wellcome Images http://creativecommons.org/licenses/by/4.0/:** Wellcome Library, London (cra, bc). **80 Wellcome Images http://creativecommons.org/licenses/by/4.0/:** Wellcome Library, London (bl). **80–81 akg-images. 81 Getty Images:** Science & Society Picture Library (br). **Rex by Shutterstock:** Paul Fievez / Associated Newspapers (cr). **82 Wellcome Images http://creativecommons.org/licenses/by/4.0/:** Wellcome Library, London. **83 Science Photo Library:** (tc). **Wellcome Images http://creativecommons.org/licenses/by/4.0/:** Wellcome Library, London (br). **84–85 akg-images:** Album / Oronoz. **86 Bridgeman Images:** Bibliotheque de la Faculte de Medecine, Paris, France / Archives Charmet (br). **Getty Images:** DEA / G. DAGLI ORTI (ca). **87 Bridgeman Images:** Bibliotheque de l'Institut d'Ophtalmologie, Paris, France / Archives Charmet. **88 Wellcome Images http://creativecommons.org/licenses/by/4.0/:** Science Museum, London (tr). **89 Mary Evans Picture Library. 90 Wellcome Images http://creativecommons.org/licenses/by/4.0/:** Wellcome Library, London (clb, r). **91 SuperStock:** Science and Society (tc). **Wellcome Images http://creativecommons.org/licenses/by/4.0/:** Wellcome Library, London (crb). **92 Alamy Stock Photo:** liszt collection (bc). **Getty Images:** Science & Society Picture Library (ca). **92–93 Wellcome Images http://creativecommons.org/licenses/by/4.0/:** Wellcome Library, London (b). **93 Wellcome Images http://creativecommons.org/licenses/by/4.0/:** Wellcome Library, London (tr). **94 Dorling Kindersley:** The Science Museum, London (tl, tc). **95 Corbis:** Inga Spence / Visuals Unlimited (br). **96 Corbis:** Scientifica (cra). **Photo Scala, Florence:** courtesy of the Ministero Beni e Att. Culturali (clb). **Wellcome Images http://creativecommons.org/licenses/by/4.0/:** Wellcome Library, London (bc). **97 TopFoto.co.uk:** PRISMA / VWPICS. **98–99 Mary Evans Picture Library:** The National Archives, London. England. **100 Alamy Stock Photo:** The Art Archive / Gianni Dagli Orti (bl). **Science Photo Library:** CCI Archives (r). **101 Science Photo Library:** Eye Of Science (bc). **Wellcome Images http://creativecommons.org/licenses/by/4.0/:** Wellcome Library, London (tr). **102 Corbis:** The Gallery Collection (clb). **102–103 Wellcome Images http://creativecommons.org/licenses/by/4.0/:** Wellcome Library, London (t, b). **103 Wellcome Images http://creativecommons.org/licenses/by/4.0/:** Wellcome Library, London (tr). **104–105 Getty Images:** Science & Society Picture Library. **106 Science Photo Library:** Sheila Terry. **107 Getty Images:** De Agostini Picture Library (tr); Science & Society Picture Library (bc). **108 Wellcome Images http://creativecommons.org/licenses/by/4.0/:** Wellcome Library, London (bl); Science Museum, London (r). **109 akg-images:** (tc). **Alamy Stock Photo:** Sputnik (br). **110–111 SuperStock:** Science and Society. **112 Corbis:** (cra). **Image courtesy of Biodiversity Heritage Library. http://www.biodiversitylibrary.org:** Taken from Anatomy, descriptive and surgical / by Henry Gray; the drawings by H.V.Carter; the dissections jointly by the author and Dr. Carter (bc). **Science Photo Library:** (c). **Wellcome Images http://creativecommons.org/licenses/by/4.0/. 113 akg-images:** (cra). **Dorling Kindersley:** Science Museum, London (cl). **Getty Images:** Science & Society Picture Library (crb). **Wellcome Images http://creativecommons.org/licenses/by/4.0/:** Wellcome Library, London (bl). **114–115 Wellcome Images http://creativecommons.org/licenses/by/4.0/:** Wellcome Library, London. **114–177 Dorling Kindersley:** Army Medical Services Museum (t). **116 Science & Society Picture Library:** Science Museum (cr). **SuperStock:** Science and Society (crb). **117 Dorling Kindersley:** (l); The Science Museum, London (cb). **Science Photo Library:** (fbr). **Science &**